# Indoor Thermal Comfort

# Indoor Thermal Comfort

Editors

**Francesca Romana d'Ambrosio Alfano**
**Boris Igor Palella**

MDPI • Basel • Beijing • Wuhan • Barcelona • Belgrade • Manchester • Tokyo • Cluj • Tianjin

*Editors*
Francesca Romana d'Ambrosio Alfano
University of Salerno
Italy

Boris Igor Palella
University of Naples Federico II
Italy

*Editorial Office*
MDPI
St. Alban-Anlage 66
4052 Basel, Switzerland

This is a reprint of articles from the Special Issue published online in the open access journal *Atmosphere* (ISSN 2073-4433) (available at: https://www.mdpi.com/journal/atmosphere/special_issues/thermal_comfort).

For citation purposes, cite each article independently as indicated on the article page online and as indicated below:

LastName, A.A.; LastName, B.B.; LastName, C.C. Article Title. *Journal Name* **Year**, *Article Number*, Page Range.

**ISBN 978-3-03943-527-2 (Hbk)**
**ISBN 978-3-03943-528-9 (PDF)**

Cover image courtesy of Francesca Romana d'Ambrosio Alfano.

# Contents

# About the Editors

**Francesca Romana d'Ambrosio Alfano** is Full Professor of Building Physics at the Department of Industrial Engineering of the University of Salerno, Co-chair of Observatory for Gender Studies and Equal Opportunities at the University of Salerno, and member of the Scientific Board of the Interdepartmental Centre of Engineering for Cultural Heritage (CIBEC) of the University Federico II of Naples. Her research interests are multidisciplinary and include indoor thermal environments, indoor air quality, energy saving, historical buildings, and the history of engineering. She is also a member of international (ICOMOS, REHVA, ASHRAE) and national (AiCARR, AISI, SIE, Associazione della Fisica Tecnica Italiana) associations. Her activity also involves the standardization field, with participation in the technical committees of ISO/CEN (thermal environments), UNI (ergonomics, cultural heritage), and CTI (heat transfer and fluid dynamics, air conditioning and refrigeration).

**Boris Igor Palella** is Associate Professor of Building Physics at Department of Industrial Engineering of the Università degli Studi di Napoli Federico II in Naples (Italy), member of the Interdepartmental Centre of Engineering for Cultural Heritage (CIBEC), and member of the Task Force on the Cultural Heritage of Federico II University. He earned his Ph.D. in Chemical Engineering from the University of Lyon Claude Bernard Lyon I in co-tutelage with the Università Degli Studi di Napoli Federico II. His research activity mostly covers the fields of building physics and building energy systems (thermal comfort, IAQ, energy-saving, and historical buildings). He is also the convenor of the Technical Committee on Indoor Environmental Quality of the Italian Society for Air Conditioning, Heating and Refrigeration (AiCARR), member of the Italian Society of Applied Thermodynamics and Heat Transfer (Associazione della Fisica Tecnica Italiana), and Treasurer of the Italian Society of the History of Engineering (AISI).

# Preface to "Indoor Thermal Comfort"

In recent years, the application of the principles of human factors has revealed the need for rethinking the whole indoor built environment design. Indoor environments should be livable, comfortable, safe, and productive, with low energy costs, and their design has to be compliant with sustainability requirements. This is also because the indoor environment has a potential impact on occupants' health and productivity, affecting their physical and psychological conditions.

In this context, the design and assessment of indoor thermal comfort, although regulated by a robust framework of standards, require a novel approach to find the best solution under the specific context every time. This implies general rules to be respected and the awareness that a project of comfort is a project that ultimately puts people and their needs at the center.

This Special Issue deals with most debated challenges in the field of thermal comfort with a special focus on design, technical, engineering, psychological, and physiological issues and, finally, potential interactions with other IEQ issues that require a holistic way to conceive the building envelope design. Covered topics include the following: fundamentals in thermal comfort and IEQ assessment; field investigations; innovative designs, systems, and/or control domains that can enhance thermal comfort; and the integration of human factors in buildings' energy performance.

We thank all authors for their submissions, and we also thank the colleagues involved in the review process of received manuscripts.

**Francesca Romana d'Ambrosio Alfano, Boris Igor Palella**
*Editors*

*Article*

# Fifty Years of PMV Model: Reliability, Implementation and Design of Software for Its Calculation

Francesca Romana d'Ambrosio Alfano [1], Bjarne Wilkens Olesen [2], Boris Igor Palella [3,*], Daniela Pepe [1] and Giuseppe Riccio [3]

[1]   DIIn Dipartimento di Ingegneria Industriale, Università degli Studi di Salerno, Via Giovanni Paolo II 132, Fisciano, 84084 Salerno, Italy; fdambrosio@unisa.it (F.R.d.A.); dpepe@unisa.it (D.P.)
[2]   Department of Civil Engineering, International Centre for Indoor Environment and Energy, Nils Koppels Alle, Building 402, DK-2800 Lyngby, Denmark; bwo@byg.dtu.dk
[3]   DII—Dipartimento di Ingegneria Industriale, Università degli Studi di Napoli Federico II, Piazzale Vincenzo Tecchio 80, 80125 Naples, Italy; riccio@unina.it
*   Correspondence: palella@unina.it; Tel.: +39-081-7682-618

Received: 11 December 2019; Accepted: 24 December 2019; Published: 29 December 2019

**Abstract:** In most countries, PMV is the reference index for the assessment of thermal comfort conditions in mechanically conditioned environments. It is also the basis to settle input values of the operative temperature for heating and cooling load calculations, sizing of equipment, and energy calculations according to EN 16798-1 and 16798-2 Standards. Over the years, great effort has been spent to study the reliability of PMV, whereas few investigations were addressed to its calculation. To study this issue, the most significant apps devoted to its calculation have been compared with a reference software compliant with EN ISO 7730 and the well-known ASHRAE Thermal Comfort Tool. It has been revealed that only few apps consider all six variables responsible for the thermal comfort. Relative air velocity is not considered by ASHRAE Thermal Comfort Tool and, finally, the correction of basic insulation values due to body movements introduced by EN ISO 7730 and EN ISO 9920 Standards has only been considered in one case. This implies that most software and apps for the calculation of PMV index should be used with special care, especially by unexperienced users. This applies to both research and application fields.

**Keywords:** PMV; comfort indices; thermal comfort; software; app; building simulation

---

## 1. Introduction

### 1.1. Background

The building stock in the world uses approximately 40% of the total energy and it is responsible for one third of the global greenhouse gases emissions [1,2]. As a consequence, achieving sustainable energy usage in buildings has received significant attention in the past years [2,3]. The requirements of high levels of Indoor Environmental Quality (IEQ) in terms of thermal, visual and acoustic comfort and indoor air quality may increase the energy demand. This means that especially thermal comfort conditions for occupants must be accurately calculated in designs of new buildings or refurbishments of existing buildings to evaluate the energy performance and safeguard the well-being of occupants [4–7].

In buildings with mechanical cooling, the basis for establishing thermal comfort criteria is the use of the PMV-PPD and local thermal discomfort indices [8–10]. PMV is also the basis for energy calculations, as underlined by European Standards EN 16798-1 [11] EN 16798-2 [12]. EN 16798-1 specifies indoor environmental input parameters for design and assessment of energy performance of buildings addressing IEQ, whereas EN 16798-2 explains how to use EN 16798-1 by specifying additional information as: (i) input parameters for building system design and energy performance calculations; (ii) methods for long term evaluation of the indoor environment; (iii) criteria for measurements which can be used if required to measure compliance by inspection; (iv) parameters to be used by monitoring and displaying the indoor environment in existing buildings. With reference to thermal comfort, EN 16798-1 suggests specific design ranges of operative temperature consistent with the desired level of environmental quality (see Table 1).

**Table 1.** Temperature ranges for hourly calculation of cooling and heating energy for some indoor environment (Category II) according to EN 16798-1 and EN 16798-2 Standards [11,12]. Resultant insulation values $I_{cl,r}$ to be used [6] are 0.5 clo (cooling) and 1.0 clo (heating).

| Type of Building Space | Operative Temperature Range for Heating (°C) | Operative Temperature Range for Cooling (°C) |
|---|---|---|
| Offices and spaces with similar activity (single offices, open plan offices, conference rooms, auditorium, cafeteria, restaurants, and classrooms). Sedentary activity: M = 1.2 met | 20.0–24.0 | 23.0–26.0 |
| Department store Standing-walking activity: M = 1.6 met | 16.0–22.0 | 21.0–25.0 |

Finally, PMV is used for the attribution of the class of risk in the prevention of stress or discomfort in thermal working conditions according to ISO 15265 Standard [13,14], and in the field of the thermal bioclimate [15,16] where more specific metrics should be applied [17,18].

*1.2. Open Issues about the Evaluation of PMV/PPD Indices*

In the past, several studies have been undertaken to highlight the limitations of PMV in predicting thermal comfort conditions in naturally ventilated buildings (and in hot and humid climates) where adaptation phenomena have to be considered. Despite this interest that has led to the formulation of modified PMV indices (e.g., ePMV [19] and aPMV [20,21]), two issues remain unresolved in the scientific debate: the effect of measurement uncertainties and its calculation.

To calculate PMV and PPD indices, the evaluation/measurement of six variables are required: the air temperature, the mean radiant temperature, the relative humidity, the air velocity, the metabolic rate and finally the clothing insulation [8]. These quantities can be measured or evaluated according to the Standards in the field of the Ergonomics of the Thermal Environment [22–25].

Although technical Standards specify methods, protocol of measurement and accuracy levels [23], due to the sensitivity of PMV to each involved quantity, the uncertainty on its final value can reach 2–3 decimals points on the PMV scale (for each single input variable) as shown in Figure 1. This phenomenon might affect the category of the environmental quality as prescribed by ISO 7730 and EN 16798-1 (See Table 2). This implies that the most accurate measurement methods for the assessment of the variables should be used [26–28].

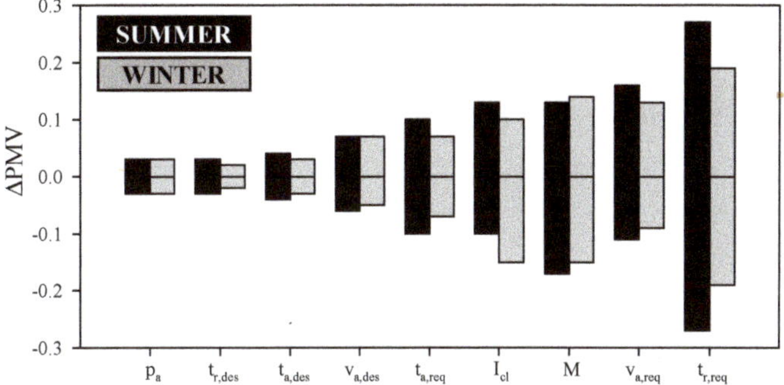



**Figure 1.** PMV sensitivity (ΔPMV) to the accuracy of each quantity required for the thermal environment assessment according to ISO 7730 Standard [9] under thermal neutrality conditions (PMV = 0). M = 1.2 met [20]. "req" and "des" subscripts are referred to the required and desired accuracy levels prescribed by ISO Standard 7726 [23].

**Table 2.** The classification proposed by ISO 7730 [9] and EN 16798-1 [11] Standards.

| Category | | Thermal State of the Body as a Whole | |
|---|---|---|---|
| **ISO 7730** | **EN 16798-1** | **Percentage of Dissatisfied (PPD), %** | **Predicted Mean Vote (PMV)** |
| A | I | <6 | −0.20 < PMV < 0.20 |
| B | II | <10 | −0.50 < PMV < 0.50 |
| C | III | <15 | −0.70 < PMV < 0.70 |
| - | IV | <25 | −1.0 < PMV < 1.0 |

For calculating the PMV, ISO 7730 Standard reports two different procedures:

- Using tables in the ANNEX E of the Standard;
- Using the computer program in BASIC in the Annex A.

To obtain reliable results both procedures (see Figure 2) require some specific conditions often not clearly reported by standards or ignored even by skilled users [29].

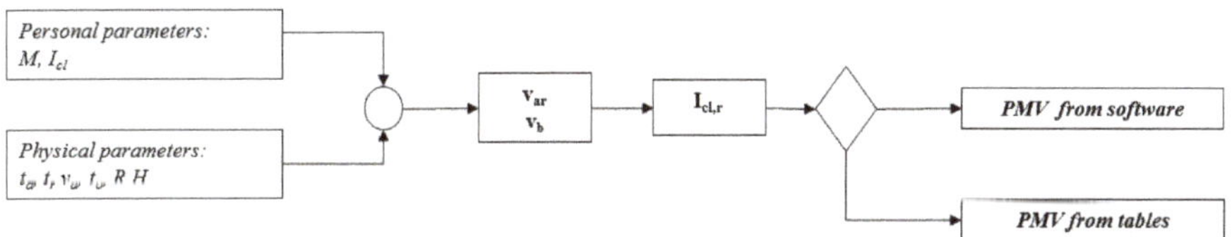



**Figure 2.** Flow chart for the calculation of PMV as required by ISO 7730 Standard [9].

ISO 7730 explicitly states that PMV values given in tables in Annex E only apply to a relative humidity of 50%. However, slight deviations from this reference value do not affect significantly the PMV, due to the relatively small influence of humidity in Fanger's thermal comfort model [30]. In addition, since the input value of tables is the operative temperature [9], the accuracy of PMV values is acceptable provided that the difference between air temperature $t_a$ and mean radiant temperature $t_r$ is less than 5 °C [6,9,31]. Although slight differences between mean radiant temperature and air temperature might result in negligible differences of PMV values, some problems occur in terms of environmental category assignment. As seen from Table 3, while keeping constant the operative temperature value, PMV index can vary as air temperature and mean radiant temperature vary. Consequently, the attribution of the environmental category becomes uncertain [9].

**Table 3.** Effect of the operative temperature on the PMV evaluation. $M$ = 1.2 met, $v_a$ = 0.10 m/s, $R.H.$ = 40% in winter (60% in summer), $I_{cl,r}$ = 1.0 clo in winter (0.5 clo in summer).

| $t_a$ (°C) | $t_r$ (°C) | $t_o$ (°C) | PMV (-) | Category |
|---|---|---|---|---|
| | | **Summer** | | |
| 24 | 28 | | 0.1 | I |
| 25 | 27 | | 0.1 | I |
| 26 | 26 | 26 | 0.2 | II |
| 27 | 25 | | 0.3 | II |
| 28 | 24 | | 0.4 | II |
| 25 | 29 | | 0.4 | II |
| 26 | 28 | | 0.5 | II |
| 27 | 27 | 27 | 0.6 | III |
| 28 | 26 | | 0.6 | III |
| 29 | 25 | | 0.8 | IV |
| | | **Winter** | | |
| 18 | 22 | | −0.8 | IV |
| 19 | 21 | | −0.7 | III |
| 20 | 20 | 20 | −0.6 | III |
| 21 | 19 | | −0.6 | III |
| 22 | 18 | | −0.5 | II |
| 19 | 23 | | −0.5 | II |
| 20 | 22 | | −0.4 | II |
| 21 | 21 | 21 | −0.4 | II |
| 22 | 20 | | −0.3 | II |
| 23 | 19 | | −0.3 | II |

Another very important issue concerns the calculation of the PMV by software. Firstly, ISO 7730 requires the correction of the basic values of the clothing insulation related to the effect of body movements with the algorithms described in ISO 9920 Standard [9,22,32,33]. This is not the case of ASHRAE Standard 55 [10], because the correction is optional and restricted only for moving occupants with the following equation:

$$I_{cl,r} = I_{cl}\left(0.6 + \frac{0.4}{M}\right) \tag{1}$$

with the metabolic rate $M$ expressed in met.

In addition, to evaluate the heat transfer coefficient by convection, ISO and ASHRAE algorithms require the relative air velocity $v_{ar}$ given by:

$$v_{ar} = v_a + v_b \tag{2}$$

where $v_b$ is given by:

$$v_b = 0.0052(M - 58.2) \tag{3}$$

In short, the BASIC program in ISO 7730 requires additional information not clearly reported in the text with unforeseeable consequences in the implementation of software devoted to the calculation of the PMV as recently discussed by our team in a short communication devoted to Fanger's equation [29]. In particular, the wrong calculation of PMV via software [29] results in uncertainties even greater than one point on the ASHRAE thermal sensation scale that is unacceptable if compared to the "physiological" uncertainty due to the measurement precision of input variables.

This applies not only to researchers and scientific studies but also to less skilled users performing the calculation of PMV during an inspection or auditing for conformity checking.

### 1.3. Aim of the Paper

50 years after Fanger's studies, the PMV index remains the most used tool for the objective assessment of indoor thermal comfort. The index is used both for evaluation in existing buildings and in building simulations for the prediction of thermal comfort levels [34]. The lack of clear information in technical Standards and wrong interpretations of the standards [29,35–37] result in increased inaccuracy in its calculation. This also applies to microclimatic dataloggers provided with built-in software.

Until a few years ago ISO and ASHRAE software ran only on Windows platform. Today, the continuous innovation of smartphones and tablets with high performances and unique portability characteristics has favored the release of web applications (web apps) and specific applications for mobile devices (apps) for thermal comfort and heat stress assessment.

Based on the above, in this investigation the reliability and the compliance with International Standards of commonly used software, web apps and apps available in the stores will be verified. This will help both professionals and researchers in the correct use of such tools, which are designed under specific conditions that are often not clearly specified. Finally, the main findings from the present study will be useful for standardization aimed at verification/certification of software.

## 2. Methods

In this study we investigated the most popular apps available on the web and apps available on the market (Apple Store and Google Play), as summarized in Table 4.

**Table 4.** Summary of web apps and apps used for the present investigation.

| Label. | OS | Details | Manufacturer | First Release | Last Update | Last Access |
|---|---|---|---|---|---|---|
| A | Web app | CBE Thermal Comfort Tool [38] | Center for the Built Environment, University of California, Berkeley (USA) | 2014 [39] | 2017 | 11.2019 |
| B | Web app | Java APPLET for ISO 7730 [40] | Lund University, Sweden | 2008 | 2008 | 11.2019 |
| C | iOS | IEQ calculator for apartment | Fishball Studio, Department of Building Services Engineering, Polytechnic University, Hong Kong | 2015 | n.a. | 05.2019 |
| D | iOS | PMV | Zantedeschi System Integrator | n.a. | 2010 | 05.2019 |
| E | iOS | PMV Simulator | Ozaki Seiichi | n.a. | 2013 | 05.2019 |
| F | Android | PMV calculator | Fishball Studio | 2011 | 2011 | 05.2019 |
| G | Android | IEQ calculator for classrooms [41,42] | Fishball Studio, Department of Building Services Engineering, Polytechnic University, Hong Kong | 2012 | n.a. | 05.2019 |

As reference for comparisons, we have used: (i) the values reported in the tables of the Annex E of ISO 7730 Standard; (ii) a software consistent with the code reported in Annex D of EN ISO 7730 Standard (TEE, Thermal Environment Evaluation) [29,43,44]; (iii) the well-known ASHRAE Thermal Comfort Tool [45] validated by ASHRAE and provided with a user-friendly interface for calculating thermal comfort parameters and making thermal comfort predictions.

The microclimatic conditions for comparisons have been based on standards EN 16798-1 and 2 [11,12] that recommend typical values of operative temperature for energy calculation for four categories of Indoor Environmental Quality (see Table 1). The air velocity value used for the investigation was $0.10 \text{ m s}^{-1}$ because it is the minimum value accepted by the ASHRAE Thermal Comfort Tool (despite in several environments lower values can be observed), whereas reference relative humidity values were 40% (60%) for heating (cooling) according to EN 16798-1 [11].

The comparison phase consists of the following steps:

1. PMV calculation
2. Comparison among obtained results
3. Analysis of inconsistent results
4. Attribution of possible causes of inconsistencies

In Table 5 are reported more specific information strictly related to input and output variables considered by each software.

**Table 5.** Input and output data for software and apps used for comparisons. (×) Included; (-) not included. [1] Input value for clothing is the resultant clothing insulation $I_{cl,r}$. [2] Input value is the air speed. [3] Input value is a "generic" clothing level or value. [4] Input value is the relative air velocity, $v_{ar}$. [5] Only specific values for the metabolic rate can be used (e.g., 1.2, 1.8 and 2.0 met). [6] Data sliders move with a random step. [7] Input value is a generic temperature (probably the air temperature). [8] Only one decimal value is accepted. [9] PMV value is rounded to one decimal place. [10] Only integer values are accepted. [11] This app returns only the value of PPD index according to and the sign of the thermal sensation. [12] Does not work on Android 4.4 and later based devices (data sliders do not appear).

| Software or App | Input Data | | | | | | Output Data | | |
| --- | --- | --- | --- | --- | --- | --- | --- | --- | --- |
| | $t_a$ | $t_r$ | RH | $v_a$ | M | $I_{cl}$ | PMV | PPD | Thermal Sensation on the ASHRAE Scale |
| TEE | × | × | × | × | × | × [1] | × | × | × |
| ASHRAE Thermal Comfort Tool 2.0 | × | × | × | × [2] | × | × [3] | × | × | × |
| A | × | × | × | × [2] | × | × [3] | × | × | × |
| B | × | × | × | × [4] | × | × | × | × | - |
| C | × | - | × | - | × [5] | - | - | - | × |
| D [6] | × [7] | - | × | × | × [8] | × [3,8] | × [9] | × | - |
| E | × [10] | × [10] | × [10] | × [4] | × [8] | × [3,8] | - | × [11] | - |
| F [12] | × | × | × | × | × | × | × | × | - |
| G | × | × | × | × | × | - | - | - | × |

It is important to emphasise that only apps A, B, E and F take into account all the variables required for the calculation of the PMV. Apps C and D do not consider the mean radiant temperature (or, probably, they assume $t_a = t_r$). This implies that they are not accurate in non-uniform environments (e.g., near windows or terminal units of HVAC systems) where the difference between air and mean radiant temperature may be significant [6]. The app C does not consider the air velocity and, similarly to the app G does not consider clothing insulation among input variables. In short, apps C, D and G are poorly designed due to the lack of one or more variable necessary to solve the heat balance equation on which PMV is based [7,9,29].

## 3. Results and Discussion

### 3.1. Uniform Environments ($t_r = t_a$)

In Table 6 are summarized the values of the PMV index calculated by means of all investigated apps for the operative temperature values in Table 1 and under homogeneous conditions ($t_a = t_r = t_o$). This hypothesis allows to investigate even apps C and D which consider only one temperature input value. No values have been reported for the app F because it crashes.

According to output data of each app, PMV values reported in Table 6 are those directly obtained only for A, B and D. In case of the app E, which returns only the PPD and the sign of the thermal sensation, the PMV has been calculated from the standard equation [7,9]:

$$PPD = 100 - 95 \cdot \exp\left(-0.3353 PMV^4 - 0.2179 PMV^2\right) \tag{4}$$

As the apps C and G return as output values only the description of the thermal state consistent with the ASHRAE 7-point thermal sensation scale [46,47], in these cases the *PMV* values were attributed by converting them into a thermal sensation vote (e.g., +1 for slightly warm, 0 for neutral and so on) [47].

**Table 6.** PMV values and comfort categories obtained with the investigated software and comparison with values from ISO 7730 tables. Relative velocity has been calculated according to Equation (2). [1] PMV value has been calculated by means of Equation (4). [2] It is not allowed changing the metabolic rate value.

| Input Data | | | | | PMV | | | | | | | | |
|---|---|---|---|---|---|---|---|---|---|---|---|---|---|
| $t_a = t_r$ (°C) | RH (%) | $v_a$ (m/s) | M (met) | $I_{cl,r}$ (clo) | ISO 7730 | TEE | ASH RAE | A | B | C | D | E [1] | G |
| 23.0 | 60 | 0.10 | 1.2 | 0.5 | −0.69 | −0.66 | −0.45 | −0.45 | −0.69 | −1 | −0.7 | −0.78 | +3 |
| 26.0 | | | 1.2 | | 0.24 | 0.31 | 0.46 | 0.46 | 0.28 | +1 | 0.2 | 0.22 | |
| 21.0 | | | 1.6 | | −0.65 | −0.58 | −0.18 | −0.18 | −0.61 | (2) | −0.6 | −0.62 | |
| 25.0 | | | 1.6 | | 0.36 | 0.44 | 0.72 | 0.72 | 0.43 | | 0.4 | 0.44 | |
| 20.0 | 40 | 0.10 | 1.2 | 1.0 | −0.47 | −0.53 | −0.39 | −0.39 | −0.55 | −1 | −0.6 | −0.58 | 0 |
| 24.0 | | | 1.2 | | 0.45 | 0.38 | 0.47 | 0.47 | 0.36 | +1 | 0.3 | 0.31 | +3 |
| 16.0 | | | 1.6 | | −0.63 | −0.65 | −0.38 | −0.37 | −0.67 | (2) | −0.7 | −0.69 | 0 |
| 22.0 | | | 1.6 | | 0.42 | 0.39 | 0.58 | 0.58 | 0.38 | | 0.4 | 0.38 | +3 |
| **Input Data** | | | | | **Thermal Environment Category** | | | | | | | | |
| 23.0 | 60 | 0.10 | 1.2 | 0.5 | III | III | II | II | III | - | IV | IV | - |
| 26.0 | | | 1.2 | | II | II | II | II | II | - | B | B | |
| 21.0 | | | 1.6 | | III | III | I | I | III | (2) | III | III | |
| 25.0 | | | 1.6 | | II | II | IV | IV | II | | II | II | |
| 20.0 | 40 | 0.10 | 1.2 | 1.0 | II | III | II | II | III | - | III | III | I |
| 24.0 | | | 1.2 | | II | II | II | II | II | - | II | II | - |
| 16.0 | | | 1.6 | | III | III | II | II | III | (2) | IV | III | I |
| 22.0 | | | 1.6 | | II | II | III | III | II | | II | II | - |

From a quick analysis of data in Table 6 it seems that only the TEE, app B, and, partially, app E return values compliant with the ISO 7730 tables, especially if the comparison is based on the agreement of the environmental category. The difference between the values obtained by the TEE and the tables is often negligible and consistent with the different values of relative humidity used for our comparison.

The ASHRAE Thermal Comfort Tool and the web app A give values similar to those obtained by using a program consistent with ISO 7730 only at low metabolic rate (M = 1.2 met). At higher metabolic rate value (1.6 met), the PMV values are about 3–4 decimal points higher than those obtained by ISO 7730 tables. A reasonable explanation of this apparent inconsistency could be the input value used by the ASHRAE Comfort Tool for the air velocity. Particularly, the ASHRAE Comfort Tool requires as input value the air speed that, according to ASHRAE Standard 55 [10] is defined as "the rate of air movement at a point without regard to direction". However, according to Fanger's model [8,9], the input value for air velocity is the air velocity relative to the person which includes body movements as expressed by Equation (2). This implies that the overestimation of PMV values at higher metabolic rate could be related to the underestimation of the heat transfer by convection which occurs when air velocity does not take into account body movements.

To verify this hypothesis, we have analysed the difference between PMV values calculated with the ASHRAE Thermal Comfort Tool ($PMV_{ASHRAE}$) and by tables of the Annex E ($PMV_{7730}$). The analysis has been carried out as a function of the operative temperature both in summer ($I_{cl,r}$ = 0.50 clo) and in winter ($I_{cl,r}$ = 1.0 clo) by using as input value the air velocity $v_a$ and the relative air velocity $v_{ar}$ calculated by means of Equation (2). Obtained results are depicted in Figure 3.

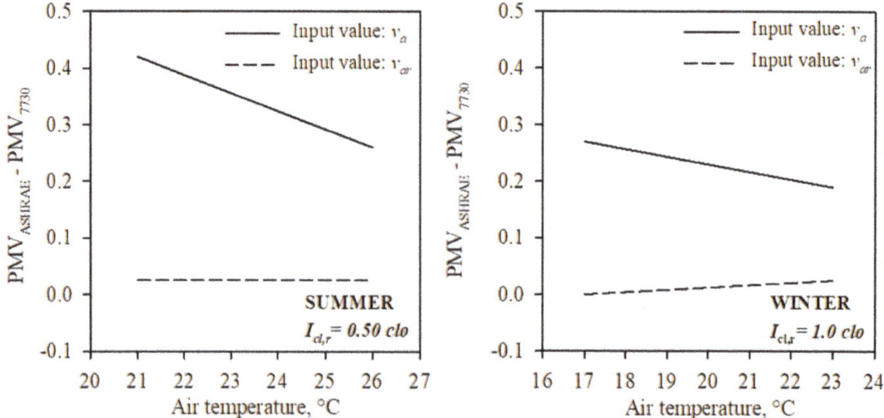

**Figure 3.** Difference between PMV values calculated by means of the ASHRAE Thermal Comfort Tool and tables reported in the Annex E of ISO 7730 Standard by using as input value air velocity (continuous lines) and relative velocity (dashed lines). $M = 1.6$ met; $t_a = t_r = t_o$; $v_a = 0.10$ m s$^{-1}$; $RH = 50\%$.

All plots reported in Figure 3 clearly demonstrate that the ASHRAE Comfort Tool is consistent with tables of ISO 7730 Standard provided that the input value for air velocity is $v_{ar}$. However, in the case of the wrong use of the input value of the air velocity, the overestimation of the PMV value varies from 0.26 to 0.42 in summer and from 0.19 to 0.27 in winter that is of the same order of magnitude of the effect of uncertainty due to the measurements of each microclimatic parameter [26,27].

### 3.2. Non-Uniform Environments ($t_r \neq t_a$)

To verify the reliability of investigated software also under non-uniform conditions ($t_r \neq t_a$), we have calculated the PMV index under the same operative temperature conditions summarized in Table 6 by applying slight differences between mean radiant temperature and air temperature (1 °C and 2 °C).

Results are summarized in Table 7 and show that only app B gives values consistent with those reported in ISO 7730 Standard and those obtained by the TEE. To the contrary, ASHRAE Thermal Comfort Tool and web app A are in agreement with each other only when the effects of the body movements are negligible as observed above (e.g., at low metabolic rate and in winter, when the contribution of the air boundary layer to the total clothing insulation is less significant).

**Table 7.** PMV values and comfort categories obtained with the investigated software and comparison with values from ISO 7730 tables for the operative temperature values in Table 6 with ($t_r$–$t_a$) values of 1 °C and 2 °C. Apps C and D were not considered as they allow only one data input value for the temperature. [(1)] PMV value has been calculated by means of Equation (4).

| Input Data | | | | | | | PMV | | | | | | |
|---|---|---|---|---|---|---|---|---|---|---|---|---|---|
| $t_o$ (°C) | $t_a$ (°C) | $t_r$ (°C) | RH (%) | $v_a$ (m/s) | M (met) | $I_{cl,r}$ (clo) | ISO 7730 | TEE | ASH RAE | A | B | E [(1)] | G |
| 23.0 | 22.0 | 24.0 | | | | | −0.69 | −0.71 | −0.49 | −0.49 | −0.74 | −0.85 | |
| | 22.5 | 23.5 | 60 | 0.10 | 1.2 | 0.5 | −0.69 | −0.71 | −0.46 | −0.46 | −0.71 | −0.75 | |
| 26.0 | 25.0 | 27.0 | | | | | 0.24 | 0.26 | 0.44 | 0.44 | 0.24 | 0.22 | |
| | 25.5 | 26.5 | | | | | 0.24 | 0.28 | 0.45 | 0.45 | 0.26 | 0.22 | +3 |
| 21.0 | 20.0 | 22.0 | | | | | −0.65 | −0.66 | −0.23 | −0.23 | −0.68 | −0.69 | |
| | 20.5 | 21.5 | 60 | 0.10 | 1.6 | 0.5 | −0.65 | −0.62 | −0.20 | −0.20 | −0.64 | −0.54 | |
| 25.0 | 24.0 | 26.0 | | | | | 0.35 | 0.37 | 0.70 | 0.70 | 0.36 | 0.31 | |
| | 24.5 | 25.5 | | | | | 0.35 | 0.41 | 0.71 | 0.71 | 0.39 | 0.54 | |
| 20.0 | 19.0 | 21.0 | | | | | −0.47 | −0.56 | −0.40 | −0.40 | −0.58 | −0.65 | 0 |
| | 19.5 | 20.5 | 40 | 0.10 | 1.2 | 1.0 | −0.47 | −0.54 | −0.39 | −0.39 | −0.56 | −0.49 | 0 |
| 24.0 | 23.0 | 25.0 | | | | | 0.45 | 0.34 | 0.46 | 0.46 | 0.33 | 0.31 | +3 |
| | 23.5 | 24.5 | | | | | 0.45 | 0.36 | 0.46 | 0.46 | 0.34 | 0.44 | +3 |
| 16.0 | 15.0 | 17.0 | | | | | −0.63 | −0.71 | −0.41 | −0.41 | −0.41 | −0.72 | 0 |
| | 15.5 | 16.5 | 40 | 0.10 | 1.6 | 1.0 | −0.63 | −0.68 | −0.39 | −0.39 | −0.39 | −0.62 | 0 |
| 22.0 | 21.0 | 23.0 | | | | | 0.42 | 0.34 | 0.57 | 0.57 | 0.57 | 0.31 | +3 |
| | 21.5 | 22.5 | | | | | 0.42 | 0.37 | 0.57 | 0.57 | 0.57 | 0.44 | +3 |

| Input Data | | | | | | | Thermal Environment Category | | | | | | |
|---|---|---|---|---|---|---|---|---|---|---|---|---|---|
| $t_o$ (°C) | $t_a$ (°C) | $t_r$ (°C) | RH (%) | $v_a$ (m/s) | M (met) | $I_{cl,r}$ (clo) | ISO 7730 | TEE | ASH RAE | A | B | E [(1)] | G |
| 23.0 | 22.0 | 24.0 | | | | | III | IV | II | II | IV | IV | |
| | 22.5 | 23.5 | 60 | 0.10 | 1.2 | 0.5 | III | IV | II | II | IV | IV | |
| 26.0 | 25.0 | 27.0 | | | | | II | II | II | II | II | B | |
| | 25.5 | 26.5 | | | | | II | II | II | II | II | B | − |
| 21.0 | 20.0 | 22.0 | | | | | III | III | II | II | III | III | |
| | 20.5 | 21.5 | 60 | 0.10 | 1.6 | 0.5 | III | III | II | II | III | III | |
| 25.0 | 24.0 | 26.0 | | | | | II | II | IV | IV | II | II | |
| | 24.5 | 25.5 | | | | | II | II | IV | IV | II | II | |
| 20.0 | 19.0 | 21.0 | | | | | II | III | III | III | III | III | I |
| | 19.5 | 20.5 | 40 | 0.10 | 1.2 | 1.0 | II | III | II | II | III | II | I |
| 24.0 | 23.0 | 25.0 | | | | | II | II | II | II | II | II | − |
| | 23.5 | 24.5 | | | | | II | II | II | II | II | II | − |
| 16.0 | 15.0 | 17.0 | | | | | III | IV | II | II | IV | III | I |
| | 15.5 | 16.5 | 40 | 0.10 | 1.6 | 1.0 | III | III | II | II | III | III | I |
| 22.0 | 21.0 | 23.0 | | | | | II | II | III | III | II | II | − |
| | 21.5 | 22.5 | | | | | II | II | III | III | II | II | − |

## 3.3. Clothing Insulation Input Value

The last issue regarding the comparison is devoted to the verification of possible effects of the input value for the clothing insulation [22,32,33]. The PMV values under the microclimatic conditions in Table 6 have been calculated by using as input value the basic clothing insulation $I_{cl}$ instead of the resultant clothing insulation $I_{cl,r}$ as specifically required by ISO 7730 [9]. Results (see Table 8) clearly prove that PMV varies only for the app B, which is the only explicitly based on the basic clothing insulation (see Table 5). This means that only the software designed by the Lund University (B) is compliant with procedures reported in ISO 7730 Standard. Unlike ISO, the ASHRAE Thermal Comfort tool—consistently with ASHRAE 55 [10]—does not take into account both the adjustment of the basic clothing insulation and the relative air velocity, and, consequently, it returns higher PMV values. However, this software can be used provided that the input value for clothing is the resultant clothing insulation and the air speed input value is the relative velocity.

**Table 8.** PMV values and comfort categories calculated with the investigated software and comparison with values from ISO 7730 tables. Relative velocity has been calculated according to Equation (2). [1] PMV value has been calculated by means of Equation (4). [2] It is not allowed changing the metabolic rate value.

| Input Data | | | | | PMV | | | | | | | | |
|---|---|---|---|---|---|---|---|---|---|---|---|---|---|
| $t_a = t_r$ (°C) | RH (%) | $v_a$ (m/s) | M (met) | $I_{cl,r}$ (clo) | ISO 7730 | TEE | ASH RAE | A | B | C | D | E [1] | G |
| 23.0 | | | 1.2 | | −0.69 | −0.66 | −0.45 | −0.45 | −0.69 | −1 | −0.7 | −0.78 | |
| 26.0 | 60 | 0.10 | | 0.5 | 0.24 | 0.31 | 0.46 | 0.46 | 0.28 | +1 | 0.2 | 0.22 | +3 |
| 21.0 | | | 1.6 | | −0.65 | −0.58 | −0.18 | −0.18 | −0.61 | (2) | −0.6 | −0.62 | |
| 25.0 | | | | | 0.36 | 0.44 | 0.72 | 0.72 | 0.43 | | 0.4 | 0.44 | |
| 20.0 | | | 1.2 | | −0.47 | −0.53 | −0.39 | −0.39 | −0.55 | −1 | −0.6 | −0.58 | 0 |
| 24.0 | 40 | 0.10 | | 1.0 | 0.45 | 0.38 | 0.47 | 0.47 | 0.36 | +1 | 0.3 | 0.31 | +3 |
| 16.0 | | | 1.6 | | −0.63 | −0.65 | −0.38 | −0.37 | −0.67 | (2) | −0.7 | −0.69 | 0 |
| 22.0 | | | | | 0.42 | 0.39 | 0.58 | 0.58 | 0.38 | | 0.4 | 0.38 | +3 |

*3.4. Final Observations*

In Table 9 precautions to be adopted when using all investigated software are briefly summarized.

**Table 9.** Summary of precautions to be adopted when using investigated software or apps.

| Software or App | Precautions |
|---|---|
| ASHRAE Thermal Comfort Tool | • Air speed is the relative air speed<br>• Clothing insulation input value is the basic clothing insulation value (ASHRAE 55) or the resultant clothing insulation (ISO 7730) |
| A | As above |
| B | No precautions |
| C | Not reliable |
| D | Reliable only in uniform environments ($t_a = t_r = t_o$) |
| E | Clothing insulation input value is the basic clothing insulation value (ASHRAE 55) or the resultant clothing or the resultant clothing insulation (ISO 7730) |
| F | Unable to be tested |
| G | Not reliable |

Unfortunately, no further discussion can be provided for the other investigated software because their algorithms and listings are not accessible to users.

These results also rise some interesting questions that cannot be easily answered:

1. Is the assessment of thermal comfort conditions easy enough to be carried out merely with a software, without experience and ergonomic skills?
2. Are smartphones suitable replacements for commercial equipment and able to measure all the needed physical variables (particularly, the air velocity and the mean radiant temperature)?
3. With accurate commercial equipment, are the standards defined completely enough and clearly enough that non experts can assess thermal comfort accurately?

## 4. Conclusions

Software for the evaluation of thermal comfort conditions by means of PMV index, presently on the market, have to be used with special care. This is also for software designed by Academics and posted on Universities web pages in the absence of clear specifications for input values (e.g., the relative air velocity and the resultant clothing insulation). In addition, most apps do not consider one or more variables affecting the thermal sensation. Regarding the mean radiant temperature this means that calculated values of PMV index are reliable only in homogeneous environments (or in the presence of very small differences between air temperature and mean radiant temperature). This is not acceptable, especially in environments with large windows or close to HVAC terminal units.

Software developers should be aware that all computer programs available in annexes of international Standards give correct results only when all specifications are considered. In particular, ISO 7730 Standard requires that basic clothing insulation values have to be preliminary adjusted for wind and body movements.

Professional and experts should be aware of the risks related to the use of unvalidated evaluation tools. This is even more important in case of cold and hot environments, where the goal of the thermal environment assessment is the safety of working conditions.

The analysis reported in this paper is not able to provide further details about the reasons why some of the investigated software fail, because their code is inaccessible to the users (closed code).

This investigation has demonstrated that the unreliability of software/apps is mainly related to a wrong interpretation of International Standards and not to inaccurate coding. This implies that International Standards should be written to avoid ambiguous or undefined input parameters (i.e., relative air velocity) resulting in inaccurate tools.

Finally, this study has emphasized that the assessment of the thermal environments requires specific skills and robust tools. Software and apps are only the final step of the objective measurement of the variables responsible for the thermal sensation. This is another crucial issue, because all measurement devices required for the objective survey are impossible to miniaturize in a smartphone. However, beyond the intrinsic limitations of the PMV model and of measurement uncertainties, it is unthinkable that the main cause of unreliable assessments might be caused by poor software implementations of the standard.

**Author Contributions:** All authors contributed in equal amounts to the whole research activity here discussed. F.R.d.A. and B.I.P. designed this research; the numerical campaign has been carried out by G.R. and D.P., finally B.W.O. supervised, reviewed and edited the manuscript. All authors have read and agreed to the published version of the manuscript.

**Funding:** This research has been carried out within the "Renovation of existing buildings in NZEB vision (nearly Zero Energy Buildings)" Project of National Interest (Progetto di Ricerca di Interesse Nazionale - PRIN) funded by the Italian Ministry of Education, Universities and Research (MIUR), Protocol no. 2015S7E247_005.

**Conflicts of Interest:** The authors declare no conflict of interest.

## Acronyms and Symbols

| | |
|---|---|
| $I_{cl}$ | Basic clothing insulation, $m^2$ K $W^{-1}$ or clo |
| $I_{cl,r}$ | Resultant clothing insulation adjusted for wind and body movements, $m^2$ K $W^{-1}$ or clo |
| M | Metabolic rate, W $m^{-2}$ or met |
| nZEB | Nearly Zero Energy Buildings |
| $p_a$ | Water vapor partial pressure, Pa |
| $RH$ | Relative humidity, % |
| PMV | Predicted Mean Vote, 1 |
| $PMV_{ASHRAE}$ | PMV value calculated by means of the ASHRAE Thermal Comfort Tool, 1 |
| $PMV_{7730}$ | PMV value calculated from tables in the Annex E of ISO 7730 Standard, 1 |
| PPD | Predicted Percentage of Dissatisfied, % |
| $t_a$ | Air temperature, °C |
| $t_{a,des}$ | Air temperature measured within desired accuracy prescribed by ISO 7726, °C |
| $t_{a,req}$ | Air temperature measured within required accuracy prescribed by ISO 7726, °C |
| $t_o$ | Operative temperature, °C |
| $t_r$ | Mean radiant temperature, °C |
| $t_{r,des}$ | Mean radiant temperature measured within desired accuracy prescribed by ISO 7726, °C |
| $t_{r,req}$ | Mean radiant temperature measured within required accuracy prescribed by ISO 7726, °C |

| | |
|---|---|
| $v_a$ | Absolute air velocity, m s$^{-1}$ |
| $v_{a,des}$ | Absolute air velocity measured within desired accuracy prescribed by ISO 7726, m s$^{-1}$ |
| $v_{ar}$ | Relative air velocity, m s$^{-1}$ |
| $v_{a,req}$ | Absolute air velocity measured within required accuracy prescribed by ISO 7726, m s$^{-1}$ |
| $v_b$ | Velocity due to body movement, m s$^{-1}$ |

## References


1. Buildings Performance Institute Europe. *Europe's Buildings under the Microscope*; Buildings Performance Institute Europe: Brussels, Belgium, 2011.
2. Ruparathna, R.; Hewage, K.; Sadiq, R. Improving the energy efficiency of the existing building stock: A critical review of commercial and institutional buildings. *Renew. Sustain. Energy Rev.* **2016**, *53*, 1032–1045. [CrossRef]
3. Parliament of the European Union. Council directive of 16 December 2002 on the energy performance of buildings (2002/91/EC). *Off. J. Eur. Commun.* **2003**, *1*, 65–71.
4. Parliament of the European Union. Directive 2018/844/EU of the European Parliament and of the council of 30 may 2018 amending directive 2010/31/EU on the energy performance of buildings and directive 2012/27/EU on energy efficiency. *Off. J. Eur. Commun.* **2018**, *156*, 75–91.
5. US Environmental Protection Agency. Available online: https://www.epa.gov/sites/production/files/2015-08/documents/vision.pdf (accessed on 25 November 2019).
6. d'Ambrosio Alfano, F.R.; Olesen, B.W.; Palella, B.I.; Riccio, G. Thermal comfort: Design and assessment of energy saving. *Energy Build.* **2014**, *81*, 326–336. [CrossRef]
7. De Santoli, L.; Garcia, D.A.; Groppi, D.; Bellia, L.; Palella, B.I.; Riccio, G.; Cuccurullo, G.; d'Ambrosio Alfano, F.R.; Stabile, L.; Dell'Isola, M.; et al. A general approach for retrofit of existing buildings towards NZEB: The windows retrofit effects on indoor air quality and the use of low temperature district heating. In Proceedings of the 2018 IEEE International Conference on Environment and Electrical Engineering and 2018 IEEE Industrial and Commercial Power Systems Europe, EEEIC/I and CPS Europe 2018, Palermo, Italy, 12–15 June 2018. [CrossRef]
8. Fanger, P.O. *Thermal Comfort*; Danish Technical Press: Copenhagen, Denmark, 1970.
9. ISO. *ISO Standard 7730. Ergonomics of the Thermal Environment—Analytical Determination and Interpretation of Thermal Comfort Using Calculation of the PMV and PPD Indices and Local Thermal Comfort*; International Organization for Standardization: Geneva, Switzerland, 2005.
10. ASHRAE. *ANSI/ASHRAE Standard 55. Thermal Environmental Conditions for Human Occupancy*; American Society of Heating, Refrigerating and Air Conditioning Engineers: Atlanta, GA, USA, 2017.
11. CEN EN Standard 16798-1. *Indoor Environmental Input Parameters for Design and Assessment of Energy Performance of Buildings Addressing Indoor Air Quality, Thermal Environment, Lighting and Acoustics—Module M1-6*; European Committee for Standardization: Brussels, Belgium, 2019.
12. CEN EN Standard 16798-2. *Energy Performance of Buildings—Ventilation for Buildings—Part. 2: Interpretation of the Requirements in EN 16798-1—Indoor Environmental Input Parameters for Design and Assessment of Energy Performance of Buildings Addressing Indoor Air Quality, Thermal Environment, Lighting and Acoustics—Module M1-6*; European Committee for Standardization: Brussels, Belgium, 2019.
13. ISO. *ISO Standard 15265. Ergonomics of the Thermal Environment—Risk Assessment Strategy for the Prevention of Stress or Discomfort in Thermal Working Conditions*; International Organization for Standardization: Geneva, Switzerland, 2004.
14. d'Ambrosio Alfano, F.R.; Palella, B.I.; Riccio, G. On the transition thermal discomfort to heat stress as a function of the PMV value. *Ind. Health* **2013**, *51*, 285–296. [CrossRef]
15. Matzarakis, A.; Rutz, F.; Mayer, H. Modelling radiation fluxes in simple and complex environments—Application of the RayMan model. *Int. J. Biometeorol.* **2007**, *51*, 323–334. [CrossRef]
16. de Freitas, C.R.; Grigorieva, E.A. A comprehensive catalogue and classification of human thermal climate indices. *Int. J. Biometeorol.* **2015**, *59*, 109–120. [CrossRef]
17. Hoppe, P. The physiological equivalent temperature—A universal index for the biometeorological assessment of the thermal environment. *Int. J. Biometeorol.* **1999**, *43*, 71–75. [CrossRef]

18. Bröde, P.; Błazejczyk, K.; Fiala, D.; Havenith, G.; Holmér, I.; Jendritzky, G.; Kuklane, K.; Kampmann, B. The universal thermal climate index UTCI compared to ergonomics standards for assessing the thermal environment. *Ind. Health* **2013**, *51*, 16–24. [CrossRef]

19. Fanger, P.O.; Toftum, J. Extension of the PMV model to non-air-conditioned buildings in warm climates. *Energy Build.* **2002**, *34*, 533–536. [CrossRef]

20. Yao, R.; Li, B.; Liu, J. A theoretical adaptive model of thermal comfort—Adaptive predicted mean vote (aPMV). *Build. Environ.* **2009**, *44*, 2089–2096. [CrossRef]

21. Carlucci, S.; Bai, L.; de Dear, R.; Yang, L. Review of adaptive thermal comfort models in built environmental regulatory documents. *Build. Environ.* **2018**, *137*, 73–89. [CrossRef]

22. ISO. *ISO Standard 9920. Ergonomics of the Thermal Environment—Estimation of the Thermal Insulation and Evaporative Resistance of a Clothing Ensemble*; International Organization for Standardization: Geneva, Switzerland, 2009.

23. ISO. *ISO Standard 7726. Ergonomics of the Thermal Environment—Instruments for Measuring Physical Quantities*; International Organization for Standardization: Geneva, Switzerland, 1998.

24. ISO. *ISO Standard 8996. Ergonomics of the Thermal Environment—Determination of Metabolic Rate*; International Organization for Standardization: Geneva, Switzerland, 2004.

25. Malchaire, J.; d'Ambrosio Alfano, F.R.; Palella, B.I. Evaluation of the metabolic rate based on the recording of the heart rate. *Ind. Health* **2017**, *55*, 219–232. [CrossRef] [PubMed]

26. d'Ambrosio Alfano, F.R.; Palella, B.I.; Riccio, G. The role of measurement accuracy on the thermal environment assessment by means of PMV index. *Build. Environ.* **2011**, *46*, 1361–1369. [CrossRef]

27. Ricciu, R.; Galatioto, A.; Desogus, G.; Besalduch, L.A. Uncertainty in the evaluation of the Predicted Mean Vote index using Monte Carlo analysis. *J. Environ. Manag.* **2018**, *223*, 16–22. [CrossRef]

28. d'Ambrosio Alfano, F.R.; Dell'Isola, M.; Palella, B.I.; Riccio, G.; Russi, A. On the measurement of the mean radiant temperature and its infuence on the indoor thermal environment assessment. *Build. Environ.* **2013**, *63*, 79–88. [CrossRef]

29. d'Ambrosio Alfano, F.R.; Palella, B.I.; Riccio, G.; Toftum, J. Fifty years of Fanger's equation: Is there anything to discover yet? *Int. J. Ind. Ergon.* **2018**, *66*, 157–160. [CrossRef]

30. Fanger, P.O. Air humidity, Comfort and Health. In Proceedings of the CISCO-ITBTP Seminar Humidity in building, Saint Remy les Chevreuse, France, November 1982; pp. 192–195.

31. Alfano, G.; Cannistraro, G.; d'Ambrosio, F.R.; Rizzo, G. Notes on the use of the tables of standard ISO 7730 for the evaluation of the PMV index. *Indoor Built Environ.* **1996**, *5*, 355–357. [CrossRef]

32. Havenith, G.; Nilsson, H. Correction of clothing insulation for movement and wind effects, a meta-analysis. *Eur. J. Appl. Physiol.* **2004**, *92*, 636–640. [CrossRef]

33. d'Ambrosio Alfano, F.R.; Palella, B.I.; Riccio, G.; Malchaire, J. On the effect of thermophysical properties of clothing on the heat strain predicted by PHS model. *Ann. Occup. Hygiene* **2016**, *60*, 231–251. [CrossRef]

34. U.S. Department of Energy's (DOE) Building Technologies Office (BTO). EnergyPlus 9.2.0. 2019. Available online: https://www.energyplus.net/sites/default/files/docs/site_v8.3.0/EMS_Application_Guide/EMS_Application_Guide/index.html (accessed on 19 November 2019).

35. Broday, E.E., de Paula Xavier, A A.; de Oliveira, R. Comparative analysis of methods for determining the metabolic rate in order to provide a balance between man and the environment. *Int. J. Ind. Ergon.* **2014**, *44*, 570–580. [CrossRef]

36. Broday, E.E.; de Paula Xavier, A.A.; de Oliveira, R. Comparative analysis of methods for determining the clothing surface temperature ($t_{cl}$) in order to provide a balance between man and the Environment. *Int. J. Ind. Ergon.* **2017**, *57*, 80–87. [CrossRef]

37. Broday, E.E.; Moreto, J.A.; Xavier, A.A.D.P.; de Oliveira, R. The approximation between thermal sensation votes (TSV) and predicted mean vote (PMV): A comparative analysis. *Int. J. Ind. Ergon.* **2019**, *69*, 1–8. [CrossRef]

38. Center for the Built Environment, University of California Berkeley. CBE Thermal Comfort Tool. Available online: http://comfort.cbe.berkeley.edu/ (accessed on 19 November 2019).

39. Schiavon, S.; Hoyt, T.; Piccioli, A. Web application for thermal comfort visualization and calculation according to ASHRAE Standard 55. *Build. Simul.* **2014**, *7*, 321–334. [CrossRef]

40. Lund University Faculty of Engineering. JAVA Applet for ISO 7730 Calculation of Predicted Mean Vote (PMV), and Predicted Percentage Dissatisfied (PPD), PMV 2008 Ver 1.0. Available online: http://www.eat.lth.se/fileadmin/eat/Termisk_miljoe/PMV-PPD.html (accessed on 19 November 2019).

41. Lee, M.C.; Mui, K.W.; Wong, L.T.; Chan, W.Y.; Lee, E.W.M.; Cheung, C.T. Student learning performance and indoor environmental quality (IEQ) in air-conditioned university teaching rooms. *Build. Environ.* **2012**, *49*, 238–244. [CrossRef]

42. Mui, K.W.; Wong, L.T.; Cheung, C.T.; Yu, H.C. An application-based indoor environmental quality (IEQ) calculator for residential buildings. *Int. J. Arch. Environ. Eng.* **2015**, *9*, 822–825.

43. d'Ambrosio Alfano, F.R.; Palella, B.I.; Riccio, G. TEE (thermal environment assessment): A friendly tool for thermal environment evaluation. In Proceedings of the 11th International Conferences on Environmental Ergonomics, Ystad, Sweden, 22–26 May 2005; pp. 503–506.

44. d'Ambrosio Alfano, F.R.; Palella, B.I.; Riccio, G. Notes on the implementation of the IREQ model for the assessment of extreme cold environments. *Ergonomics* **2013**, *56*, 707–724. [CrossRef]

45. American Society of Heating, Refrigerating and Air Conditioning Engineers. *ASHRAE Thermal Comfort Tool Version 2*; American Society of Heating, Refrigerating and Air Conditioning Engineers: Atlanta, GA, USA, 2013.

46. ISO. *ISO Standard 10551. Ergonomics of the Thermal Environment—Subjective Judgement Scales for Assessing Physical Environments*; International Organization for Standardization: Geneva, Switzerland, 2019.

47. American Society of Heating, Refrigerating and Air Conditioning Engineers. Thermal comfort. In *ASHRAE Handbook of Fundamentals*; American Society of Heating, Refrigerating and Air Conditioning Engineers: Atlanta, GA, USA, 2017.

 *atmosphere*

Review

# An Extensive Collection of Evaluation Indicators to Assess Occupants' Health and Comfort in Indoor Environment

**Fabio Fantozzi and Michele Rocca ***

Department of Energy, Systems, Territory and Constructions Engineering, University of Pisa, 56122 Pisa, Italy; f.fantozzi@ing.unipi.it
* Correspondence: michele.rocca.au@gmail.com; Tel.: +39-050-221-7104

Received: 26 November 2019; Accepted: 4 January 2020; Published: 12 January 2020

**Abstract:** Today, the effects of the indoor environment on occupants' health and comfort represent a very important topic and requires a holistic approach in which the four main environmental factors (thermal comfort, air quality, acoustics, and lighting) should be simultaneously assessed. The present paper shows the results of a literature survey that aimed to collect the indicators for the evaluation of occupants' health and comfort in indoor environmental quality evaluations. A broad number of papers that propose the indicators of a specific environmental factor is available in the scientific literature, but a review that collects the indicators of all four factors is lacking. In this review paper, the difference between indicators for the evaluation of risk for human health and for comfort evaluation is clarified. For each environmental factor, the risk for human health indicators are proposed with the relative threshold values, and the human comfort indicators are grouped into categories according to the number of parameters included, or the specific field of application for which they are proposed. Furthermore, the differences between human health and comfort indicators are highlighted.

**Keywords:** health and comfort; evaluation indicators; work environments; indoor environmental quality; indoor comfort; human health

---

## 1. Introduction

Nowadays, the awareness of the relevance of obtaining buildings with high performance to reduce both energy consumption and the impact on the environment ($CO_2$ emissions into the atmosphere), have been collectively reached. The challenge is now to guarantee a high level of housing quality while maintaining low energy consumption. People spend almost 90% of their life in indoor environments (houses, schools, work environments, etc.) and the effects of indoor conditions on human health cannot be ignored. The interest in buildings that guarantee high levels of occupant health is increasing both in research and in professional practice. In different areas (i.e., health, economy, etc.), comfort, well-being, and quality of life are becoming increasingly important [1,2]. The built environment can be interpreted as a physical and social environment that must guarantee the environmental conditions to promote wellbeing, health, productivity, and interactions between people. To this aim, at the European level, new building certification assets have already been developed; they include not only the assessment of energy performance, but also the levels of comfort and well-being that can be achieved within the indoor environment [3–6]. Health is defined by the World Health Organization (WHO) as "a state of complete physical, mental, and social well-being and not merely the absence of disease or infirmity" [7]. Corresponding to this definition, there is a change of the public health focus from life expectancy to health expectancy, which no longer includes mortality in favor of aspects of quality of life [8,9].

Indoor environmental quality (IEQ) assessments, with particular reference to work environments, should combine the evaluation of all the possible factors that can have negative effects on health with the evaluation of the perceived levels of comfort [10]. New holistic approaches in which building energy performance is combined with the indoor environmental conditions, are increasingly used [11–14].

The health risks evaluation is generally carried out using "pass/fail criteria" purely quantitative that generally involve comparisons between the measured or calculated parameters and the related limits (threshold values).

The comfort evaluation includes quantitative, qualitative, and subjective investigations. In this field, the human perception is particularly relevant and cannot be neglected. Comfort assessment is generally performed through the evaluation of the IEQ, that is based on four environmental factors that simultaneously affect the human perception, namely: thermal environment, indoor air quality, acoustical environment, and visual environment [1,15–18].

Concerning the effects of the indoor environment on human health, in addition to the building-related illnesses caused by specific exposures in indoor environments (e.g., rhinitis, asthma, and hypersensitivity pneumonitis, etc.), sick building syndrome (SBS) consists of a group of mucosal, skin, and general symptoms that are temporally related to working in particular buildings [19,20]. The SBS is characterized by "non-specific" effects and its signs can be highly variable, affecting diverse parts of the human body, and correlating them to SBS could be challenging in the first place [21]. The SBS was introduced the first time by the WHO in 1983 [22], and it is well demonstrated that sick buildings can also induce stress, anxiety, and aggression [23,24]. These negative effects can further result in an increase in the possibility of hazardous events in workplaces [21].

Another important point is the selection of evaluation indicators. There are numerous indicators in the scientific literature, but the distinction between health and comfort indicators is not always clear. Concerning the thermal environment among the huge amount of indicators proposed in the literature, only five are normalized: Wet Bulb Globe Temperature (WBGT), required clothing insulation (IREQ), Predicted Heat Strain (PHS), and Predicted Mean Vote (PMV) and Predicted Percentage of Dissatisfied (PPD) for the thermal comfort [25,26]. On the contrary, concerning the visual environment, specific indicators for the assessment of the risk arising from a non-adequate lighting exposure do not exist, and the same indicators are used for both visual comfort and improper light exposure assessments. Moreover, only some indicators (e.g., PMV, PPD, etc.) can be used alone because they combine different physical parameters and provide overall information on the human perception of a specific environmental factor. The other indicators (e.g., illuminance, reverberation time, etc.) provide detailed information and it is necessary to combine the results of more indicators to obtain an overall evaluation on environmental factor.

In the literature, articles are available that collect indicators of individual environmental factors (i.e., [27–31]); however, there is no review to collect indicators of multiple environmental factors. For this reason, and given the growing interest in overall assessments, the authors decided to collect the indicators of all four main IEQ environmental factors: thermal environment, indoor air quality, acoustics and lighting. The objective of this paper is to create a comprehensive overview of indicators for assessing human health and comfort in the indoor environment. This collection could also be a useful starting point for those who must approach global assessments of the indoor environment. The present paper is composed of a first section on health and comfort evaluations where some general indications are provided, four further sections, one for each main environmental factor, and the conclusive remarks. The four sections on the environmental factors are developed following the same structure. The first part is related to the risk for human health evaluation and it is composed of an "Overview", a framework of "Guideline and legislative outline", and a collection of the "Indicators". The second part is related to the comfort aspects and it is composed of an "Overview", a collection of "Indicators", and some information on the "Current research trends". For both risk and comfort, the most commonly used indicators are briefly described, and the main aspects discussed. With regard to the risk assessment for human health, together with the most important risk indicators, the reference limit values (minimum/threshold values) that must be guarantee at international level are reported. As regards the assessment of comfort, the indicators are grouped into categories according to the number of parameters included, or the specific field of application for which they were proposed.

## 2. Literature Search

The literature research started by separately search in the scientific databases "ScienceDirect", "MDPI", "Web of Science" (WOS) and "Google Scholar" the terms related to health and comfort evaluation of the four main environmental factors: "thermal comfort", "thermal stress" "indoor air quality", "indoor air pollution", acoustics comfort", "noise exposure", "visual comfort", "visual fatigue". The summary of the results of this search are reported in Table 1. It is possible to observe that "thermal comfort" and "indoor air quality" are the most studied topics, while "acoustic comfort" is the subject on which it was less published. However, it should be considered that, although the terms "thermal comfort" and "indoor air quality" are the specific terms uniquely used to refer to the relative factor, "visual comfort" and "acoustic comfort" are not always the only terms used. In fact, the term "acoustic comfort" is sometimes replaced by "aural comfort" or "sound quality".

**Table 1.** Results of general literature search in different scientific databases.

| | Google Scholar | Science Direct | WOS | MDPI |
|---|---|---|---|---|
| Search in | All (not optional) | Title, abstract, keywords | Title, abstract, keywords | Title, keywords |
| Sort type | Relevance (not optional) | Relevance | Time cited | Time cited |
| Meaning of classification | Publisher, authors, number of citations, recent citations | Highest occurrence of search item | Highest number of citation | Highest number of citation |
| keywords | Number of results (reviews) | | | |
| "Thermal comfort" | 185,000 | 20,831 (1201) | 12,830 (495) | 466 |
| "Thermal stress" | 901,000 | 75,807 (2549) | 20,113 (401) | 254 |
| "Indoor air quality" | 271,000 | 14,551 (797) | 8046 (346) | 290 |
| "Indoor air pollution" | 94,800 | 5871 (603) | 3086 (286) | 64 |
| "Acoustic comfort" | 9100 | 881 (62) | 375 (9) | 21 |
| "Noise exposure" | 122,000 | 6897 (500) | 4962 (253) | 103 |
| "Visual comfort" | 25,000 | 2646 (187) | 1290 (52) | 43 |
| "Visual fatigue" | 18,100 | 1145 (36) | 825 (16) | 9 |

Concerning the databases, it can be noted that a higher number of papers was obtained from Google Scholar, and this is also due to the impossibility to set the search option. In Science Direct and Web of Science (WOS) it was also possible to distinguish the paper type and it is was useful to easily identify the review papers.

Among these papers, the most "relevant" and the most "cited" (results sort type) were selected and reviewed by reading title, abstract and keywords. Special attention was paid on the review articles (numbers in brackets in Table 1) because they provided already organized information. Then the bibliographies of such papers were examined, in order to identify possible other interesting researches not present in the previous results. Subsequently, using the references of the consulted papers, a search was carried out aimed at identifying the original papers in which these indicators were presented for the first time.

At the same time, the websites and publications (guidelines) on these issues of the main international organizations such as the World Health Organization (WHO) and the European Agency for Safety and Health and Work (EU-OSHA) were consulted, as well as the European Directives and international standards. Overall, around 300 documents were collected which were the subject of this review article.

## 3. Health and Comfort Evaluation

Nowadays, it has become essential to guarantee high standards of occupants' health and comfort in indoor environments (especially in the workplaces) [9,32–35]. To achieve high levels of IEQ, it is necessary to consider that the human perception of the indoor environment is affected by four basic environmental factors: thermal environment, air, acoustics, and lighting. The evaluation of each factor should be different in relation to the purpose of the investigation. Indeed, if it is necessary to consider that two different approach are generally followed for the assessment of the risk for human health and the evaluation of the human comfort.

The assessment of human health risks regards the human exposure to potentially dangerous conditions; its evaluation must include the containment measures and the risk prevention activities that are considered primary interventions. The safety of occupants represents an essential need for all environments. The aspects involved in the occupants' safety must be never overlooked. The evaluation of the risk for human health shall be generally carried out measuring some specific parameters and comparing the obtained values with the threshold values provided by the national or international standards and laws. The comparison (pass/fail) is purely quantitative.

In the ASHRAE TC 1.6 (Terminology), the IEQ is defined as "a perceived indoor experience about the building indoor environment that includes aspects of design, analysis, and operation of energy efficient, healthy, and comfortable buildings" [36]. The effects of problems related to poor IEQ levels on occupant comfort, health, well-being, and productivity were clearly demonstrated by the scientific literature [37–39]. These problems can also have consequences on the lifecycle costs and on the additional energy consumption due to the attempts to compensate for the non-adequate environmental conditions [40].

Although to guarantee a risk exempt environment is a basic need, it is not sufficient to guarantee a high level of well-being that instead can be achieved only considering the comfort perceived by the occupants, too. Figure 1 represents a map for reading the rest of the paper: each of the next sections is dedicated to an environmental factor, and for each of them, the indicators relating to comfort and risk to human health are reported and classified.

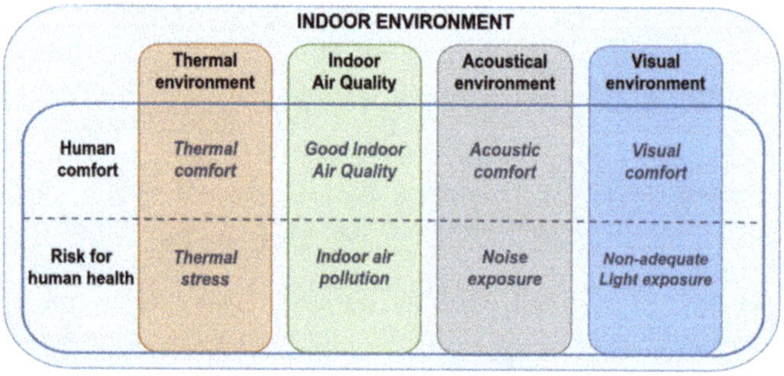

**INDOOR ENVIRONMENT**

| | Thermal environment | Indoor Air Quality | Acoustical environment | Visual environment |
|---|---|---|---|---|
| **Human comfort** | *Thermal comfort* | *Good Indoor Air Quality* | *Acoustic comfort* | *Visual comfort* |
| **Risk for human health** | *Thermal stress* | *Indoor air pollution* | *Noise exposure* | *Non-adequate Light exposure* |

**Figure 1.** Main aspects related to the comfort and to the risk for human health for the four IEQ environmental factors.

With regards to comfort and health, in the ASHRAE guideline 10-2016 "Interactions Affecting the Achievement of Acceptable Indoor Environments" [1], the four main environmental factors and the interactions among environmental conditions are treated, highlighting the importance to not limit the analysis to the single environmental factor. Furthermore, "in order to provide an acceptable indoor environment, it is necessary not only that each aspect of the environment be at a satisfactory level but also that the adverse impact of interactions between these aspects is limited" [1]. The environmental factors do not necessarily have equal importance (weight), but in any case, if an environmental factor is rated as least satisfactory, it is also likely to be rated as the most relevant one. It is therefore essential that all aspects involved in the four environmental factors should be considered satisfactory in order not adversely affect the overall satisfaction. In the overall evaluations of the indoor environment, it is therefore important not to limit the assessments to quantitative evaluations only, but also to include qualitative considerations.

## 4. Thermal Environment

The thermal environment involves such different environmental aspects as air temperature, surrounding surface temperatures, air speed and humidity. Furthermore, the human perception of thermal comfort depends not only on physical factors, but also on metabolic activity, clothing, and personal preference. As stated by Parsons [41]: "The challenge for every person is to successfully interact with his or her local environment. The human body responds to environmental variables in a dynamic interaction that can lead to death if the response is inappropriate, or if energy levels are beyond survivable limits, and it determines the strain on the body as it uses its resources to maintain an optimum state. In the case of the thermal environment this will determine whether a person is too hot, too cold or in thermal comfort. Air temperature, radiant temperature, humidity, and air movement are the four basic environmental variables that affect human response to thermal environments. Combined with the metabolic heat generated by human activity and clothing worn by a person, they provide the six fundamental factors (sometimes called the six basic parameters as they vary in space and time but fixed representative values are often used in analysis) that define human thermal environments".

### 4.1. Thermal Stress

### 4.1.1. Overview

Thermal stress can severely affect human health, but it can also reduce attention span, increasing the risk of injury. Exposure to inadequate thermal conditions can reduce productivity and tolerance to other environmental risks. Thermal stress represents a relevant factor in many industrial situations, sports events and military scenarios [42]. In the industrial context, when people are exposed to extremely hot or cold environments, thermal discomfort is considered as one of the major causes of dissatisfaction [43]. People who work in severe hot or cold environments may lose the ability to make decisions and/or perform manual operations, and therefore the likelihood that they will behave unsafely increases. When it becomes difficult to control internal body temperature, heat stress occurs.

### 4.1.2. Guidelines and Legislative Outline

The control of thermal risks in workplaces is not included in any specific EU Directives, although such risks are covered by the general provisions of the Framework Directive (89/391/EEC) [44,45]. In any case, general recommendations on the protection of workers from thermal risks and against thermal hazards are provided in the guidelines of Workplace Safety and Health Services of the occupational national health authorities [46]. All these guidelines suggest that the temperature in the work environments should be adequate for the human body during all the work activities taking into account the physical effort of the workers. It is not sufficient consider only the effects of the temperature, but is necessary to take into consideration also the influence of humidity and ventilation in the perception of the work environment. In fact, in addition to the air temperature, other factors can help to minimize both the heat stress and the thermal discomfort. When it is not convenient to modify the temperature of the entire environment, workers must be protected against too high or too low temperatures by means of collective or individual protection devices [47]. Furthermore, Malchaire et al. [48] developed a strategy to assess the risks of heat disturbance in any work situation. This strategy is based on the three highest phases of the SOBANE strategy: "observation" to improve the thermal conditions of work; "Analysis" to assess the extent of the problem and optimize the choice of solutions; "Expert", when necessary, for an in-depth analysis of the work situation [49,50].

### 4.1.3. Indicators

The creation of a single indicator that allows the evaluation of the perceived thermal stress in a wide range of environmental conditions and physiological activities required numerous attempts [42]. In the Table 2, according to [27,42,51], the most important thermal stress indicators proposed since 1945 are summarized and divided according to their application in hot or cold environments.

Among the thermal stress assessment indicators for hot environments, the most common is the wet bulb globe temperature (WBGT). The WBGT was introduced by the US Navy within a study on accidents related to heat during military training [52], and it was derived from the "corrected effective temperature" (CET). The WBGT comprises the weighed combination of the dry bulb (dry-bulb) (Ta), the wet bulb (wet-bulb) (Tw), and the black-globe (Tg) temperatures. Even if the WBGT has no physiological correlations, weighted coefficients were obtained empirically. The reference values of the WBGT are reported in Table 3.

**Table 2.** A collection of thermal stress indicators.

| Reference | Year | Index | Symbol | Application |
|---|---|---|---|---|
| [53] | 1945 | Wind Chill Index | WCI | |
| [54] | 1984 | Required Clothing Insulation | IREQ | |
| [55] | 1987 | Survival time outdoor in extreme cold | STOEC | Cold Environment |
| [28] | 1988 | Wind-Chill Effective Temperature | WET | |
| [56] | 2005 | Wind Chill equivalent Temperature | WCT | |
| [57] | 2007 | Wind Chill Temperature | $t_{wc}$ | |
| [58] | 1945 | Index of physiological effect | $E_p$ | |
| [59] | 1947 | Predicted 4-h sweat rate | P4SR | |
| [60] | 1955 | Heat stress index | HIS | |
| [52] | 1957 | Wet bulb globe temperature | WBGT | |
| [61] | 1958 | Thermal strain index | TSI | |
| [62] | 1960 | Index of physiological Strain | Is | |
| [63] | 1962 | Index of Thermal Stress | ITS | |
| [64] | 1966 | Heat Strain Index corrected | HIS | |
| [65] | 1966 | Prediction of heart rate | HR | |
| [66] | 1970 | Prescriptive zone | | |
| [67] | 1971 | Wet Globe Temperature | WGT | |
| [68] | 1972 | Skin wettedness | | |
| [69] | 1974 | Predicted heart rate | PHR | |
| [70] | 1978 | Skin wittedness | | |
| [71] | 1979 | Fighter Index of Thermal Stress | FITS | Hot Environment |
| [72] | 1981 | Effective Heat Strain Index | EHSI | |
| [73] | 1982 | Predicted sweat loss | msw | |
| [74] | 1984 | Munich energy balance model for individuals | MEMI | |
| [74] | 1985 | Skin Temperature energy balance index | STEBIDEX | |
| [75] | 1985 | Heat Budget index | HEBIDEX | |
| [76] | 1996 | Cumulative Heat Strain Index | CHSI | |
| [77] | 1998 | Physiological Strain Index | PSI | |
| [78] | 1989 | Required sweating | SWreq | |
| [79] | 2001 | Predicted Heat strain | PHS | |
| [80] | 2001 | Environmental stress index | ESI | |
| [81] | 2005 | Wet bulb dry temperature | WBDT | |
| [82] | 2005 | Relative Humidity Dry Temperature | RHDT | |
| [83,84] | 2007 | Overheating Risk | OR | |
| [85] | 2010 | Thermal work limit | TWL | |
| [86] | 2013 | Equivalent Wet Bulb Globe Temperature | eWBGT | |

Both ISO and ACGIH (American Conference of Governmental Industrial Hygienists) recommend considering the WBGT index as a screening tool, while the PHS (Predicted Heat Strain) approach must be used to investigate more severe working conditions in the heat. Based upon a rational approach, PHS it is the only index allowing the assessment of the work limit duration [87].

Concerning the cold environment, the most used indicator is the required clothing insulation (IREQ). The IREQ model was proposed in 1984 by Holmer [54] for the assessment of thermal stress during work in cold environments. The IREQ combines the effects of different thermo-hygrometric parameters (i.e., air temperature, mean radiant temperature, relative humidity, air velocity) with and metabolic activity and, in order to maintain its thermal equilibrium, it specified the required clothing insulation [88]. The IREQ is created upon rational basis and it is recommended by ISO 11079 [82,89].

**Table 3.** Example of WBGT reference values.

| Metabolic Rate Class | Metabolic Rate W | Reference Value of WBGT | |
|---|---|---|---|
| | | PERSONS Acclimatized to Heat °C | Persons Unacclimatized to Heat °C |
| 0 Resting metabolic rate | 115 | 33 | 32 |
| 1 Low metabolic rate | 180 | 30 | 29 |
| 2 Moderate metabolic rate | 300 | 28 | 26 |
| 3 High metabolic rate | 415 | 26 | 23 |
| 4 Very high metabolic rate | 520 | 25 | 20 |

## 4.2. Thermal Comfort

### 4.2.1. Overview

The American Society of Heating, Refrigerating and Air-Conditioning Engineers (ASHRAE) defined thermal comfort as "the condition of the mind in which satisfaction is expressed with the thermal environment" [90]. In addition to environmental conditions, the perception of thermal comfort can be influenced by personal differences in mood, culture and other social factors. In the ASHRAE definition, thermal comfort is not only a state condition, but rather a state of mind (a cognitive process that involves many inputs influenced by physical, physiological and other factors). "Satisfaction with the thermal environment is a complex subjective response to several interacting and less tangible variables" [29]. An absolute thermal comfort condition does not exist; the sensation of thermal comfort occurs when the body temperatures fall in rather narrow intervals, the moisture on the skin is low, and the physiological effort is minimal [29].

### 4.2.2. Indicators

Nowadays, for the evaluation of the thermal comfort, two main approaches are proposed: the rational or heat balance approach, and the adaptive approach. The first uses data from climate chamber studies to support its theory, the second uses data from field studies of people in buildings.

The heat balance approach is based on the results of the experiments carried out by Fanger in 1970 [91]. Using a steady-state heat transfer model, Fanger studied the response of a sample of 126 Danish student who underwent specific tests in a climate chamber [92]. Then, in order to determine range of comfort temperature in which the occupant feel comfortable, he combines theories on heat balance with the physiology of the thermoregulation. In order to maintain a balance between the heat produced by metabolism and the heat lost from the body, the human body is engaged in physiological processes (e.g., sweating, shivering, regulating blood flow to the skin). Consequently, the condition in which this thermal equilibrium is maintained can be considered the first condition to obtain a neutral thermal sensation [29]. Then, the tests were repeated until they involved a sample of 1296 participants, and a comfort equation was obtained that could predict when the occupants feel thermally neutral.

With the Fanger comfort equation it is possible to calculate all combinations of the environmental variables (air temperature, air humidity, mean radiant temperature and relative air velocity) which will create optimal thermal comfort for any activity level and any clothing insulation [92]. The comfort equation related thermal conditions to the seven-point ASHRAE thermal sensation scale and became known as the predicted mean vote (PMV) indicator. Furthermore, to the PMV, the predicted percentage of dissatisfied (PPD) was added in order to obtain a direct information of the percentage number of people who may experience discomfort. Using the Predicted Mean Vote and the PDD indicators it is possible to assess the thermal sensation perceived by occupants of indoor environments. It is important to note that the PMV should be calculated taking into account the range of application and the limitations, as it is very often improperly used without taking into account all the input variables, or calculated outside its range of validity [93,94].

The adaptive thermal comfort approach is based on the consideration that occupants are passive subjects, but they constantly interacting with the environment and adapting to it [95]. Using this approach, thermal perception is determined in relation to the indoor operating temperature and the outdoor air temperature. The adaptive thermal comfort approach takes into account that there are different factors that affects the thermal perception. These factors can include such different aspects as: demographics (gender, age, economic status), context (building design, building function, season, climate, social conditioning), and cognition (attitude, preference, and expectations) [96]. The adaptation can be divided into three different categories: behavioral, psychological, and physiological.

Behavioral adaptation is the most important factor and it offers the opportunity to adjust the body's heat balance to maintain thermal comfort by changing the clothing levels, opening or closing the windows, and switching on/off the fans. Psychological adaptation describes how habits and expectations can change perceptions of the thermal environment and, more specifically, it refers to the effects of the cognitive, social, and cultural variables. Physiological adaptation or acclimatization is the less relevant of the three aspects in the moderate range of more common indoor environmental conditions [97].

Although the Fanger indicators (PMV and PPD) are the most widely used [15–17], in the literature, other indicators for the evaluation of thermal comfort were proposed, among which some of these were designed to better evaluate the thermal sensation in particular climatic conditions. A collection of the indicators found in the literature with the reference of the scientific publications in which they were proposed, is reported in Table 4. These indicators were classified in relation to the number of parameters necessary for their determination. In particular, they were classified in: "single parameters based" (if directly obtained from one parameter), "double parameters based" (if obtained from the combination of two parameters), and in "multiple parameters based" (if obtained from the combination of three or more parameters).

4.2.3. Current Research Trends

Climate change, the urgency to decarbonize the built environment, and the reduction of energy consumption are still the topics of greatest interest. The increased policy attention on these topics has increased research and development investment. The demand for increasingly high-performance buildings has led to the research of new strategies to mitigate energy consumption and environmental impact. The new challenge is to minimize energy consumption by guaranteeing high housing standards. For this reason, the thermal comfort indicators suitable for evaluating air-conditioned environments and occupant interaction with plant systems are still studied.

The use of adaptive thermal comfort models can be particularly well combined with climate-based software for the dynamic simulation of the buildings behavior that allow to study the variation of internal conditions in relation to the external temperature changes. The comfort simulation methods based on easily accessible numerical simulation tools of the built environment are becoming more popular. A possible critical aspect of this method concerns the risk of moving too far away from the actual behavior of the users, which is very often difficult to predict. Practitioners' confidence in comfort models (design phase) and in engineering calculations, as well as conformity assessments for existing buildings, is widespread and perfectly reasonable. However, subjective surveys of thermal comfort are a superior contribution to knowledge, with more lasting value for the research community than simulated comfort assessments coming from a comfort model [98].

**Table 4.** A collection of thermal comfort indicators.

| Reference | Index | Symbol | Parameters | Year | Category |
|---|---|---|---|---|---|
| [99] | Wet bulb temperature | $T_w$ | | 1905 | Single parameter based |
| [100] | Katathermometer | | | 1914 | |
| [101] | Globe-Thermometer Temperature | GtT | | 1930 | |
| [102] | Effective Radiant Field | ERF | | 1967 | |
| [103] | Humiture or Heat Index | HI | Air temperature, humidity. | 1937 | Double parameters based |
| [104] | Craig Index | I | Heart rate, predicted body temperature | 1950 | |
| [105] | Oxford index | WD | Wet-bulb temperature, Dry-bulb temperature | 1957 | |
| [106] | Discomfort index | DI | Wet-bulb temperature, Dry-bulb temperature | 1957 | |
| [107] | Cumulative Discomfort Index (**) | CumDI | Wet-bulb temperature, Dry-bulb temperature | 1961 | |
| [108] | Humiture revisited | | Air temperature, humidity. | 1960 | |
| [109] | Relative Strain Index | RSI | Air temperature, water vapor pressure. | 1963 | |
| [110] | Temperature-Humidity Index | THI | Air temperature, Relative Humidity | 1977 | |
| [111] | Humidex | H | Air Temperature, water vapor pressure | 1979 | |
| [112] | Humisery | | Air temperature, Relative Humidity | 1982 | |
| [113] | Summer simmer Index | SSI | Air temperature, Relative Humidity | 1987 | |
| [114] | Modified discomfort index | MDI | Wet-bulb temperature, Dry-bulb temperature | 1998 | |
| [115] | New Summer Simmer Index | newSSI | Air temperature, Relative Humidity | 2000 | |
| [116] | Degree-hour criterion (**) | | Air temperature, mean radiant temperature | 2007 | |
| [117] | EsConTer Index | | Wet-bulb temperature, Dry-bulb temperature | 2009 | |
| [118] | Exceedance$_M$ (**) | | Air temperature, mean radiant temperature | 2010 | |
| [119] | Effective temperature | ET | Dry-bulb temperature, air temperature, humidity, radiant conditions, air movements. | 1923 | Multi-parameters based |
| [120] | Equivalent temperature | $T_{eq}$ | Dry-bulb temperature, air temperature, radiant temperature, humidity. | 1929 | |
| [121] | Corrected effective temperature | CET | Dry-bulb temperature, air temperature, humidity, radiant conditions, air movements. | 1932 | |
| [122] | Standard Operative Temperature | SOpT | Air temperature, air velocity, solar radiation, body temperature. | 1937 | |
| [123] | Operative temperature | $T_o$ | Air temperature, mean radiant temperature, air velocity | 1937 | |
| [124] | Thermal acceptance ratio | TAR | Vapor pressure, metabolic heat, skin temperature, evaporation, convection and radiation constants. | 1945 | |
| [125] | Corrected Effective Temperature | CET | Dry-bulb temperature, humidity, radiant conditions, air movements. | 1946 | |
| [126] | Resultant temperature | RT | Dry-bulb temperature, air temperature, humidity, radiant conditions, air movements. | 1948 | |

**Table 4.** *Cont.*

| Reference | Index | Symbol | Parameters | Year | Category |
|---|---|---|---|---|---|
| [52] | Effective Temperature including radiantion | ETR | Dry-bulb temperature, air temperature, humidity, globe thermometer temperature, air movements. | 1957 | |
| [127] | Equatorial comfort index | ECI | Wet-bulb temperature, Dry-bulb temperature, air velocity, geographic coordinates | 1959 | |
| [128] | Cumulative Effective Temperature (**) | CumET | Dry-bulb temperature, air temperature, humidity, radiant conditions, air movements. | 1962 | |
| [91] | Predicted mean vote | PMV | Air temperature, relative humidity, mean radiant temperature and relative air velocity, metabolic activity, clothing. | 1970 | |
| [129] | New effective temperature | ET * | Dry-bulb temperature, air temperature, humidity, radiant conditions, air movements. | 1971 | |
| [130] | Humid operative temperature | $T_{OH}$ | Air temperature, mean radiant temperature, dew point temperature, skin temperature, air movements, clothing. | 1971 | Multi-parameters based |
| [131] | Apparent Temperature | AT | Air temperature, relative humidity, air velocity. | 1971 | |
| [132] | Standard effective temperature | SET | Dry-bulb temperature, air temperature, humidity, radiant conditions, air movements. | 1973 | |
| [133] | Apparent Temperature (revised) | AT | Air temperature, relative humidity, air velocity. | 1979 | |
| [134] | Equivalent Uniform Temperature | EUT | Air temperature, relative humidity, mean radiant temperature and relative air velocity, metabolic activity, clothing. | 1980 | |
| [135] | Predicted mean vote modified | PMV * | Air temperature, relative humidity, mean radiant temperature and relative air velocity, metabolic activity, clothing. | 1986 | |
| [136] | Tropical Summer Index | TSI | Wet-bulb temperature, globe temperature, Relative Humidity, | 1987 | |
| [137] | CIBSE Guide J-criterion | | Dry-bulb temperature, mean radiant temperature, air velocity. | 2002 | |
| [138] | PPD weighted criterion (**) | | Air temperature, relative humidity, mean radiant temperature and relative air velocity, metabolic activity, clothing. | 2005 | |
| [138] | Average PPD (**) | | Air temperature, relative humidity, mean radiant temperature and relative air velocity, metabolic activity, clothing. | 2005 | |
| [138] | Cumulative PPD (**) | | Air temperature, relative humidity, mean radiant temperature and relative air velocity, metabolic activity, clothing. | 2005 | |
| [116] | CIBSE Guide A-criterion | | Dry-bulb temperature, mean radiant temperature, air velocity. | 2006 | |

Note (**): indicators that require monitoring in pre-established time intervals and the determination of the percentage of time in which a certain parameter falls outside the relative reference value.

## 5. Indoor Air Quality

Clean air is considered a basic requirement of human health, well-being, comfort and productivity [139]. Indoor air quality is affected by different factors as the interactions between building materials, building services, location, climate, contaminant sources, and occupancy [140–143]. Indoor air quality can be approached from three points of view: the human, the air of the indoor space, and the sources contributing to indoor air pollution [136].

From the human point of view, indoor air quality is the physical effect of people's exposures to substances in the indoor air. From the indoor air point of view, indoor air quality is often expressed with the ventilation rate or with the concentrations for specific compounds. From the sources point of view, air pollution levels may be higher near specific air pollution sources such as roads; the protection of people from air pollution may require specific measures to maintain the pollution levels below the threshold values indicated in the national and international guidelines [139].

### 5.1. Indoor Air Pollution

### 5.1.1. Overview

Considering that especially in urban areas, people spend long periods of time in indoor environments, the international scientific community deals with air pollution in living environments because indoor pollution can cause side effects ranging from discomfort to serious health consequences. In this regard, illness such as sick building syndrome (SBS) [144,145] and building related illness (BRI) are referred to [145]. It is important to note that a study conducted by the United States Environmental Protection Agency (EPA) in 1998 has estimated that indoor concentrations and indoor exposures are generally 1–5 times, and 10–50 times higher than outdoor ones, respectively [146]. In 2000, the World Health Organization (WHO), with the document "The Right to Healthy Indoor Air", recognized a healthy indoor air as a fundamental human right [147].

### 5.1.2. Guidelines and Legislative Outline

The WHO, with "the health for all in the 21st century", provided a policy framework for the European region [148]. The target was that, by the year 2015, people in the EU region had to live in a safer physical environment, with contaminants exposure lower than internationally agreed thresholds. This target has consequently resulted in the introduction of legislative measures for the regulation of the surveillance and control of the air quality in both indoor and outdoor environments.

The healthiness of the indoor air is influenced by such different factors as undesirable substances emitted by construction, furniture materials, and human being, as well as ventilation and heating systems if not adequately cleaned. Concerning the indoor air quality, it is important to consider individual exposure to pollutants that can be defined as the air pollutants concentration over time. This depends upon several aspects: the emission rate of a pollutant ($\mu g/s$ or $\mu g/s$ per $m^2$ surface area of source); the ventilation rate of the space in which the pollutants are produced ($m^3/h$ or $L/s$); the pollutants concentration in the ventilation air (ppm or $\mu g/m^3$).

In Italy, a clear legislative framework for the control and the maintenance of the air quality in the indoor environment does not exist. In the national law 81/2008 [47], the only limit values provided are referred to the concentrations of $CO_2$, radon, and bacteria allowed in the indoor environment during the daily activities. For example, the maximum permissible level of CO2 is equal to 5000 ppm in a time interval of eight working hours.

### 5.1.3. Air Pollutants

At the international level, the exposure limit values of a wide range of pollutants were defined by the EU and the WHO [44]. Since 1958, when the first report was published, the WHO has been working on the potential adverse effects on health of air quality [149–151]. In the first edition of the WHO air quality guidelines (AQGs) [152], 28 air pollutants (organic and inorganic) were classified, and different approaches to deal with carcinogenic (i.e., unit risk factors) and non-carcinogenic-health (i.e., LOAEL and protection factors) aspects were used. Furthermore, $SO_2$ and particulate matter (PM) were considered jointly. The second edition of the WHO AQGs was published in 2000 [153], following the evidence of health effects occurring at lower levels of exposure, and it was considered as a starting point for the EU Air Quality Directive and the definition of legally binding limit values [154]. In this AQGs 32 air pollutants are concerned (see Table 5), moreover the assessments for three organic air pollutants (butadiene, polychlorinated biphenyls and polychlorinated dibenzodioxins and dibenzofurans) and a specific section for indoor air pollutants (radon, environmental tobacco smoke, and man-made vitreous fibers) are included.

The latest WHO AQGs "WHO Air Quality Guidelines, Global Update 2005" [155] concerned the policy development and risk reduction application of AQGs, as well as a comprehensive risk assessment for the four classical air pollutants: PM, $O_3$, $NO_2$, and $SO_2$. In addition to numerical guidelines, it also proposed interim targets above the guideline value to promote steady progress in different world regions [155]. Moreover, the WHO has published a series of indoor AQGs on dampness and mould, selected pollutants, and household fuel combustion [156–158]. In Table 6 are shown the limit values of the air pollutants with the latest version of the relative WHO Air quality guidelines (WHO AQGs).

**Table 5.** Summary of the most important air pollutants.

| Organic Pollutants | Inorganic Pollutants | Classical Pollutants |
|---|---|---|
| Acrylonitrile; Butadiene; Benzene; Carbon disulfide; Carbon monoxide; 1.2 Dichloromethane; Dichloromethane; Formaldehyde; Polycyclic aromatic hydrocarbons; Polychlorinated biphenyls; Styrene; Tetrachloroethylene; Toluene; Trichloroethylene; Vinyl chloride; Polychlorinated dibenzodioxins and dibenzofurans. | Arsenic; Asbestos; Cadmium; Chromium; Fluoride; Hydrogen sulfide; Lead; Manganese; Mercury; Nickel; Platinum; Vanadium | Particulate matter; Ozone and other photochemical oxidants; Nitrogen dioxide; Sulfur dioxide |
|  |  | **Indoor Air Pollutants** |
|  |  | Man-made vitreous fibers; Environmental tobacco smoke; Radon. |

**Table 6.** Example of limit values for the air pollutants.

| Pollutant | Symbol | Limit Values | Reference Period | Year of Latest WHO AQGs |
|---|---|---|---|---|
| *Organic Pollutants* | | | | |
| Carbon disulphide | $CS_2$ | 100 µg/m$^3$<br>20 µg/m$^3$ | 24 h<br>30 min | 1987 |
| 1,2-Dichloroethane | EDC | 0.7 mg/m$^3$ | 24 h | 2000 |
| Dichloromethane | DCM | 3.0 mg/m$^3$ | 24 h | 2000 |
| Formaldehyde | | 100 µg/m$^3$ | 30 min | 2010 |
| Styrene | | 0.26 mg/m$^3$<br>70 µg/m$^3$ | Weekly<br>30 min | 2000 |
| Tetrachloroethylene | PERC | 0.25 mg/m$^3$<br>8 mg/m$^3$ | Annual<br>30 min | 2010 |
| Toluene | | 0.26 mg/m$^3$<br>1 mg/m$^3$ | Weekly<br>30 min | 2000 |
| *Inorganic Pollutants* | | | | |
| Cadmium | Cd | 100 mg/m$^3$ | 30 min | 2000 |
| Hydrogen sulphide | $H_2S$ | 150 µg/m$^3$<br>7 µg/m$^3$ | 24 h<br>30 min | 2000 |
| Lead | Pb | 0.5 µg/m$^3$ | Annual | 2000 |
| Manganese | Mn | 0.15 µg/m$^3$ | Annual | 2000 |
| Mercury | Hg | 1 µg/m$^3$ | Annual | 2000 |
| Vanadium pentoxide | | 1 µg/m$^3$ | 24 h | 1987 |
| *Classical Pollutants* | | | | |
| Particulate matter | PM2.5 | 10 µg/m$^3$<br>25 µg/m$^3$ | Annual<br>24 h | 2006 |
| | PM10 | 20 µg/m$^3$<br>50 µg/m$^3$ | Annual<br>24 h | |
| Ozone | $O_3$ | 100 µg/m$^3$ | 8 h | 2006 |
| Nitrogen dioxide | $NO_2$ | 40 µg/m$^3$<br>200 µg/m$^3$ | Annual<br>1 h | 2010 |
| Sulphur dioxide | $SO_2$ | 20 µg/m$^3$<br>500 µg/m$^3$ | 24 h<br>10 min | 2006 |

## 5.2. Good Indoor Air Quality

### 5.2.1. Overview

Indoor air quality (IAQ), in terms of comfort, is very important because it affects building occupants and their ability to conduct activities, in addition to creating positive or negative impressions on customers, clients, and other visitors.

When IAQ is not adequate, building owners and managers must devote considerable resources to resolve occupant complaints or deal with long periods of building closure, major repair costs, and expensive legal actions. On the contrary, when IAQ is good, building are more desirable places to work, to learn, to conduct business, and to rent [159,160].

High levels of IAQ are achieved by providing air in which there are low contaminant concentrations and no conditions that can be associated with occupant health or comfort complaints. The current knowledge on the health and comfort impacts of specific contaminants and contaminant mixture in nonindustrial environments, does not allow for the development of a single IAQ metric able to provide a summary measure of IAQ in buildings [159]. Considering the difficulties associated with the air quality assessment in confined environments, to date has not possible to formulate a universally accepted air quality definition. At present, for buildings, the definition of the ASHRAE considers acceptable the indoor air quality when "it does not contain known contaminants in harmful concentrations, as established by the competent authorities, and for which a substantial majority of people exposed (80% or more) does not express dissatisfaction" [161].

### 5.2.2. Indicators

Regarding the air quality assessment indicators, it is certainly important to mention the 1987 Fanger study [162]. Fanger, similarly to what was previously done for the evaluation of thermal comfort, introduced a subjective indicator called DECIPOL. The DECIPOL is defined as "pollution perceived in the presence of a normal subject (1 olf) in an environment with ventilation equal to 10 L/s of clean air". In the DECIPOL definition, 1 olf represents the main quantity and it is defined as the "amount of bio-effluents emitted from a standard source consisting of a subject that performs sedentary activities in conditions of thermal well-being with a hygienic standard of 0.7 baths/d" [162].

Later, different IAQ indicators were proposed for the evaluation of IAQ in buildings [163–177]. Two different approaches are commonly used to construct IAQ indicators: subjective surveys and field measurements [178].

The IAQ indicators based on subjective surveys include the administration of questionnaires on the perception of IAQ and indoor comfort, (e.g., the ABCD tool proposed in the Netherlands [179]) or the compilation of checklists describing building facilities (e.g., Indoor airPLUS proposed by the USA Environmental Protection Agency [180]).

The IAQ indicators based on field measurements, such as the BILGA index [163,164] and the IAQ Certification [165], are more common, and generally they can be calculated using equations, e.g., the indoor environmental index (IEI) proposed in the USA [176–178]. According to [181], the IAQ indicators can be classified according to used approaches in: "one indicator per single pollutant"; "simple aggregation"; "aggregation according to the sources of pollutants and/or types of pollutants"; "aggregation accounting for the IAQ of the building stock"; "aggregation by simple addition of health impacts". Among the more recent studies, it is possible to find comfort indicators built on the basis of the PMV proposed by Fanger for the thermal comfort. In particular, Zhu and Li [182] tried to connect the objective environment parameters with the subjective comfort perception introducing the indoor air quality indicator $PMV_{IAQ}$. Such an indicator is defined as the maximum value of the PMV indicators related to three different pollutant concentrations: CO2 ($PMV_{CO2}$), $PM_{10}$ ($PMV_{PM10}$), and the formaldehyde ($PMV_{HCHO}$).

The main indoor air quality indicators proposed in literature are collected in Table 7.

**Table 7.** A collection of indoor air quality indicators.

| Reference | Symbol | Pollutants | Category | Year |
|---|---|---|---|---|
| [169] | | Any | One index per single pollutant * | |
| | | $CO_2$, TVOC | | 2016 |
| [163,164] | BILGA | Any | | 1996 |
| [164,170] | CLIM2000 | CO, $CO_2$, $NO_2$, formaldehyde | | 1996 |
| [164,170] | LHVP | CO, $CO_2$, bacteria | Simple aggregation | 1998 |
| [166–168] | IEI | CO, $CO_2$, formaldehyde, TVOC, PM10 | | 2002 |
| [165] | IAQC | CO, $CO_2$, formaldehyde, TVOC, PM10 | | 2003 |
| [172] | QUAD | Group A: $CO_2$<br>Group B: $NO_2$, $SO_2$, $O_3$<br>Group C: CO, formaldehyde, acetaldehyde, ethylbenzene, styrene, toluene, o-xylene, acetone<br>Group D: PM2.5, PM10 | Aggregation according the sources of pollutants and/or type of pollutants | 2012 |
| [173] | IAPI | CO, $CO_2$, formaldehyde, TVOC, PM2.5, PM10, fungi, bacteria | Aggregation accounting for the IAQ in the building stock | 2003 |
| [174] | DALY | Any | Aggregation by simple addition of health impacts | 2011 |
| [182] | $PMV_{IAQ}$ | $CO_2$; PM10; HCHO | One index per single pollutant; and an overall index | 2017 |

* One index per single indicators are commonly obtained with the ratio between the concentration of the pollutants and the relative exposure limit value.

### 5.2.3. Current Research Trends

Despite the existing air quality indicators, the lack of metrics which quantitatively describe the IAQ can be regarded as one of the most relevant complications for the achievement of the integration of energy and IAQ strategies in indoor environment design. Such an indicator could allow the analysis and comparisons of different plans for achieving high IAQ levels, high energy performance, and low greenhouse gas emissions [183]. Recently, the International Energy Agency (IEA) in agreement with the Energy in Buildings and Communities (EBC) defined a project (IAE EBC annex 68) to provide a guideline for the design and control strategy of high energy efficiency residential buildings. One of the first steps of this project is to define an indicator that will have to include the additional energy consumption necessary to improve IAQ (compared to standard practices), e.g., an increased consumption induced by higher air change rates [181,183].

## 6. Acoustical Environment

People are continuously exposed to noise, even when they sleep. Consequently, for workers, whether alone in a private office or among a large number of colleagues in an industrial setting, a complete absence of noise never occurs. A good acoustical environment should not involve any physical, physiological or psychological effects on the human body that could negatively affect health. Furthermore, the acoustical environment should allow a person to be in the most suitable state of mind for a specific activity [184]. On the contrary, exposure to noise can affect quality of life and, in the worst case, can lead to health problems. A workplace with good acoustics allows for confidential conversations between collaborators without affecting those engaged in individual work. The internal environment must protect from excessive noise pollution from internal and external sources and encourage the performance of the planned activities, specifically in all those environments where verbal communication and correct listening take priority (i.e., schools, conference rooms, etc.).

*6.1. Noise Exposure*

6.1.1. Overview

The effects that caused by excessive noise exposure are now well known and they can even lead to hearing loss (e.g., hypoacusis). Noise exposure can also cause effects on other organs such as the cardiovascular, the endocrine, and the central nervous systems, and it involves different effects such as fatigue, interference with sleep and rest, and reduction of work performance. Another possible effect on safety that cannot be neglected is that noise can produce masking effects that can disturbs the verbal communication with other persons and impedes the correct perception of warning signals. Such an effect can increase the risk of accidents at work. However, it is not acceptable either a completely silent environment because it can cause a sense of estrangement [185]. For these reasons, human exposure to noise should be evaluated in the field of safety in the workplace, but also in the domestic and private sphere, using indicators that take into account noise exposure even during the rest hours.

6.1.2. Guidelines and Legislative Outline

The legislative framework in the field of the human exposure to noise in work environments is very clear today. At the international level, the recommendations are given in the Physical Agents (Noise) Directives [186], where the minimum requirements for the protection of workers from risks arising from exposure to noise are provided. Such a directive was implemented by the European Union member states, including Italy with the law 81/2008 [47].

The assessment procedure of noise human exposure is indicated in the Technical Standard ISO 9612 [187] and is composed by the following phases: work analysis, selection of measurement strategy, measurements, error handling and uncertainty evaluations, equivalent exposure levels calculation, and presentation of results. In these procedures the choice of the most appropriate measurement strategy is extremely relevant in order to actually detect the effective human exposure and to estimate the results uncertainty. The indicators for this type of assessment are the daily exposure and the peak noise. Their action levels and limit values are given in [186] to ensure that appropriate actions are taken to guarantee the protection of workers. Personal hearing protection must be used when other controls cannot adequately reduce the noise exposures [188]. Hearing protection must be selected to reduce noise exposure without isolating the worker (overprotection), which may compromise safety.

In the Physical Agents (Noise) Directives [186] two exposure action values (EAV) and an exposure limit value are defined (Table 8). The two EAVs are the "upper exposure actions", and the "lower exposure actions" values. To the first correspond the noise exposure level, above which the employer must require the use of the personal protective equipment, and to the second correspond the noise exposure level above which the employer must provide the workers with the personal protective equipment (but it is not required to oblige their use). The exposure limit value represents the noise exposure level which must not be exceeded.

6.1.3. Indicators

It is important to note that sometimes, despite the continuous equivalent levels being lower than the limit values, exposure for a long time to noise characterized by particular frequency emissions (for example continuous noise at low frequencies) can still cause health problems. In this case, there is no risk of hearing loss, but rather a series of side effects (extra-auditory effects) such as headaches, dizziness, problems with the gastrointestinal system, increased respiratory rate, etc. The psychosocial effects are reflected on the modification of interpersonal relationships and on social relations. In addition to the Physical Agents Directives [186], international level guidelines (e.g., [189,190]) were published, in which indicators and the relative reference values, above which the first effects of noise on human health can begin, are provided (Table 8).

**Table 8.** A collection of noise exposure indicators.

| Symbol | Indicator | Time Constant | Threshold [2] |
|---|---|---|---|
| $L_{EX,8h}$ | Daily noise exposure level | 8 h | 80/85/87 dB(A) [3] |
| $P_{peak}$ | Peak sound pressure | 125 ms | 135/137/140 dB(C) [3] |
| $L_{Amax}$ | Maximum sound pressure level occurring in an interval | 125 ms | 35 dB(A) |
| SEL | Sound exposure level or sound pressure level over an interval normalised to 1 s | 1 s | 53 dB(A) |
| $L_{day}$ | Average sound pressure level over a 1 day | 12 or 16 h | |
| $L_{night}$ [1] | Average sound pressure level over 1 night | 8 h | 42 dB(A) |
| $L_{24h}$ | Average sound pressure level over a whole day | 24 h | |
| $L_{dn}$ | Average sound pressure level over a whole day getting to the night values a penalty of 10 dB | 24 h | |
| $L_{den}$ [1] | Average sound pressure level over all days, evening and nights in a year getting a penalty of 5 dB to the evening values and a penalty of 10 dB for the night values | Year | 42 dB(A) |

Note [1]: referred to outside exposure levels. Note [2]: Levels above which the first health effects start to occur. Note [3]: Limit values related to the lower exposure action values, the upper exposure action values, and exposure limit values, respectively.

## 6.2. Acoustic Comfort

### 6.2.1. Overview

The term "acoustic comfort" is not commonly used; a good acoustic environment is mainly associated with preventing the occurrence of discomfort. An acoustically comfortable building is able to isolate occupants from internal and external noise and at the same time to guarantee an environment acoustically appropriate for the activities carried out within it [191]. It can be defined as "a state of contentment with acoustic conditions" [192]. The acoustic sensation does not depend only on the sound pressure level but also on the sound frequency composition. The acoustic quality of an environment therefore depends on how the sound is propagated inside it (how the sound is absorbed and reflected by the surfaces), how the envelope structure is isolated from external noise, and how the sound coming from inside sources is treated. The acoustic comfort evaluations in workplaces are often undervalued, but especially in environments like offices where annoyance due to the noise can often cause mental stress and concentration loss, it is very important [44]. Very often, the acoustic comfort is considered adequate when the building envelope is sufficiently isolated from the noise coming from outside or adjacent environments (e.g., residential buildings). However, in many other cases, noise insulation is not sufficient to consider an environment acoustically comfortable and more detailed parameters should be used to obtain important information on the correct perception of sounds.

### 6.2.2. Indicators

In the literature, several indicators for the evaluation of the acoustic comfort in indoor environment are introduced. Such indicators can be divided into three wide groups: sound pressure levels, architectural acoustics, and building acoustics (Table 9).

The first group is composed of indicators related to the sound pressure levels established in an environment due to noise produced by a source located outside or inside the analyzed room.

The second group is composed of indicators which describe the acoustic transient and the quality of sound perception within the analyzed environment.

The third group is composed of indicators that do not describe directly the sound perception, but describe the sound insulation of buildings from noise coming from outside and adjacent environments and the levels of noise produced by the building services.

Among the sound pressure levels indicators, the most widely used is the A-weighted equivalent sound pressure level $L_{eqA}$. This takes into account the duration, the variation in time, and the frequency composition of the sound. [193]. However, this indicator may not always be the most suitable, while other more specific indicators can be preferable. Among the architectural acoustic indicators, the most common is the reverberation time ($RT_{60}$), although in literature many other indicators for detailed evaluations of the acoustic quality of rooms are present [194].

### 6.2.3. Current Research Trends

The evaluation of acoustic quality, especially in schools, is a very topical subject. One of the problems related to the evaluation of acoustic quality is that most indicators are obtained from field measurements or advanced simulations, and only a few indicators can be calculated in the design phase with simplified formulas. For this reason, topics of research concern the determination of equations based on an empirical basis (generally obtained using field measurement results) to correlate the indicators with the reverberation time, which can be calculated with simple and well-known equations. For example, in Italy, for the educational buildings, the speech transmission index (STI) must be verified during the design stage, but the calculus of STI requires field measurements or advanced simulations that are incompatible with the level of detail of the preliminary stage [195,196].

**Table 9.** A collection of indicators for the evaluation of the acoustic comfort in indoor environment.

| Reference | Acoustic Comfort Metric | Symbol | Category |
|---|---|---|---|
| | A-weighted equivalent sound pressure level | $L_{eqA}$ | |
| | A-weighted statistical levels | $L_{A90}, L_{A10}, L_{A5}$ | |
| | Linear equivalent sound pressure level | $L_{eqlin}$ | |
| [197] | Balanced noise criterion | NCB | |
| [198] | Combined noise index | CNI | |
| [199] | Preferred noise criterion | PNC | |
| [200] | Noise climate | $(L_{A10}-L_{A90})$ | |
| [201] | Noise criterion curves | NC | |
| [202] | Noise pollution level | LNP | Sound pressure level |
| [203] | Noise rating curves | NR | |
| [204] | Office noise index | ONI | |
| [205] | Quality assessment index | QAI | |
| [206] | Room criterion | RC | |
| [207,208] | Room Criterion Mark II | $RC_{markII}$ | |
| [209] | Speech interference level | SIL | |
| [210] | Stevens' loudness level | LLS | |
| [211] | Zwicker's loudness level | LLZ | |
| [212] | Articulation index | AI | |
| [213] | Articulation loss of consonants | ALcons | |
| [214] | Clarity index | $C_{80}$ | |
| [215] | Definition index | $D_{50}$ | |
| [216,217] | Early decay time | EDT | |
| [218] | Late arriving sound/ Strength of the late arriving | $G_{late}$ | |
| [219–222] | Reverberation time | $RT_{30}, RT_{60}$ | |
| [223] | Room Acoustic Speech Transmission Index | RASTI | Architectural |
| [224] | Room gain | $GR_{G,0.5-2kHz}$ | acoustic |
| [225] | Signal to noise ratio | S/N | |
| [214] | Speech clarity | $C_{50}$ | |
| [226] | Speech Intelligibility Index | SII | |
| [227] | Speech Transmission Index | STI | |
| [223,228] | Speech Transmission Index for Public Access | STIPA | |
| [229] | Strength value | G | |
| [230] | Voice support | $ST_{V,0.5-2kHz}$ | |
| [231] | Acoustic Performance index | AP | |
| [232] | Noise level produced by discontinuous service equipment | $L_{ic}$ | |
| [232] | Noise level produced by continuous service equipment | $L_{id}$ | Building acoustic |
| [233] | Weighted normalized impact sound pressure level of floor | $L'_{nw}$ | |
| [234] | Weighted sound reduction index | $R'_{w}$ | |
| [235] | Weighted standardized sound level difference of facade | $D_{2m,nt,w}$ | |

A second current research topic concerns the proposal of assessment methods of the overall acoustic quality that allow to evaluate the existing environments and to identify all the critical aspects that need improvements. These methods are generally based on Multicriteria analysis in order to combine the results of different indicators and obtain an overall rating [236,237]. In acoustics, only single parameter-based indicators exist, and for this reason, the multi-criteria approach is very interesting. This represents a potential overall assessment tool taking into account more indicators simultaneously. Until now, in fact, in acoustics, the individual indicators are determined and evaluated separately, and the combination of results is decided by the evaluator on the basis of subjective personal knowledge and sensitivity.

## 7. Visual Environment

Lighting has significant relevance on comfort and human life [238], with direct and indirect (non-visual) effects on the human body. The direct effects are related to visual performance (visual task activity) and visual comfort; the indirect effects are related to the possible consequences on safety and health. Traditionally, lighting assessment is limited to the assessment of illuminance in the task areas, however, there is a growing awareness of the effects of light on both health and the quality of human life. This is confirmed by new certification systems as the building certification system, developed by the International WELL Building Institute with the aim of "measure, monitor, and certify the performance of building features that impact health and well-being" [239].

### 7.1. Non-Adequate Light Exposure

#### 7.1.1. Overview

Properly lighting is a basic need for visual performance and safety. Non-adequate lighting systems can cause problems with visual fatigue, as well as causing errors or possible accidents. In work environment special attention is focused on the video display terminal (VDT) workstations: electric lighting must guarantee adequate levels of illuminance for the reading and writing activities, but they should also favor a satisfactory contrast of luminance between the main object of view (the display) and the background in order to reduce visual fatigue [240,241]. For the video-terminal workstation, the asthenopic complaints are considered signs of visual discomfort. The most common visual discomfort markers are: red eyes, burning eyes, double vision, eye fatigue, blurred vision, headache, etc. [241–243]. Such effects together with inadequate lighting conditions can be related to the computer vision syndrome (CVS) [244]. At the end of 1980s, Bergqvist [245] revealed that eye discomfort and hand/wrist problems were associated with the work at the VDT workstation. Subsequently, several researches have further investigated the consequences of incorrect postures during VDT use [240–254] and the psychological effects of computer use [248]. Exposure to light is studied in relation to its photobiological effects, in particular the consequences of artificial optical radiations (AOR) emitted by light sources are evaluated. If in the past general lighting was considered risk exempt, recently, the advent of new LED sources has rekindled interest in this field, especially in the blue light hazard assessment [255]. In any case, the assessment of exposure to artificial optical radiation (not only produced by general lighting sources) is part of the physical agent, and it is included in the workers risk assessment procedures.

#### 7.1.2. Guidelines and Legislative Outline

At the international level, Directive 89/654/EEC [256] includes an indication about lighting, and provides the requirements to guarantee health and safety in the workplaces. In [257] it is indicated that: "the workplace must have sufficient daylight and be equipped with devices that allow artificial lighting adequate to ensure the safety, health and welfare of workers". This directive was included in the Italian Law 81/2008 [47], where only rather general instructions (without specific limit values) are provided.

Lighting not only affects visual performance and visual comfort, but also human health with so called "non-visual effects". In particular, it is necessary to evaluate human exposure to artificial optical radiation (AOR) emitted by light sources. Optical radiation regards the electromagnetic radiation in the wavelength range between 100 nm and 1 mm. They are divided in: "ultraviolet" (UV), "infrared" (IR), and "visible light" (VIS) or more simply "light" radiation [258]. Biological effects of AOR affect the skin and eyes but systemic effects may also occur [259,260]. Regarding the optical radiation, the Directive 2006/25/CE [257] represents the basic guideline. This directive is adopted in the Italian Law 81/2008 [47], and for a series of wavelength ranges specifies indicators and limit values for the evaluation of the AOR workers exposure. On these issues, important documents are those published by the International Commission on Non-Ionizing Radiation Protection (ICNIRP) [261,262], and the recent international standards [263,264].

### 7.1.3. Indicators

As indicated in the previous section, the national and international legislation does not provide limit values, and for this reason, the evaluations are generally qualitative and it is necessary to use the same indicators used for the evaluation of visual comfort (see Section 7.2).

The only field in which specific indicators are provided by the legislation is the artificial optical radiation (AOR). The AOR are included in the physical agents and it is required the evaluation for the workers' risk assessment. In Table 10, for each wavelength range, the exposure limit values, the type of hazard, and the action spectrum that must be used are indicated.

**Table 10.** Example of exposure limit values related to the artificial optical radiation.

| Wavelength. Range (nm) | Exposure Limit Values | Units | Action Spectrum | Hazard | Effects |
|---|---|---|---|---|---|
| 180 ÷ 400 (UVA, UVB, UVC) | $H_{EFF} = 30$ Daily value (8 h) | $J/m^2$ | $S(\lambda)$ | Actinic UV | Eye: Photokeratitis; Conjunctivitis; Cataractogenesis. Skin: Elastosis |
| 315 ÷ 400 (UVA) | $H_{UVA} = 10^4$ Daily value (8 h) | $J/m^2$ | | UVA | Eye: Cataractogenesis |
| 300 ÷ 700 (Blue Light) | $L_B = 10^6 \cdot t^{-1}$ for $t \leq 10{,}000$ s; $L_B = 100$ for $t > 10{,}000$ s | $W/m^2 \cdot sr$ | $B(\lambda)$ | Retinal Blue-Light small source | Eye: Photoretinitis |
| | $E_B = 100 \cdot t^{-1}$ for $t \leq 10{,}000$ s; $E_B = 0.01$ for $t > 10{,}000$ s | $W/m^2$ | $B(\lambda)$ | Retinal Blue-Light small source | |
| 380 ÷ 1400 (Visible, IRA) | $L_R = (2.8 \cdot 10^7) \cdot C_\alpha^{-1}$ for $t > 10$ s; $L_R = (5 \times 10^7) \cdot C_\alpha^{-1} \cdot t^{-0.25}$ for $10\ \mu s \leq t \leq 10$ s; $L_R = (8.89 \times 10^8) \cdot C_\alpha^{-1}$ for $t < 10\ \mu s$ | $W/m^2 \cdot sr$ | $R(\lambda)$ | Retinal thermal | Eye: Retinal burn |
| 780 ÷ 1400 (IRA) | $L_R = (6 \times 10^6)\ C_\alpha^{-1}$ for $t > 10$ s; $L_R = (5 \times 10^7) \cdot C_\alpha^{-1} \cdot t^{-0.25}$ for $10\ \mu s \leq t \leq 10$ s; $L_R = (8.89 \times 10^8)\ C_\alpha^{-1}$ for $t < 10\ \mu s$ | | $R(\lambda)$ | Retinal thermal-weak visual stimulus | Eye: Retinal burn |
| 780 ÷ 3000 (IRA, IRB) | $E_{IR} = 18{,}000 \cdot t^{-0.75}$ for $t \leq 1000$ s; $E_{IR} = 100$ for $t > 1000$ s | $W/m^2$ | | Infrared radiation eye | Eye: Cornea burn; Cataractogenesis |
| 380 ÷ 3000 (Visible, IRA, IRB) | $H_{SKIN} = 20{,}000 \cdot t^{0.25}$ for $t < 10$ s | $J/m^2$ | | Thermal skin | Skin: Skin burn |

### 7.2. Visual Comfort

#### 7.2.1. Overview

In EN 12665, visual comfort is defined as "a subjective condition of visual well-being induced by the visual environment" [265]. Lighting, daylighting, and the use of colors have a significant impact on the perception of the environment, and can affect both physical and mental well-being. Visual comfort was commonly studied through the evaluation of some factors that characterize the relationship between human needs and the enlightened environment [30].

In inadequate lighting conditions, visual discomfort may not be immediately perceived due to the effect of adaptation of the visual apparatus, but may affect work performance or lead to visual fatigue.

#### 7.2.2. Indicators

The indicators for the visual comfort evaluation can be separated in four groups: amount of light, color rendering, daylight availability, and glare. For each group, a collection of indicators is proposed as reported in Table 11.

**Table 11.** A collection of visual comfort indicators.

| Reference | Visual Comfort Metric | Symbol | Category |
|---|---|---|---|
| [266] | Illuminance | E (lx) | |
| [266] | Illuminance uniformity | U | |
| [266] | Luminance | L (cd/m$^2$) | Amount of light |
| [266] | Luminance ratio | LR | |
| [267] | Scotopic/Photopic ratio | S/P ratio | |
| [268] | Color Discrimination Index | CDI | |
| [269] | Color Preference Index | CPI | |
| [270] | Color Rendering Capacity | CRC | |
| [266] | Color Rendering Index | CRI | |
| [271] | Color Quality Scale | CQS | |
| [272] | Feeling of contrast Index | FCI | Color rendition |
| [273] | Flattery Index | | |
| [274] | Gamut Area Index | GAI | |
| [275] | Memory color quality metric | S(a) | |
| [276] | Pointer's new color rendering index | | |
| [277] | Annual Sunlight Exposure | ASE | |
| [278] | Continuous Daylight Autonomy | cDA | |
| [279] | Daylight Autonomy | DA | |
| [280] | Daylight Factor | DF | Daylight availability |
| [281] | Frequency of Visual Comfort | FVC | |
| [281] | Intensity Visual Discomfort | IVD | |
| [277] | Spatial Daylight Autonomy | sDA | |
| [282] | Useful Daylight Illuminance | UDI | |
| [283] | British Glare Index | | |
| [284] | CIE Glare Index | CGI | |
| [285] | Discomfort Glare Index | DGI | |
| [286] | Discomfort Glare Probability | DGP | |
| [287] | Enhanced simplified Discomfort Glare Probability | eDGPs | |
| [284,288] | Great-room Glare Rating | GGR | |
| [289] | Hviid simplified Discomfort Glare Probability | DGPs$_{Hviid}$ | Glare |
| [290] | J-index | J | |
| [291] | New Discomfort Glare Index | DGI$_N$ | |
| [292] | Predicted Glare Sensation Vote | PGSV | |
| [293,294] | Unified Glare Rating | UGR | |
| [284,288] | Unified Glare Rating for small sources | UGR$_{small}$ | |
| [295] | Visual Comfort Probability | VCP | |
| [296] | Wienold Simplified Discomfort Glare Probability | DGPs$_{Wienold}$ | |

Amount of light concerns the quantitative lighting parameters commonly used for the lighting design. They are generally used to describe the performance of electric lighting systems, or in other words, if the lighting systems allow creating sufficient light conditions. In this group, the most used indicator is the illuminance over the task areas.

Color rendition concerns the property of a light source to return the colors correctly. In this group the most used indicator is the color rendering index (CRI). Nevertheless, with the diffusion of LED sources, this indicator is no longer considered the most reliable. New indicators are now under investigation, as for example the color quality scale [271] and the memory color quality metric [275].

Daylight availability concerns the amount of daylight that can be exploited in an environment. This group includes quantitative indicators aimed at assessing the penetration of daylight into the environment, while some other indicators make it possible to carry out assessments on extended time profiles and to estimate the reduction in yearly electricity consumption due to the lower use of electric lighting. In this group, the most used indicator is the daylight factor (DF). However the improvement of simulation software allows to evaluate the daylight availability day by day in real condition (Climate Based daylight modelling software). With the development of such new software the creation of new indicators (e.g., useful daylight illuminance, daylight autonomy, etc.) was possible. These new indicators have great potentialities, because they allow also for optimizing the use of building automation control systems (BACS) and their use is rapidly becoming more and more common.

Glare concerns the conditions of vision in which, caused by an unsuitable luminance distribution, there is discomfort, annoyance or a reduction in visual performance and visibility. In this group, the most used indicator is the unified glare rating (UGR); however, in the literature, many indicators for the evaluation of glare from different sources were proposed.

In lighting, as in acoustics, overall indicators do not exist, and visual comfort must be evaluated as a synergic combination of the results of more indicators properly chosen for the analyzed environment. In these evaluations, at least one indicator is generally used for each category, while the benchmark values depend on the specific visual tasks carried out.

### 7.2.3. Current Research Trends

The discovery of new photoreceptors inside the human vision system able to influence the human body physiological functions has opened a new branch [297]. These new photoreceptors, that are called intrinsically photoreceptive retinal ganglion cells (ipRGCs) serve no visual function [298].

It is currently recognized that exposure to insufficient or inappropriate lighting scenarios can bother the standard human rhythms, which can have negative consequences for performance but also for health [299]. The development of new circadian metrics is based on scientific information and expert judgments related to the duration of the exposures, time of the day, intensity, light spectrum, and history of light exposures. [300]. In order to evaluate the potential effects of the light sources on the circadian rhythms, it was necessary to quantify the light exposure in biologically meaningful units [301–307]. Until now, a consensus on the appropriate minimum light exposure threshold to ensure effective circadian stimulus in buildings has not yet obtained. The WELL Building Standard's "Circadian Lighting Design" implements a minimum threshold of 250 Equivalent Melanopic Lux (EML), which must be available for at least 4 h each day. Such requirement can be met with daylight, artificial light (exclusively), or a combination of both sources [239,298,306]. Figueiro et al. [308] recommend exposure to a "circadian stimulus (CS) of 0.3 or greater at the eye for at least 1 h in the early part of the day (equivalent to 180 lux, D65)".

In any case, the relationships between spectral distribution, duration, timing, and intensity of light exposure for optimal circadian health should be further clarified by the research community.

## 8. Conclusive Remarks

The main aim of this review is to create a comprehensive framework on the indicators for the health and comfort evaluations in indoor environments that is missing today in the scientific literature. Indeed, until now, only review papers that collect indicators of single environmental factors were published.

Occupants' health and comfort represent two basic needs in indoor environment and, although their assessments are characterized by two very different approaches, in some cases, the selection of the proper indicators is not always so obvious. In fact, for the thermal environment, indoor air quality, and lighting, selection is not so obvious and some indicators are commonly proposed for both health risk and comfort evaluation.

Concerning the thermal environment, in the literature there are numerous indicators but the distinction between comfort and thermal stress indicators is not always clear. This means that, in this field, the boundary between health and comfort is particularly subtle and it is not always easy to distinguish the two approaches. Among the thermal indicators, many can be considered "overall" indicators because they are based on two or more parameters and provide direct information about the human thermal perception. In this paper, 35 heat stress indicators and 46 thermal comfort indicators were collected. Despite the large number of indicators, the most used are historically the PMV for the evaluation of thermal comfort, and the WBGT for the evaluation of thermal stress. Today, the use of adaptive comfort models combined with software simulations that can implement climate-based model are much more frequent, however there are still doubts about how these systems can estimate the actual behavior of the users (strength of the post-occupancy evaluation methods).

Concerning the air quality, few indicators were proposed, however, it is very common to evaluate the single air pollutant concentration and compare the results with threshold limit values more or less restrictive depending on the type of evaluation (i.e., minimum safety evaluation or high levels of air quality). Although the indoor environments are often characterized by pollutants concentrations higher than the external environments (where detailed monitoring takes place), only a small number of indoor pollutants can be commonly measured. In this review, nine indicators for the air quality assessment and the limit values of 17 air pollutants are reported. Given the difficulty of measuring all the indoor air pollutants and linking them to human perception, overall indicators similar to those used for the assessment of thermal comfort, were recently proposed. Such indicators are based on subjective evaluations, but at an international level, it is not yet a consensus on which indicator is most correct to use.

Concerning acoustics, the separation between health and comfort is more evident, because human exposure to noise in work environment is clearly defined in the physical agent risk assessment directive. However, it is important to observe that there are many cases where, despite compliance with the limit values, the noise exposure can lead to a series of extra-hearing side effects that must be analyzed in relation to other typical aspects of work-related stress. In addition, people are continuously exposed to noise, even when they sleep; therefore, also the noise level exposures during the whole day and/or the night-time are relevant. Regarding acoustic comfort, 39 indicators classified into three different groups (sound pressure levels, architectural acoustics, and building acoustics) are collected. Among these, overall indicators based on more parameters do not exist, so it is not possible to directly obtain information about acoustic comfort perception. For this reason, for a complete acoustic comfort assessment, it is necessary to use sets of indicators suitably chosen according to the type of environment analyzed.

Concerning lighting, it must be considered that visual and non-visual effects are related to the use of lighting systems. The visual effects are directly related to the visual performance and they are the first analyzed when the lighting of an indoor environment is evaluated. The non-visual effects should be considered because they can affect the occupants health and well-being. For the assessment of occupants' health, the literature focused on two aspects: the asthenopic effects due to the use of video terminal equipment and the exposure to artificial optical radiation. For the evaluation of the asthenopic effects, there are no limit values or specific indicators, but qualitative assessments are necessary. Artificial optical radiation is included in the physical agents assessment and characterized by a series of indicators according to the different wavelength ranges. For the assessment of visual comfort, 37 indicators are identified and divided into 4 groups (amount of light, glare, color rendition, daylight availability). In lighting, especially for the evaluation of the daylight availability, a rapid evolution connected to the development of software that allows simulation in real climate conditions is under investigation. New daylight indicators allow to combine the evaluation of lighting with the related lighting systems energy consumption, and to estimate the energy consumption reduction due to the optimization of daylight availability.

The growing interest in the IEQ, which is also reflected in the launch of new building certification systems, has led to a growing need for new overall indicators, sets of indicators, or standardized assessment procedures approved at international level. These indicators can allow to evaluate the human perception of the single environmental factor, as well as to carry out overall evaluations in which the four environmental factors are simultaneously taken into consideration. In this paper, the authors have tried to collect and organize a wide range of indicators for assessing the health and comfort of indoor environments. These indicators are not often clearly identified in the evaluation processes, but the interest in these aspects is increasing and clear information to allow the selection of the most appropriate indicators is necessary.

The present review paper represents the state of the art of the indicators of indoor environmental factors. It should be pointed out that the overall assessments of IEQ could include a selection of the indicators proposed in the literature or the proposal of new indicators that take into account several factors simultaneously. In any case, as suggested in [1], it is necessary to verify that all the aspects analyzed reach satisfactory levels, but also that the adverse impact of the interaction between these aspects is as limited as possible. In any case, when global assessments are carried out, it must be taken into consideration that the global acceptance of an environment does not depend exclusively on the achievement of adequate levels of comfort related to the four environmental factors. This is also influenced by other aspects, such as the level of expectation, the adaptation capacity, personal subjective preferences, as well as the psychological and physiological conditions of the occupants.

**Author Contributions:** All the authors contributed in equal parts to the research activity and to the paper writing. All authors have read and agreed to the published version of the manuscript.

**Funding:** This research received no external funding.

**Conflicts of Interest:** The authors declare no conflict of interest.

## References

1. American Society of Heating, Refrigerating and Air-Conditioning Engineering (ASHRAE). *Guideline 10P, Interactions Affecting the Achievement of Acceptable Indoor Environments*; ASHRAE: Atlanta, GA, USA, 2016.
2. Pinto, S.; Fumincelli, L.; Mazzo, A.; Caldeira, S.; Martins, J.C. Comfort, well-being and quality of life: Discussion of the differences and similarities among the concepts. *Porto Biomed. J.* **2017**, *2*, 6–12. [CrossRef]
3. Directive (EU). *2018/844 of the European Parliament and of the Council of 30 May 2018 amending Directive 2010/31/EU on the Energy Performance of Buildings and Directive 2012/27/EU on Energy Efficiency*; EU: Brussels, Belgium, 2018.
4. Directive 2012/27/EU of the European Parliament and of the Council of 25 October 2012 on Energy Efficiency, Amending Directives 2009/125/EC and 2010/30/EU and Repealing Directives 2004/8/EC and 2006/32/EC. Available online: https://eur-lex.europa.eu/legal-content/EN/TXT/?uri=celex%3A32012L0027 (accessed on 7 January 2020).
5. Directive 2010/31/EU of the European Parliament and of the Council of 19 May 2010 on The Energy Performance of Buildings. Available online: https://eur-lex.europa.eu/legal-content/EN/TXT/?uri=CELEX%3A32010L0031 (accessed on 7 January 2020).
6. Directive 2002/91/EC of the European Parliament and of the Council of 16 December 2002 on the energy performance of buildings. Available online: https://eur-lex.europa.eu/legal-content/EN/TXT/?uri=LEGISSUM%3Al27042 (accessed on 7 January 2020).
7. Bluyssen, P.M. Towards new methods and ways to create healthy and comfortable buildings. *Build. Environ.* **2010**, *45*, 808–818. [CrossRef]
8. World Health Organization (WHO). Definition of Health. Available online: http://www.who.int/about/definition/en/print.html (accessed on 12 March 2019).

9.   Commission of the European Communities. *White Paper—Together for Health. A Strategic Approach for the EU 2008–2013*; Commission of the European Communities: Brussels, Belgium, 2007; Available online: https://ec.europa.eu/health/ph_overview/Documents/strategy_wp_en.pdf (accessed on 6 June 2019).

10.  Tang, H.; Ding, Y.; Singer, B. Interactions and comprehensive effect of indoor environmental quality factors on occupant satisfaction. *Build. Environ.* **2020**, *167*, 106462. [CrossRef]

11.  Shan, A.; Nissa Melina, A.; Yang, E.H. Impact of indoor environmental quality on students' wellbeing and performance in educational building through life cycle costing perspective. *J. Clean. Prod.* **2018**, *204*, 298–309. [CrossRef]

12.  Lee, E. Indoor environmental quality (IEQ) of LEED-certified home: Importance-performance analysis (IPA). *Build. Environ.* **2019**, *149*, 571–581. [CrossRef]

13.  Geng, Y.; Lin, B.; Yu, J.; Zhou, H.; Ji, W.; Chen, H.; Zhang, Z.; Zhu, Y. Indoor environmental quality of green office buildings in China: Large-scale and long-term measurement. *Build. Environ.* **2019**, *150*, 266–280. [CrossRef]

14.  Anand, P.; Cheong, D.; Sekhar, C.; Santamouris, M.; Kondepudi, S. Energy saving estimation for plug and lighting load using occupancy analysis. *Renew. Energy* **2019**, *143*, 1143–1161. [CrossRef]

15.  European Committee for Standardization. *EN 15251: Indoor Environmental Input Parameters for Design and Assessment of Energy Performance of Buildings Addressing Indoor Air Quality, Thermal Environment, Lighting and Acoustics*; European Committee for Standardization: Brussels, Belgium, 2007.

16.  European Committee for Standardization. Part 1: Indoor Environmental Input Parameters for Design and Assessment of Energy Performance of Buildings Addressing Indoor Air Quality, Thermal Environment Lighting and Acoustics. Module M1-6. In *EN 16798-1 Energy Performance of Buildings. Ventilation for Buildings*; European Committee for Standardization: Brussels, Belgium, 2019.

17.  European Committee for Standardization. *PD CEN/TR 16798-2 Energy Performance of Buildings. In Ventilation for Buildings; Part 2: Interpretation of the Requirements in EN 16798-1. Indoor Environmental Input Parameters for Design and Assessment of Energy Performance of Buildings Addressing Indoor Air Quality, Thermal Environment, Lighting and Acoustics. Module M1-6*; European Committee for Standardization: Brussels, Belgium, 2019.

18.  Anand, P.; Deb, C.; Alur, R. A simplified tool for building layout design based on thermal comfort simulations. *Front. Archit. Res.* **2017**, *6*, 218–230. [CrossRef]

19.  Redlich, C.A.; Sparer, J.; Cullen, M.R. Sick-building syndrome. *Lancet* **1997**, *349*, 1013–1016. [CrossRef]

20.  Burge, P.S. Sick Building Syndrome. *Occup. Environ. Med.* **2004**, *61*, 185–190. [CrossRef]

21.  Ghaffarianhoseini, A.; AlWaer, H.; Omrany, H.; Ghaffarianhoseini, A.; Alalouch, C.; Clements-Croome, D.; Tookey, J. Sick building syndrome: Are we doing enough? *Archit. Sci. Rev.* **2018**, *61*, 99–121. [CrossRef]

22.  World Health Organization (WHO). *Indoor Air Pollutants: Exposure and Health Effects*; EURO Reports and Studies 78: 1–42.; WHO: Geneva, Switzerland, 1983.

23.  Kamaruzzaman, S.; Sabrani, N.A. The Effect of Indoor Air Quality (IAQ) Towards Occupants' Psychological Performance in Office Buildings. *J. Rekabentuk Dan Binaan* **2011**, *4*, 49–61.

24.  Runeson-Broberg, R.; Norbäck, D. Sick Building Syndrome (SBS) and Sick House Syndrome (SHS) in Relation to Psychosocial Stress at Work in the Swedish Workforce. *Int. Arch. Occup. Environ. Health* **2013**, *86*, 915–922. [CrossRef] [PubMed]

25.  De Freitas, C.R.; Grigorieva, E.A. A comprehensive catalogue and classification of human thermal climate indices. *Int. J. Biometeorol.* **2015**, *59*, 109–120. [CrossRef]

26.  D'Ambrosio Alfano, F.R.; Palella, B.I.; Riccio, G. Thermal Environment Assessment Reliability Using Temperature-Humidity Indices. *Ind. Health* **2011**, *49*, 95–106. [CrossRef]

27.  Carlucci, S.; Pagliano, L. A review of indices for the long-term evaluation of the general thermal comfort conditions in buildings. *Energy Build.* **2012**, *53*, 194–205. [CrossRef]

28.  Beshir, M.Y.; Ramsey, J.D. Heat stress indices: A review paper. *Int. J. Ind. Ergon.* **1988**, *3*, 89–102. [CrossRef]

29.  Djongyang, N.; Tchinda, R.; Njomo, D. Thermal comfort: A review paper. *Renew. Sustain. Energy Rev.* **2010**, *14*, 2626–2640. [CrossRef]

30.  Carlucci, S.; Causone, F.; De Rosa, F.; Pagliano, L. A review of indices for assessing visual comfort with a view to their use in optimization processes to support building integrated design. *Renew. Sustain. Energy Rev.* **2015**, *47*, 1016–1033. [CrossRef]

31.  Sundell, J. On the history of indoor air quality and health. *Indoor Air* **2004**, *14*, 51–58. [CrossRef]

32. European Agency of Safety and Health at Work (EU-OSHA). *Well-Being at Work: Creating a Positive Work Environment*; EU-OSHA: Luxembourg, 2013.

33. Institution of Occupational Safety and Health (IOSH). *Working Well. Guidance on Promoting Health and Wellbeing at Work*; IOSH: Leicestershire, UK, 2018.

34. Public Health Agency (HSC). *Health and Wellbeing at Work: A Resource Guide*; HSC: Belfast, Northern Ireland, 2017.

35. Anand, P.; Sekhar, C.; Cheong, D.; Santamouris, M.; Kondepudi, S. Occupancy-based zone-level VAV system control implications on thermal comfort, ventilation, indoor air quality and building energy efficiency. *Energy Build.* **2019**, *204*, 109473. [CrossRef]

36. American Society of Heating, Refrigerating and Air-Conditioning Engineering (ASHRAE). "TC 1.6 Terminology", Atlanta, USA. Available online: http://tc0106.ashraetcs.org/ (accessed on 21 January 2019).

37. Fisk, W.J. Health and productivity gains from better indoor environments and their relationship with building energy efficiency. *Annu. Rev. Energy Environ.* **2000**, *25*, 537–566. [CrossRef]

38. Andersson, J.; Boerstra, A.; Clements-Croome, D.; Fitzner, K.; Hanssen, S.O. *Indoor Climates and Productivity in Offices*; Wargocki, P., Seppänen, O., Eds.; REHVA: Ixelles, Belgium, 2006; Volume 6.

39. Seppanen, O.A. Association of ventilation rates and CO2 concentrations with health and other responses in commercial and institutional buildings. *Indoor Air* **1999**, *9*, 226–252. [CrossRef]

40. De Giuli, V.; Da Pos, O.; De Carli, M. Indoor environmental quality and pupil perception in Italian primary schools. *Build. Environ.* **2012**, *56*, 335–345. [CrossRef]

41. Parsons, K.C. *Human Thermal Environments: The Effects of Hot, Moderate, and Cold Environments on Human Health, Comfort and Performance*, 3nd ed.; Taylor and Francis: London, UK, 2014.

42. Epstein, Y.; Moras, D.S. Thermal Comfort and the Heat Stress Indices. *Ind. Health* **2006**, *44*, 388–398. [CrossRef] [PubMed]

43. Morgado, M.; Talaia, M.; Teixeira, L. A new simplified model for evaluating thermal environment and thermal sensation: An approach to avoid occupational disorders. *Int. J. Ind. Ergon.* **2017**, *60*, 3–13. [CrossRef]

44. European Agency for Occupational Health and Safety at Work (EU-OSHA)—OSHwiki. Available online: http:$\delimiter"026E30F$$\delimiter"026E30F$oshwiki.eu (accessed on 12 March 2019).

45. Council Directive 89/391/EEC of 12 June 1989 on the Introduction of Measures to Encourage Improvements in the Safety and Health of Workers at Work. Available online: https://eur-lex.europa.eu/legal-content/EN/TXT/?uri=CELEX:52006PC0390 (accessed on 7 January 2020).

46. Workplace Safety & Health Division of the Manitoba Government. *Guideline for Thermal Stress*; Workplace Safety & Health Division of the Manitoba Government: Winnipeg, MB, Canada, 2007. Available online: https://www.gov.mb.ca/labour/safety/ (accessed on 21 January 2019).

47. Italian Legislative Decree, No. 81/2008 "Code on Health and Safety Protection of Employees in the workplace", implementation of Article 1 of Law 3 August 2007, No 123 on Health and Safety in the workplaces. Available online: https://www.ispettorato.gov.it/it-it/Documenti-Norme/Documents/Testo-Unico-Dlgs-81-08-edizione-di-luglio-2018.pdf (accessed on 7 January 2020).

48. Malchaire, J.B.; Gebhardt, H.J.; Piette, A. Strategy for evaluation and prevention of risk due to work in thermal environment. *Ann. Occup. Hyg.* **1999**, *43*, 367–376. [CrossRef]

49. Malchaire, J.B. Occupational heat stress assessment by the Predicted Heat Strain model. *Ind. Health* **2006**, *44*, 380–387. [CrossRef]

50. Palella, B.I.; Quaranta, F.; Riccio, G. On the management and prevention of heat stress for crews on board ships. *Ocean Eng.* **2016**, *112*, 277–286. [CrossRef]

51. Taleghani, M.; Tenpierik, M.; Kurvers, S.; van den Dobbelsteen, A. A review into thermal comfort in buildings. *Renew. Sustain. Energy Rev.* **2013**, *26*, 201–215. [CrossRef]

52. Yaglou, C.P.; Minard, D. Control of heat casualties at military training centers. *Am. Med. Assoc. Arch. Ind. Health* **1957**, *16*, 302–316.

53. Siple, P.A.; Passel, C.F. Measurement of dry atmospheric cooling in subfreezing temperatures. *Proc. Am. Phil. Soc.* **1945**, *89*, 177–199. [CrossRef]

54. Holmer, I. Required Clothing Insulation (IREQ) as an Analytical Index of the Cold Stress. *Ashrae Trans.* **1984**, *90*, 1116–1128.

55. De Freitas, C.R.; Symon, L.V. A bioclimatic index of human survival times in the Antarctic. *Polar Rec.* **1987**, *23*, 651–659. [CrossRef]

56. Shitzer, A. Wind-chill-equivalent temperatures: Regarding the impact due to the variability of the environmental convective heat transfer coefficient. *Int. J. Biometeorol.* **2006**, *50*, 224–232. [CrossRef]

57. International Organization for Standardization. *ISO 11079. Ergonomics of the Thermal Environment—Determination and Interpretation of Cold Stress When Using Required Clothing Insulation (IREQ) and Local Cooling Effects*; International Organization for Standardization: Geneve, Switzerland, 2007.

58. Robinson, S.; Turrell, E.S.; Gerking, S.D. Physiologically equivalent conditions of air temperature and humidity. *Am. J. Physiol.* **1945**, *143*, 21–32. [CrossRef]

59. McArdle, B.; Dunham, W.; Holling, H.E.; Ladel, W.S.S.; Scott, J.W.; Thomson, M.L.; Weiner, J.S. The Prediction of the Physiological Effects of Warm and Hot Environments. Available online: https://www.scienceopen.com/document?vid=f0cf60ca-b22b-4615-be54-548ad606fd56 (accessed on 7 January 2020).

60. Belding, H.S.; Hatch, T.F. Index for evaluating heat stress in terms of resulting physiological strain. *Heat. Pip. Air Condit.* **1955**, *27*, 129–136.

61. Lee, D.H.K. Proprioclimates of Man and Domestic Animals. Climatology, Arid Zone Research—X, UNESCO. 1958, p. 102. Available online: https://unesdoc.unesco.org/ark:/48223/pf0000179751 (accessed on 6 June 2019).

62. Lally, V.E.; Watson, B.F. Humiture revisited. *Weatherwise* **1960**, *13*, 254–256. [CrossRef]

63. Givoni, B. The influence of work and environmental conditions on the physiological responses and thermal equilibrium of man. *Arid Zone Res.* **1962**, *24*, 199–204.

64. McKarns, J.S.; Brief, R.S. Nomographs give refined estimate of heat stress index. *J. Occup. Med.* **1966**, *8*, 557.

65. Fuller, F.H.; Brouha, L. New engineering methods for evaluating the job environment. *Ashrae J.* **1966**, *8*, 39–52.

66. Lind, A.R. Effect of individual variation on upper limit of perspective zone of climates. *J. Appl. Physiol.* **1970**, *28*, 57–62. [CrossRef] [PubMed]

67. Botsford, J.H. A wet globe thermometer for environmental heat measurement. *Am. Ind. Hyg. Assoc. J.* **1971**, *32*, 1–10. [CrossRef] [PubMed]

68. Kerslake, D.M. *The Stress of Hot Environment*; Cambridge University Press: Cambridge, UK, 1972.

69. Givoni, B.; Pandolf, R.R. Predicting heart rate response to work, environment and clothing. *J. Appl. Physiol.* **1973**, *34*, 201–204. [CrossRef] [PubMed]

70. Gonzalez, R.R.; Bergulnd, L.G.; Gagge, A.P. Indices of thermoregulatory strain for moderate exercise in the heat. *J. Appl. Physiol.* **1978**, *44*, 889–899. [CrossRef] [PubMed]

71. Nunneley, S.H.; Stribley, F. Fighter index of thermal stress (FITS): Guidance for hot-weather aircraft operations. *Aviat. Space Environ. Med.* **1979**, *50*, 639–642. [PubMed]

72. Kamon, E.; Ryan, C. Effective heat strain index using pocket computer. *Am. Ind. Hyg. Assoc. J.* **1981**, *42*, 611–615. [CrossRef]

73. Shapiro, Y.; Pandolf, K.B.; Goldman, R.F. Predicting sweat loss response to exercise, environment and clothing. *Eur. J. Appl. Physiol. Occup. Physiol.* **1982**, *48*, 83–96. [CrossRef]

74. Höppe, P. Die Energiebilanz des Menschen. Ph.D. Thesis, University of München, Munich, Germany, 1984.

75. De Freitas, C.R. Assessment of human bioclimate based on thermal response. *Int. J. Biometeorol.* **1985**, *29*, 97–119. [CrossRef]

76. Frank, A.; Moran, D.; Epstein, Y.; Belokopytov, M.; Shapiro, Y. The Estimation of Heat Tolerance by a New Cumulative Heat Strain Index. Available online: https://www.lboro.ac.uk/microsites/lds/EEC/ICEE/textsearch/1996/Frank-1996.pdf (accessed on 7 January 2020).

77. Moran, D.S.; Shitzer, A.; Pandolf, K.B. A physiological strain index to evaluate heat stress. *Am. J. Physiol.* **1998**, *275*, 129–134. [CrossRef]

78. International Standard Organization (ISO). *Hot Environments—Analytical Determination and Interpretation of Thermal Stress Using Calculation of Required Sweat Rate*; ISO: Geneva, Switzerland, 1989.

79. Malchaire, J.; Piette, A.; Kampmann, B.; Mehnert, P.; Gebhard, H.; Havenith, G.; Den hartog, E.; Holmer, I.; Parsons, K.; Alfano, G.; et al. Development and Validation of the Predicted Heat Strain Model. *Ann. Occup. Hyg.* **2001**, *45*, 123–135. [CrossRef]

80. Moran, D.S.; Pandolf, K.B.; Shapiro, Y.; Heled, Y.; Shani, Y.; Matthew, W.T.; Gonzales, R.R. An environmental stress index (ESI) as a substitute for the wet bulb globe temperature (WBGT). *J. Therm. Biol.* **2001**, *26*, 427–431. [CrossRef]

81. Wallace, R.F.; Kriebel, D.; Punnett, L.; Wegman, D.H.; Wenger, C.B.; Gardner, J.W.; Gonzalez, R.R. The effects of continuous hot weather training on risk of exertional heat illness. *Med. Sci. Sports Exerc.* **2005**, *37*, 84–90. [CrossRef] [PubMed]

82. European Committee for Standardization. *EN ISO 11079. Ergonomics of the Thermal Environment—Determination and Interpretation of Cold Stress When Using Required Clothing Insulation (IREQ) and Local Cooling Effects (ISO 11079:2007)*; European Committee for Standardization: Brussel, Belgium, 2007.

83. Robinson, D.; Haldi, F. Model to predict overheating risk based on an electrical capacitor analogy. *Energy Build.* **2007**, *40*, 1240–1245. [CrossRef]

84. Nicol, F.; Hacker, J.; Spires, B.; Davies, H. Suggestion for new approach to overheating diagnostics. In *Proceedings of Air Conditioning and the Low Carbon Cooling Challenge*; Cumberland Lodge: Windsor, UK, 2008.

85. Brake, D.J.; Bates, G.P. Limiting Metabolic Rate (Thermal Work Limit) as an Index of Thermal Stress. *Appl. Occup. Environ. Hyg.* **2010**, *17*, 176–186. [CrossRef]

86. Sakoi, T.; Mochida, T. Concept of the equivalent wet bulb globe temperature index for indicating safe thermal occupational environments. *Build. Environ.* **2013**, *67*, 167–178. [CrossRef]

87. D'Ambrosio Alfano, F.R.; Malchaire, J.; Palella, B.I.; Riccio, G. The WBGT index revisited after 60 years of use. *Ann. Occup. Hyg.* **2014**, *58*, 955–970.

88. Olesen, B.W.; d'Ambrosio, F.R.; Parsons, K.; Palella, B.I. The creation of a single indicator that allows the evaluation of the perceived thermal stress in a wide range of environmental conditions and physiological activities required numerous attempts. In *Proceeding of the 9th Windsor Conference on "Making Comfort Relevant"*; Cumberland Lodge: Windsor, UK, 2016.

89. D'Ambrosio Alfano, F.R.; Palella, B.I.; Riccio, G. Notes on the Implementation of the IREQ Model for the Assessment of Extreme Cold Environments. *Ergonomics* **2013**, *56*, 707–724. [CrossRef]

90. American Society of Heating, Refrigerating and Air-Conditioning Engineering (ASHRAE). *Thermal Environmental Conditions for Human Occupancy*; Standard 55-2004; ASHRAE: Atlanta, GA, USA, 2017.

91. Fanger, P.O. *Thermal Comfort*; Danish Technical Press: Copenhagen, Denmark, 1970.

92. Hamdi, M.; Lachiver, G.; Michaud, F. A new predictive thermal sensation index of human response. *Energy Build.* **1999**, *29*, 167–178. [CrossRef]

93. D'Ambrosio Alfano, F.R.; Palella, B.I.; Riccio, G.; Toftum, J. Fifty years of Fanger's equation: Is there anything to discover yet? *Int. J. Ind. Ergon.* **2018**, *66*, 157–160. [CrossRef]

94. D'Ambrosio Alfano, F.R.; Olesen, B.W.; Palella, B.I. Povl Ole Fanger's Impact Ten Years Later. *Energy Build.* **2017**, *152*, 243–249. [CrossRef]

95. Yang, L.; Yan, H.; Lam, J.C. Thermal comfort and building energy consumption implications—A review. *Appl. Energy* **2014**, *115*, 164–173. [CrossRef]

96. De Dear, R.; Brager, G.; Cooper, D. *Developing an Adaptive Model of Thermal Comfort and Preference*; RP-884, Final Report; ASHRAE: Berkeley, CA, USA, 1997.

97. Beccali, M.; Nucara, A.; Rizzo, G. Thermal Comfort. In *The Encyclopaedia of Energy*; Cutler, J.C., Ed.; Elsevier Science: Amsterdam, The Netherlands, 2004; pp. 55–64. ISBN 0-12-176480-X.

98. De Dear, R.J.; Akimoto, T.; Arens, E.A.; Brager, G.; Candido, C.; Cheong, K.W.D.; Li, B.; Nishihara, N.; Sekhar, S.C.; Tanabe, S. Progress in Thermal Comfort Research over the Last Twenty Years. *Indoor Air* **2013**, *23*, 442–461. [CrossRef] [PubMed]

99. Haldane, J.S. The influence of high air temperature. *Hygiene* **1905**, *5*, 494–513.

100. Hill, L.; Griffith, O.W.; Flack, M. The measurement of the rate of heat-loss at body temperature by convection, radiation, and evaporation. *Philos. Trans. Res. Soc. Lond. B* **1916**, *207*, 183–220. [CrossRef]

101. Vernon, H.M. The measurements of radiant heat in relation to human comfort. Proceeding of the Physiological Society. *J. Physiol.* **1930**, *70*. [CrossRef]

102. Gagge, A.P.; Rapp, G.M.; Hardy, J.D. Effective radiant field and operative temperature necessary for comfort with radiant heating. *Ashrae Trans.* **1967**, *73*, 2.1–2.9.

103. Hevener, O.F. All about humiture. *Weatherwise* **1959**, *12*, 56–85. [CrossRef]

104. Craig, F.N. Relation between heat balance and physiological strain in walking men clad in ventilated impermeable. *Fed. Proc.* **1950**, *9*, 26.

105. Lind, A.R.; Hallon, R.F. Assessment of physiologic severity of hot climate. *J. Appl. Physiol.* **1957**, *11*, 35–40. [CrossRef]

106. Thom, E.C. The discomfort index. *Weatherwise* **1959**, *12*, 57–61. [CrossRef]

107. Tennenbaum, J.; Sohar, E.; Adar, R.; Gilat, T.; Yaski, D. The physiological significance of the cumulative discomfort index (Cum DI). *Harefuah* **1961**, *60*, 315–319. [PubMed]

108. Hall, J.F.K.; Polte, W. Physiological index of strain and body heat storage in hyperthermia. *J. Appl. Physiol.* **1960**, *15*, 1027–1030. [CrossRef] [PubMed]

109. Lee, D.H.K.; Henschel, A. *Evaluation of Thermal Environment in Shelters, Department of Health, Education and Welfare*; Public Health Service, Division of Occupational Health: Washington, DC, USA, 1963.

110. Nieuwolt, S. *Tropical Climatology: An Introduction to the Climates of the Low Latitudes*; John Wiley and Sons: London, UK, 1977.

111. Masterton, J.M.; Richardson, F.A. *Humidex, a Method of Quantifying Human Discomfort Due to Excessive Heat and Humidity*; Environment Canada: Ottawa, ON, Canada, 1979.

112. Weiss, M. The humisery and other measures of summer discomfort. *Nat. Wea. Dig.* **1982**, *7*, 10–18.

113. Pepi, W.J. The Summer Simmer Index. *Weatherwise* **1987**, 40. [CrossRef]

114. Moran, D.S.; Shapiro, Y.; Epstein, Y.; Matthew, W.; Pandolf, K.B. A Modified Discomfort Index (MDI) as an Alternative to the Wet Bulb Globe Temperature (WBGT). In *Environmental Ergonomics VIII, Proceedings of the 8th International Conference on Environmental Ergonomics, 18–23 October 1998, San Diego, CA, USA*; Naval Health Research Center and San Diego State University: San Diego, CA, USA, 1998; pp. 77–80.

115. Pepi, W.J. The New Summer Simmer Index. In Proceedings of the 80th Annual Meeting of the AMS, Long Beach, CA, USA, 9–14 January 2000.

116. Chartered Institution of Building Services Engineers. *CIBSE Guide a Chapter 1: Environmental Criteria for Design*; CIBSE: London, UK, 2006.

117. Talaia, M.; Simoes, H. EsConTer: Um índice de avaliaçao de ambiente termico. In *Proceedings of the V Congresso Cubano de Meteorologia, Somet-Cuba*; Sociedade de Meteorologia de Cuba: La Havana, Cuba, 2009; pp. 1612–1626.

118. Borgeson, S.; Brager, G.S. Comfort standards and variations in exceedance for mixed-mode buildings. *Build. Res. Inf.* **2010**, *39*, 118–133. [CrossRef]

119. Houghton, F.C.; Yaglou, C.P. Determining equal comfort lines. *Am. Soc. Heat. Vent. Eng.* **1923**, *29*, 165–176.

120. Dufton, A.F. The eupatheostat. *Sci. Instrum.* **1929**, *6*, 249–251. [CrossRef]

121. Vernon, H.M.; Warner, C.G. The influence of the humidity of the air on capacity for work at high temperature. *Hygiene* **1932**, *32*, 431–463. [CrossRef]

122. Gagge, A.P. Man, his environment, his comfort. *Heat. Pip. Air Cond.* **1969**, *41*, 209–224.

123. Winslow, C.E.A.; Herrington, L.P.; Gagge, A.P. Physiological reactions and sensations of pleasantness under varying atmospheric conditions. *Trans. Ashve* **1937**, *44*, 179–196.

124. Ionides, M.; Plummer, J.; Siple, P.A. The thermal acceptance ratio. Interm. report No. 1, Climatology and envelope. *Fed. Proc.* **1945**, *9*, 26.

125. Bedford, T. *Environmental Warmth and Its Measurement*; H.M.S.O: London, UK, 1946.

126. Missenard, A. A thermique des ambiences: Équivalences de passage, équivalences de séjours. *Chal. Indust* **1948**, *276*, 159–172.

127. Webb, C.G. An analysis of some observations of thermal comfort in an equatorial climate. *Brit. J. Ind. Med.* **1959**, *16*, 297–310. [CrossRef] [PubMed]

128. Sohar, E.; Tennenbaum, D.J.; Robinson, N. *A Comparison of the Cumulative Discomfort Index (Cum DI) and Cumulative Effective Temperature (Cum ET), as Obtained by Meteorological Data*; Tromp, S.W., Ed.; Biometeorology, Pergamon Press: Oxford, UK, 1962; pp. 395–400.

129. Gagge, A.P.; Stolwijk, A.; Nishi, Y. An effective temperature scale based on a simple model of human physiological regulatory response. *Ashrae Trans.* **1971**, *77*, 247–257.

130. Nishi, Y.; Gagge, A.P. Humid operative temperature. A biophysical index of thermal sensation and discomfort. *J. Physiol.* **1971**, *63*, 365–368.

131. Steadman, R.G. Indices of wind chill of clothed persons. *J. Appl. Meteor.* **1971**, *10*, 674–683. [CrossRef]

132. Gonzalez, R.R.; Nishi, Y.; Gagge, A.P. Experimental evaluation of standard effective temperature: A new biometeorological index of man's thermal discomfort. *Int. J. Biometeorol.* **1974**, *18*, 1–15. [CrossRef]

133. Steadman, R.G. The Assessment of Sultriness. Part I: A Temperature-Humidity Index Based on Human Physiology and Clothing Science. *J. Appl. Meteor.* **1979**, *18*, 861–873. [CrossRef]

134. Wray, W.O. A simple procedure for assessing thermal comfort in passive solar heated buildings. *Sol. Energy* **1980**, *25*, 327–333. [CrossRef]

135. Gagge, A.P.; Fobelets, A.P.; Berglund, L.G. A standard predictive index of human response to the thermal environment. *Ashrae Trans.* **1986**, *92*, 709–731.

136. Bureau of Indian Standards. *Handbook of Functional Requirements of Buildings (Other than Industrial Buildings)*; Bureau of Indian Standards: New Delhi, India, 1987.

137. Guide, J. *Weather, Solar and Illuminance Data, Chartered Institution of Building Services Engineers*; CIBSE: London, UK, 2002; Volume 8.

138. International Standard Organization (ISO). *7730: Ergonomics of the Thermal Environment—Analytical Determination and Interpretation of Thermal Comfort Using Calculation of the PMV and PPD Indices and Local Thermal Comfort Criteria*, 3rd ed.; ISO: Geneva, Switzerland, 2005.

139. World Health Organizzation (WHO). *WHO Air Quality Guidelines for Particulate Matter, Ozone, Nitrogen Dioxide and Sulfur Dioxide*; Summary of Risk Assessment; WHO: Geneva, Switzerland, 2005.

140. Norhidayah, A.; Lee, C.K.; Azhar, M.K.; Nurulwahida, S. Indoor air quality and sick building syndrome in three selected buildings. *Procedia Eng.* **2013**, *53*, 93–98. [CrossRef]

141. Yang, J.; Pantazaras, A.; Chaturvedi, K.A.; Chandran, A.K.; Santamouris, M.; Lee, S.E.; Tham, K.W. Comparison of different occupancy counting methods for single system-single zone applications. *Energy Build.* **2018**, *172*, 221–234. [CrossRef]

142. Sekhar, C.; Anand, P.; Schiavon, S.; Tham, K.W.; Cheong, D.; Saber, E.M. Adaptable cooling coil performance during part loads in the tropics—A computational evaluation. *Energy Build.* **2018**, *159*, 148–163. [CrossRef]

143. Bluyssen, P.M. Towards an integrative approach of improving indoor air quality. *Build. Environ.* **2009**, *44*, 1980–1989. [CrossRef]

144. Jansz, J. Sick Building Syndrome. In *International Encyclopedia of Public Health*, 2nd ed.; Academic Press: Cambridge, MA, USA, 2017; pp. 502–505.

145. Crook, B.; Burton, N.C. Indoor moulds, Sick Building Syndrome and building related illness. *Fungal Biol. Rev.* **2010**, *24*, 106–113. [CrossRef]

146. U.S. Environmental Protection Agency (EPA). Guidelines for Ecological Risk Assessment. Federal Register 63. pp. 26846–26924. Available online: https://www.epa.gov/sites/production/files/2014-11/documents/eco_risk_assessment1998.pdf (accessed on 16 February 2019).

147. World Health Organization. The Right to Healthy Indoor Air. In Proceedings of the WHO Meeting, Bilthoven, The Netherlands, 15–17 May 2000.

148. World Health Organization. *Health 21—An Introduction to the Health for all Policy Framework for the WHO European Region: European Health for all Series No 5*; World Health Organization Europe Regional Office: Copenhagen, Denmark, 2000.

149. World Health Organization (WHO). *Air Pollution: Fifth Report of the Expert Committee on Environmental Sanitation*; WHO Technical Report Series, No. 157; World Health Organization: Geneva, Switzerland, 1958; Available online: http://apps.who.int/iris/handle/10665/40416 (accessed on 6 December 2018).

150. World Health Organization (WHO). *Atmospheric Pollutants: Report of a WHO Expert Committee*; WHO Technical Report Series, No. 271; World Health Organization: Geneva, Switzerland, 1964; Available online: http://apps.who.int/iris/handle/10665/40578 (accessed on 6 December 2018).

151. World Health Organization (WHO). *Air Quality Criteria and Guides for Urban Air Pollutants: Report of a WHO Expert Committee*; WHO Technical Report Series, No. 506; World Health Organization: Geneva, Switzerland, 1972; Available online: http://apps.who.int/iris/handle/10665/40989 (accessed on 6 December 2018).

152. World Health Organization (WHO). *Regional Office for Europe*; Air Quality Guidelines for Europe; World Health Organization Europe Regional Office: Copenhagen, Denmark, 1987.

153. World Health Organization (WHO). *Air Quality Guidelines for Europe. Second Edition*; European Health for all Series No 91; World Health Organization Europe Regional Office: Copenhagen, Denmark, 2000.

154. European Commission. Council directive 96/62/EC of 27 September 1996 on Ambient Air Quality Assessment and Management 96/62/EC. Available online: https://eur-lex.europa.eu/legal-content/EN/TXT/?uri=CELEX%3A31996L0062 (accessed on 10 January 2020).

155. World Health Organization (WHO) Regional Office for Europe. *Air Quality Guidelines*; Global Update; WHO: Geneva, Switzerland, 2005.

156. World Health Organization (WHO) Regional Office for Europe. *WHO Guidelines for Indoor Air Quality: Dampness and Mould*; World Health Organization Europe Regional Office: Copenhagen, Denmark, 2009.

157. World Health Organization (WHO) Regional office for Europe. *Selected Pollutants: Who Guideline for Indoor Air Quality*; World Health Organization Europe Regional Office: Copenhagen, Denmark, 2010.

158. World Health Organization (WHO) Regional office for Europe. *Indoor Air Quality Guidelines: Household Fuel Combustion*; World Health Organization Europe Regional Office: Copenhagen, Denmark, 2014.

159. American Society of Heating, Refrigerating and Air-Conditioning Engineering (ASHRAE). *Indoor Air Guide*; Environmental Protection Agency: Atlanta, GA, USA, 2009.

160. Anand, P. Development of a Data-Driven Occupancy-Based Building Energy and IEQ Model. Ph.D. Thesis, National University of Singapore, Singapore, 2019.

161. American Society of Heating, Refrigerating and Air-Conditioning Engineering (ASHRAE). *Ventilation and Acceptable Indoor Air Quality*; ANSI/ASHRAE standard 62.1; ASHRAE: Atlanta, GA, USA, 2013.

162. Fanger, P.O. Introduction of the olf and the decipol Units to Quantify Air Pollution Perceived by Humans Indoors and Outdoors. *Energy Build.* **1987**, *12*, 1–6. [CrossRef]

163. Cohas, M. *Ventilation et Qualité de l'Air dans l'Habitat*, 5th ed.; Les éditions Parisiennes (EDIPA): Paris, France, 1996.

164. Kirchner, S.; Jedor, B.; Mandin, C. Elaboration D'indices de la Qualite de L'Air Interieur: Phase 1: Inventaire des Indices Disponibles. 2015. Available online: http://www.oqai.fr/userdata/documents/483_Inventaire_OQAI_Indices_2006.pdf (accessed on 9 February 2019).

165. Indoor Air Quality Management Group. A Guide on Indoor Air Quality Certification Scheme for Offices and Public Places. 2015. Available online: http://www.iaq.gov.hk/en/publications-and-references/guidance-notes.aspx (accessed on 9 February 2019).

166. Chiang, C.M.; Lai, C.M. A study on the comprehensive indicator of indoor environment assessment for occupants' health in Taiwan. *Build. Environ.* **2002**, *37*, 387–392. [CrossRef]

167. Moschandreas, D.; Sofuoglu, S.C. The indoor air pollution index. *Indoor Air* **1999**, *99*, 261–266.

168. Moschandreas, D.J.; Sofuoglu, S.C. The indoor environmental index and its relationship with symptoms of office building occupants. *J. Air Waste Manag. Assoc.* **2004**, *54*, 1440–1451. [CrossRef]

169. Rojas, G.; Pfluger, R.; Feist, W. Ventilation concepts for energy efficient housing in Central European climate—A simulation study comparing IAQ, mold risk and ventilation losses. In Proceedings of the 14th International Conference on Indoor Air Quality and Climate, Indoor Air 2016, Ghent, Belgium, 3–8 July 2016.

170. Gadeau, A.L. Assessment of ventilation strategies using an air quality index introduced in CLIM 2000 software, Espoo, Finland. *Proc. Healthy Build.* **1996**, *4*, 23–24.

171. Castanet, S. Contribution À L'étude de la Ventilation Et de la Qualité de L'air Intérieur des Locaux. Ph.D. Thesis, INSA, Lyon, France, 1998.

172. QUAD-BBC. *Choix de Paramètres de Suivi de la Qualité de L'air Intérieur*; Report T1.2 for the project QUAD-BBC; International Energy Agency: Paris, France, 2012.

173. Sofuoglu, S.C.; Moschandreas, D.J. The link between symptoms of office building occupants and in-office air pollution: The Indoor Air Pollution Index. *Indoor Air* **2003**, *13*, 332–343. [CrossRef] [PubMed]

174. Marchand, D.; Belair, F.; Kirchner, S. Indices de qualité d'air intérieur: Vers une culture du risque sanitaire. *Environnement Risques & Santé* **2008**, *7*, 341–347.

175. Teichman, K.; Howard-Reed, C.; Persily, A.; Emmerich, S. *Characterizing Indoor Air Quality Performance Using a Graphical Approach*; NIST: Gaithersburg, MD, USA, 2016.

176. Logue, J.M.; Price, P.N.; Sherman, M.H.; Singer, B.C. A method to estimate the chronic health impact of air pollutants in U.S. residences. *Environ. Health Perspect.* **2010**, *120*, 216–222. [CrossRef]

177. Luo, N.; Weng, W.; Xu, X.; Hong, T.; Fu, M.; Su, K. Assessment of occupant-behavior based indoor air quality and its impacts on human exposure risk: A case study based on the wildfires in Northern California. *Sci. Total Environ.* **2019**, *686*, 1251–1261. [CrossRef]

178. Wei, W.; Ramalho, O.; Derbez, M.; Riberon, J.; Kirchner, S.; Mandin, C. Applicability and relevance of six indoor air quality indexes. *Build. Environ.* **2016**, *109*, 42–49.

179. van Dijken, F.; Boerstra, A.C. The ABCD Tool for schools. In Proceedings of the 9th International Conference and Exhibition on Healthy Buildings 2009, Syracuse, NY, USA, 13–17 September 2009.

180. Environment Protection Agency. Indoor AirPLUS Construction Specifications Version 1 (Rev 03) USA, 2015. Available online: http://www2.epa.gov/sites/production/files/2015-10/documents/construction_specification_rev_3_508.pdf (accessed on 16 February 2019).

181. Cony, L.R.S.; Abadie, M.; Wargocki, P.; Rode, C. Towards the definition of indicators for assessment of indoor air quality and energy performance in low-energy residential buildings. *Energy Build.* **2017**, *152*, 492–502. [CrossRef]

182. Zhu, C.; Li, N. Study on indoor air quality evaluation index based on comfort evaluation experiment. *Energy Procedia* **2017**, *205*, 2246–2253. [CrossRef]

183. Energy in Building and Community programme (EBC) and International Energy Agency (IEA) Indoor Air Quality Design and Control in Low Energy Residential Buildings IEA EBC Annex 68 2015. Available online: http://www.iea-ebc-annex68.org/about_annex-68 (accessed on 8 November 2018).

184. Reinten, J.; Braat-Eggen, P.E.; Hornikx, M.; Kort, H.S.M.; Kohlrausch, A. The indoor sound environment and human task performance: A literature review on the role of room acoustics. *Build. Environ.* **2017**, *123*, 315–332. [CrossRef]

185. Health and Safety Executive (HSE). *Controlling Noise at Work. The Control of Noise at Work*, 2nd ed.; HSE Books: Sudbury, Suffolk, UK, 2005; ISBN 978-0-7176-6164-4.

186. Directive 2003/10/EC of the European Parliament and of the Council of 6 February 2003 on the Minimum Health and Safety Requirements Regarding the Exposure of Workers to the Risks Arising from Physical Agents (Noise) (Seventeenth Individual Directive within the Meaning of Article 16 of Directive 89/391/EEC), OJ L 42, 15.2.2003. Available online: https://eur-lex.europa.eu/legal-content/EN/ALL/?uri=CELEX%3A32003L0010 (accessed on 10 January 2020).

187. International Organization for Standardization ISO 9612. *Acoustics-Determination of Occupational Noise Exposure-Engineering Method*; ISO: Geneva, Switzerland, 2009.

188. Leccese, F.; Salvadori, G.; Rocca, M.; Spinelli, N. Risk assessment of noise exposure in a machine shop and choice of hearing protection equipment. *G Ital. Med. Lav. Ergon.* **2016**, *38*, 5–13.

189. European Environment Agency. *Good Practice Guide on Noise Exposure and Potential Health Effects*; EEA Technical Report no. 11/2010; European Environment Agency: Copenhagen, Denmark, 2010.

190. World Health Organization. *Night Noise Exposure Guidelines for Europe*; WHO Regional Office for Europe: Copenhagen, Denmark, 2009; Available online: http://www.euro.who.int/__data/assets/pdf_file/0017/43316/E92845.pdf (accessed on 6 June 2019).

191. Al horr, Y.; Arif, M.; Katafygiotou, M.; Mazroei, A.; Kaushik, A.; Elsarrag, E. Impact of indoor environmental quality on occupant well-being and comfort: A review of the literature. *Int. J. Sustain. Built Environ.* **2016**, *5*, 1–11. [CrossRef]

192. Navai, M.; Veitch, J.A. *Acoustic Satisfaction in Open-Plan Offices: Review and Recommendations, Research Report RR-151*; Institute for Research in Construction, National Research Council Canada: Ottawa, ON, Canada, 2003; Available online: http://www.nrc-cnrc.gc.ca/obj/irc/doc/pubs/rr/rr151/rr151.pdf (accessed on 12 February 2019).

193. Choi, Y.J. Effect of occupancy on acoustical conditions in university classrooms. *Appl. Acoust.* **2016**, *114*, 36–43. [CrossRef]

194. Bistafa, S.R.; Bradley, J.S. *A Comparative study of Speech Intelligibility Metrics and the Derivation of Optimum Reverberation Time and Maximum Background-Noise Level for Classrooms*; Institute of Research in Construction, National Research Council Canada: Ottawa, ON, Canada, 1999.

195. Nowoświat, A.; Olechowska, M. Fast estimation of speech transmission index using the reverberation time. *Appl. Acoust.* **2016**, *102*, 55–61. [CrossRef]

196. Leccese, F.; Rocca, M.; Salvadori, G. Fast estimation of Speech Transmission Index using Reverberation Time: Comparison between predictive equations for educational rooms of different sizes. *Appl. Acoust.* **2018**, *140*, 143–149. [CrossRef]

197. Beranek, L.L. Balanced noise criterion (NCB) curves. *J. Acoust. Soc. Am.* **1989**, *86*, 650–664. [CrossRef]

198. Ayr, U.; Cirillo, E.; Martellotta, F. A new parameter for the assessment of noise annoyance in air-conditioned buildings. In Proceedings of the Clima 2000/Naples 2001 World Congress, Napoli, Italy, 15–18 September 2001.

199. Beranek, L.L.; Blazier, W.E.; Figwer, J.J. Preferred noise criterion (PNC) curves and their application to rooms. *J. Acoust. Soc. Am.* **1971**, *50*, 1223–1228. [CrossRef]

200. Kryter, K.D. *The Effects of Noise on Man*; Academic Press: New York, NY, USA, 1970.

201. Beranek, L.L. Criteria for office quieting based on questionnaire rating studies. *J. Acoust. Soc. Am.* **1956**, *28*, 833–852. [CrossRef]

202. Robinson, D.W. Towards a unified system of noise assessment. *J. Acoust. Soc. Am.* **1971**, *14*, 279–298. [CrossRef]

203. International Organization for Standardization. *ISO 1996-1: Acoustics—Description, Measurement and Assessment of Environmental Noise, Part 1: Basic Quantities and Assessment Procedures*; ISO: Geneva, Switzerland, 2016.

204. Hay, B.; Kemp, K.F. Measurement of noise in air-conditioned, landscaped offices. *J. Sound Vib.* **1972**, *23*, 363–373. [CrossRef]

205. Ayr, U.; Cirillo, E.; Martellotta, F. Further investigations on the definition of a new parameter to assess noise annoyance in air conditioned offices. *Energy Build.* **2002**, *34*, 765–774. [CrossRef]

206. Blazier, W.E. Revised noise criterion for application in the acoustical design and rating of HVAC systems. *Noise Control Eng. J.* **1981**, *162*, 64–73. [CrossRef]

207. Blazier, W.E. Sound quality consideration in rating noise from heating, ventilating and air-conditioning (HVAC) systems in buildings. *Noise Control Eng. J.* **1995**, *43*, 53–63. [CrossRef]

208. Blazier, W.E. RC Mark II: A refined procedure for rating the noise of heating, ventilating, and air-conditioning (HVAC) systems in buildings. *Noise Control Eng. J.* **1997**, *45*, 243–250. [CrossRef]

209. Beranek, L.L. The Design of Speech Communication Systems. In *Proceedings of the IRE*; IEEE: Piscataway, NJ, USA, 1947; Volume 35, pp. 880–890.

210. Stevens, S. Procedure for calculating loudness: Mark IV. *J. Acoust. Am. Soc.* **1983**, *33*, 1577–1585. [CrossRef]

211. International Organization for Standardization. *ISO 532-1: Acoustics—Methods for Calculation Loudness, Part 1— Zwicker method*; ISO: Geneva, Switzerland, 2017.

212. French, N.R.; Steinberg, J.C. Factors governing the intelligibility of speech sounds. *J. Acoust. Am. Soc.* **1947**, *19*, 90–119. [CrossRef]

213. Peutz, V.M. A Articulation Loss of Consonants as a Criterion for Speech Transmission in a Room. *J. Audio Eng. Soc.* **1971**, *19*, 915–919.

214. Reichardt, W.; Abdel Alim, O.; Schmidt, W. Definition und Messgrundlage eines objektiven Masses zur Ermittlung der Grenze zwischen brauchbarer und unbrauchbarer Durchsichtigkeit beim Musikdarbietung. *Acustica* **1975**, *32*, 126–137.

215. Thiele, R. Die Richtungsverteilung und Zeitfolge der Schallrückwürfe in Räumen (Directional distribution and time sequence of sound reflections in rooms). *Acta Acust. United Acust.* **1953**, *3*, 291–302.

216. Jordan, V.L. A comprehensive Musical Criterion: The inversion Index. In Proceedings of the 47th Audio Engineering Society Convention, Copenhagen, Denmark, 26–29 March 1974.

217. Cremer, L.; Muller, H. *Principles and Applications of Room Acoustics vol. 1 & 2*; English translation with additions by T. Schultz.; Applied Science Publishers: New York, NY, USA, 1982.

218. Dammerud, J.J.; Barron, M.; Kahle, E. Objective Assessment of Acoustic Conditions for Symphony Orchestras. *Build. Acoust.* **2011**, *18*, 207–219. [CrossRef]

219. Sabine, W.C. *Collected Papers on Acoustics*; Reprinted by Dover; University Press Harvard: New York, NY, USA, 1964.

220. Eyring, C.F. Reverberation time in "Dead" rooms. *J. Acoust. Soc. Am.* **1930**, *26*, 217–241. [CrossRef]

221. Arau-Puchades, H. An Improved Reverberation Formula. *Acta Acust.* **1988**, *65*, 163–180.

222. Millington, G. A modified formula for reverberation. *J. Acoust. Soc. Am.* **1932**, *4*, 69–82. [CrossRef]

223. IEC 60268-16. In *Sound System Equipment. Part 16: Objective Rating of Speech Intelligibility by Speech Transmission Index*, 4th ed.; IEC: Geneva, Switzerland, 2011.

224. Pelegrin Garcia, D.; Brunkoq, J.; Lyberg-Ahlander, V.; Lofquist, A. Measurement and prediction of voice support and room gain in school classroom. *J. Acoust. Soc. Am.* **2012**, *131*, 194–204. [CrossRef] [PubMed]

225. Latham, H.G. The signal-to-noise ratio for speech intelligibility—An auditorium acoustics design index. *Appl. Acoust.* **1979**, *12*, 253–320. [CrossRef]

226. American National Standards Institute (ANSI). *Methods for Calculation of the Speech Intelligibility Index*; ANSI/ASA S3.5-1997 (R2017); ANSI: Washington, DC, USA, 2017.

227. Houtgast, T.; Steeneken, H.J.M.; Plomp, R. Predicting speech intelligibility in rooms from the modulation transfer function. I. General room acoustics. *Acustica* **1980**, *46*, 60–72.

228. Van Wijngaarden, S.; Verhave, J.; Steeneken, H. The speech transmission index after four decades of development. *Acoust. Aust.* **2012**, *40*, 134–138.

229. Barron, M.; Lee, L.J. Energy relations in concert auditoriums. I J. *Acoust. Soc. Am.* **1988**, *84*, 618–628. [CrossRef]

230. Pelegrín-García, D. Comment on 'Increase in voice level and speaker comfort in lecture rooms'. *J. Acoust. Soc. Am.* **2011**, *129*, 1161.

231. Cotana, F.; Goreti, M. Acoustic classification of Buildings: Impact of acoustic performances of high acoustic performance of a high energy-efficient building on quality and sustainability indicators. In Proceedings of the 20th International Congress on Acoustics, ICA 2010, Sydney, Australia, 23–27 August 2010.

232. European Committee for Standardization. *EN 12354-5 Building Acoustics—Estimation of Acoustic Performance of Buildings from the Performance of Elements—Part 5: Sounds Levels Due to the Service Equipment*; European Committee for Standardization: Brussels, Belgium, 2000.

233. European Committee for Standardization. *EN 12354-2 Building Acoustics—Estimation of Acoustic Performance of Buildings from the Performance of Elements—Part 2: Impact Sound Insulation Between Rooms*; European Committee for Standardization: Brussels, Belgium, 2000.

234. European Committee for Standardization. *EN 12354-1 Building Acoustics—Estimation of Acoustic Performance of Buildings from the Performance of Elements—Part 1: Airborne Sound Insulation Between Rooms*; European Committee for Standardization: Brussels, Belgium, 2000.

235. European Committee for Standardization. *EN 12354-3 Building Acoustics—Estimation of Acoustic Performance of Buildings from the Performance of Elements—Part 3: Airborne Sound Insulation Against Outdoor Sound*; European Committee for Standardization: Brussels, Belgium, 2000.

236. Yang, D.; Mak, C.M. An assessment model of classroom acoustical environment based on fuzzy comprehensive evaluation method. *Appl. Acoust.* **2017**, *127*, 292–296. [CrossRef]

237. Madbouly, A.I.; Noaman, A.Y.; Ragab, A.H.M.; Khedra, A.; Fayoumi, A.G. Assessment model of classroom acoustics criteria for enhancing speech intelligibility and learning quality. *Appl. Acoust.* **2016**, *114*, 147–158. [CrossRef]

238. Boyce, P.R. *Human Factors in Lighting*, 2nd ed.; Taylor and Francis: London, UK, 1981.

239. International Well Building Institute. WELL Building Standard. Available online: https://www.wellcertified.com/ (accessed on 6 June 2019).

240. Heiden, M.; Zetterberg, C.; Lindberg, P.; Nylen, P.; Hemphala, H. Validity of a computer-based risk assessment method for visual ergonomics. *Int. J. Ind. Ergon.* **2019**, *72*, 180–187. [CrossRef]

241. Leccese, F.; Salvadori, G.; Rocca, M. Visual discomfort among university students who use CAD workstations. *Work* **2016**, *55*, 171–180. [CrossRef] [PubMed]

242. Taino, G.; Ferrari, M.; Mestad, I.J.; Fabris, F.; Imbriani, M. Astenophia and work at VDT: Study on a population of 191 workers exposed to risk through administering of anamnestic questionnaire focused on ophthalmic assessment. *G. Ital. Med. Lav. Ergon.* **2006**, *34*, 487–497.

243. Leccese, F.; Salvadori, G.; Montagnani, C.; Ciconi, A.; Rocca, M. Lighting assessment of ergonomic workstation for radio diagnostic reporting. *Int. J. Ind. Ergon.* **2017**, *57*, 42–54. [CrossRef]

244. Yan, Z.; Hu, L.; Chen, H.; Lu, F. Computer Vision Syndrome: A widely spreading but largely unknown epidemic among computer users. *Comput. Hum. Behav.* **2008**, *24*, 2026–2042. [CrossRef]

245. Bergqvist, U. Visual Display Terminal work—A perspective on long-term changes and discomfort. *Int. J. Ind. Ergon.* **1995**, *16*, 201–209. [CrossRef]

246. Turville, K.L.; Psihogios, J.P.; Ulmer, T.R.; Mirka, G.A. The effects of video display terminal height on the operator: A comparison of the 15° and 40° recommendations. *Appl. Ergon.* **1998**, *29*, 239–246. [CrossRef]

247. Pillastrini, P.; Mugnaia, R.; Bertozzi, L.; Costi, S.; Curti, S.; Guccione, A.; Mattioli, S.; Violante, F.S. Effectiveness of an ergonomic intervention on work-related posture and low back pain in video display terminal operators: A 3 year cross-over trial. *Appl. Ergon.* **2010**, *41*, 436–443. [CrossRef]

248. Mirmohammadi, S.J.; Mehrparvar, A.H.; Olia, M.B.; Mirmohammadi, M. Effects of training intervention on non-ergonomic positions among video display terminals (VDT) users. *Work* **2012**, *42*, 429–433. [CrossRef]

249. Powell, A.L. Computer anxiety: Comparison of research from the 1990s and 2000s. *Comput. Hum. Behav.* **2013**, *29*, 2337–2381. [CrossRef]

250. Leccese, F.; Salvadori, G.; Rocca, M. Visual ergonomics of video-display-terminal workstations: Field measurements of luminance for various display settings. *Displays* **2016**, *42*, 9–18. [CrossRef]

251. Lin, Y.T.; Lin, P.H.; Hwang, S.L.; Jeng, S.C.; Liao, C.C. Investigation of legibility and visual fatigue for simulated flexible electronic paper under various surface treatments and ambient illumination conditions. *Appl. Ergon.* **2009**, *40*, 922–928. [CrossRef] [PubMed]

252. Blehm, C.; Vishnu, S.; Khattak, A.; Mitra, S.; Yee, R.W. Computer vision syndrome: A review. *Surv. Ophthalmol.* **2005**, *50*, 253–262. [CrossRef] [PubMed]

253. Helland, M.; Horgen, G.; Kvikstad, T.R.; Garthus, T.; Bruenech, J.R.; Aaras, A. Musculoskeletal, visual and psychosocial stress in VDU operators after moving to an ergonomically designed office landscape. *Appl. Ergon.* **2008**, *39*, 284–295. [CrossRef]

254. Wilkins, A.J.; Evans, B. Visual stress, its treatment with special filters, and its relationship to visually induced motion sickness. *Appl. Ergon.* **2010**, *41*, 509–515. [CrossRef]

255. Pinto, I.; Brogi, A.; Piccolo, F.; Stacchini, N.; Buonocore, G.; Bellieni, C. Blue Light and Ultraviolet Radiation Exposure from infant Phototherapy Equipment. *J. Occup. Environ. Hyg.* **2015**, *12*, 603–610. [CrossRef]

256. Directive 89/654/EEC of the European Parliament and of the Council of 30 November 1989 Concerning the Minimum safety and Health Requirements for the Workplace. Available online: https://eur-lex.europa.eu/legal-content/EN/TXT/?uri=CELEX%3A31989L0654 (accessed on 10 January 2020).

257. Directive 2006/25/EC of the European Parliament and of the Council of 5 April 2006 on the Minimum Health and Safety Requirements regarding the Exposure of Workers to Risks Arising from Physical Agents (Artificial Optical Radiation). Available online: https://eur-lex.europa.eu/legal-content/EN/TXT/?uri=CELEX%3A02006L0025-20140101 (accessed on 10 January 2020).

258. Leccese, F.; Salvadori, G.; Casini, M.; Bertozzi, M. Analysis and measurements of artificial optical radiation (AOR) emitted by lighting sources found in offices. *Sustainability* **2014**, *6*, 5941–5954. [CrossRef]

259. European Commission. Council directive 96/62/EC of 27 September 1996 on ambient air quality assessment and management 96/62/EC. Available online: https://osha.europa.eu/en/legislation/guidelines/non-binding-guide-to-good-practice-for-implementing-directive-2006-25-ec-201aartificial-optical-radiation2019 (accessed on 7 January 2020).

260. Leccese, F.; Vandelanotte, V.; Salvadori, G.; Rocca, M. Blue light hazard and risk group classification of 8 W LED tubes, replacing fluorescent tubes, through optical radiation measurements. *Sustainability* **2015**, *7*, 13454–13468. [CrossRef]

261. International Commission on Non-Ionizing Radiation Protection. ICNIRP Guidelines on limits of exposure to incoherent visible and infrared radiation. *Health Phys.* **2013**, *105*, 74–96.

262. International Commission on Non-Ionizing Radiation Protection. ICNIRP Guidelines on limits of exposure to ultraviolet of wavelength between 180 nm and 400 nm (incoherent optical radiation). *Health Phys.* **2004**, *87*, 171–186. [CrossRef]

263. European Committee for Standardization. *Photobiological Safety of Lamps and Lamp Systems*; EN 62471; European Committee for Standardization: Bruxelles, Belgium, 2009.

264. International Electrotechnical Commission. *Application of IEC 62471 for the Assessment of Blue Light Hazard to Light Sources and Luminaires*; IEC/TR 62778:2012; International Electrotechnical Commission: Geneva, Switzerland, 2014.

265. European Committee for Electrotechnical Standardization. *Light and Lighting—Basic Terms and Criteria for Specifying Lighting Requirements*; EN 12665; European Committee for Electrotechnical Standardization: Bruxelles, Belgium, 2018.

266. European Committee for Standardization. *Light and Lighting—Lighting of Work Places—Part 1: Indoor Work Places*; EN 12464-1; European Committee for Standardization: Brussels, Belgium, 2011.

267. Berman, S.M. Energy Efficiency Consequences of Scotopic Sensitivity. *J. Illum. Eng. Soc.* **1992**, *21*, 3–14. [CrossRef]

268. Thornton, W.A. Color-discrimination index. *J. Opt. Soc. Am.* **1972**, *62*, 191–194. [CrossRef] [PubMed]

269. Thornton, W.A. A validation of the color preference index. *J. Illum. Eng. Soc.* **1974**, *4*, 48–52. [CrossRef]

270. Xu, H. Color-rendering capacity of illumination. *J. Opt. Soc. Am.* **1983**, *73*, 1709–1713. [CrossRef]

271. Davis, W.; Ohno, Y. Color quality scale. *Opt. Eng.* **2010**, *49*, 336021–3360216. [CrossRef]

272. Hashimoto, K.; Yano, T.; Nayatani, Y. Proposal of practical method for calculating and indexing feeling of contrast for a light source. *Illum. Eng. Inst. Jpn.* **2000**, *84*, 843–849. [CrossRef]

273. Judd, D.B. A flattery index for artificial illuminants. *Illum. Eng.* **1967**, *62*, 593–598.

274. Freyssinier-Nova, J.P.; Rea, M.S. A two-metric proposal to specify the color-rendering properties of light sources for retail lighting, in Tenth International Conference of Solid-State Lighting. In Proceedings of the SPIE, San Diego, CA, USA, 23 August 2010.

275. Smet, K.G.; Ryckaert, W.R.; Pointer, M.R.; Deconick, G.; Hanseler, P. Memory color quality metric for white light sources. *Energy Build.* **2012**, *49*, 216–225. [CrossRef]

276. Pointer, M.R. Measuring color rendering—A new approach. *Light Res. Technol.* **1986**, *18*, 175–184. [CrossRef]

277. Illuminating Engineering Society (IES). *Approved Method: IES Spatial Daylight Autonomy (sDA) and Annual Sunlight Exposure (ASE) (IES-LM-83-12)*; IES: Washington DC, USA, 2012.

278. Rogers, Z.; Goldman, D. *Daylighting Metric Development Using Daylight Autonomy Calculations in the Sensor Placement Optimization Tool—Development Report and Case Studies*; Architectural Energy Corporation: Boulder, CO, USA, 2006.

279. Reinhart, C.; Walkenhorst, O. Validation of dynamic RADIANCE-based daylight simulations for a test office with external blinds. *Energy Build* **2001**, *33*, 683–697. [CrossRef]

280. Walsh, J.W.T. The early years of Illuminating engineering in Great Britain. *Trans. Illum. Eng. Soc.* **1951**, *15*, 49–60. [CrossRef]

281. Sicurella, F.; Evola, G.; Wurtz, E. A statistical approach for the evaluation of thermal and visual comfort in free-running buildings. *Energy Build* **2012**, *47*, 402–410. [CrossRef]

282. Nabil, A.; Mardaljevic, J. Useful daylight illuminances: A replacement for daylight factors. *Energy Build* **2006**, *38*, 905–913. [CrossRef]

283. Einhorn, H.D. Discomfort glare: A formula to bridge differences. *Lighting Res. Technol.* **1979**, *11*, 90–94. [CrossRef]

284. International Commission on Illumination. *Glare from Small, Large and Complex Sources (CIE 147)*; International Commission on Illumination: Vienna, Austria, 2002.

285. Guth, S.K. A method for the evaluation of discomfort glare. *Illum. Eng.* **1963**, *57*, 351–364.

286. Nazzal, A. A new evaluation method for daylight discomfort glare. *Int. J. Ind. Ergon.* **2005**, *35*, 295–306. [CrossRef]

287. Wienold, J. Dynamic daylight glare evaluation. In Proceedings of the 11th International IBPSA Conference, Building Simulation, Glasgow, UK, 27–30 July 2009; pp. 944–951.

288. International Commission on Illumination. *Equations for Disability Glare (CIE 146)*; International Commission on Illumination: Vienna, Austria, 2002.

289. Hviid, C.; Nielsen, T.; Svendsen, S. Simple tool to evaluate the impact of daylight on building energy consumption. *Sol. Energy* **2008**, *82*, 787–798. [CrossRef]

290. Meyer, J.J. Visual discomfort: Evaluation after introducing modulated light equipment. In Proceedings of the 2nd European Conference on Energy-Efficient Lighting, in Right Light, Arnhem, NL, 26–29 September 1993; pp. 348–357.

291. Chauvel, P.; Collins, J.B.; Dogniaux, R.; Longmore, J. Glare from windows: Current views of the problem. *Light. Res. Technol.* **1982**, *14*, 31–46. [CrossRef]

292. Tokura, M.; Iwata, T.; Shukuya, M. Experimental study on discomfort glare caused by windows; development of a method for evaluating discomfort glare from a large light source. *J. Arch. Plan Envrion. Eng.* **1996**, *489*, 17–25.

293. International Commission on Illumination. *Discomfort Glare in Interior Lighting (CIE 117)*; International Commission on Illumination: Vienna, Austria, 1995.

294. Ente Italiano di Normazione. *Luce e Illuminazione-Illuminazione di Interni-Valutazione Dell'abbagliamento Molesto con il Metodo UGR (UNI 11165)*; Ente Italiano di Normazione: Milan, Italy, 2005.

295. Petherbridge, P.; Hopkinson, R.G. Discomfort glare and the lighting of buildings. *Trans Illum. Eng. Soc.* **1950**, *15*, 29–79. [CrossRef]

296. Wienold, J.; Christoffersen, J. Towards a new daylight glare rating. In Proceedings of the 10th European Lighting Conference Lux Europa, Berlin, Germany, 19–21 September 2005; pp. 157–161.

297. Bellia, L.; Bisegna, F.; Spada, G. Lighting in indoor environments: Visual and non-visual effects of light sources with different spectral power distributions. *Build. Environ.* **2010**, *46*, 1984–1992. [CrossRef]

298. Zeitzer, J.M.; Dijk, D.J.; Kronauer, R.E.; Brown, E.N.; Czeisler, C.A. Sensitivity of the human circadian pacemaker to nocturnal light: Melatonin phase resetting and suppression. *J. Physiol.* **2000**, *526*, 695–702. [CrossRef] [PubMed]

299. Bisegna, F.; Burattini, C.; Li Rosi, O.; Blaso, L.; Fumagalli, S. Non visual effects of light: An overview and an Italian experience. *Energy Procedia* **2015**, *78*, 723–728.

300. Rea, M.S. Lighting simply made better: Providing a full range of benefits without much fuss. *Build. Environ.* **2018**, *144*, 57–65. [CrossRef]

301. Enezi, J.; Revell, V.; Brown, T.; Wynne, J.; Schlangen, L.; Lucas, R. A "Melanopic" Spectral Efficiency Function Predicts the Sensitivity of Melanopsin Photoreceptors to Polychromatic Lights. *J. Biol. Rhythm.* **2011**, *26*, 314–323. [CrossRef]

302. Lucas, R.; Peirson, S.; Berson, D.; Brown, T.; Cooper, H.; Czeisler, C.; Figueiro, M.; Gamlin, P.; Lockley, S.; O'Hagan, J.; et al. Measuring and using light in the melanopsin age. *Trends Neurosci.* **2014**, *37*, 1–9. [CrossRef]

303. Lucas Group. Excel-Based Melanopic Illuminance Calculator. University of Manchester. Available online: http://lucasgroup.lab.ls.manchester.ac.uk/research/measuringmelanopicilluminance/ (accessed on 25 February 2019).

304. Rea, M.S.; Figueiro, M.; Bierman, A.; Hamner, R. Modelling the spectral sensitivity of the human circadian system. *Light. Res. Technol.* **2012**, *44*, 386–396. [CrossRef]

305. Lighting Research Center. Circadian Stimulus Calculator. Available online: http://www.lrc.rpi.edu/programs/lightHealth/ (accessed on 25 February 2019).

306. Konis, K. A novel circadian daylight metric for building design and evaluation. *Build. Environ.* **2017**, *113*, 22–38. [CrossRef]

307. Chang, A.M.; Santhi, N.; St Hilaire, M.; Gronfier, C.; Bradstreet, D.S.; Duffy, J.F.; Lockley, S.W.; Kronauer, R.E.; Czeisler, C.A. Human responses to bright light of different durations. *J. Physiol.* **2012**, *590*, 3103–3112. [CrossRef]

308. Figueiro, M.G.; Gonzales, K.; Pedler, D. Designing with Circadian Stimulus. *Des. Appl.* **2016**, *46*, 31–34. Available online: https://www.lrc.rpi.edu/programs/lightHealth/index.asp (accessed on 6 June 2019).

*Article*

# A Comparative Study on Cooling Period Thermal Comfort Assessment in Modern Open Office Landscape in Estonia

**Martin Kiil [1,*], Raimo Simson [1], Martin Thalfeldt [1] and Jarek Kurnitski [1,2]**

[1]  Department of Civil Engineering and Architecture, Tallinn University of Technology, Ehitajate tee 5, 19086 Tallinn, Estonia; raimo.simson@taltech.ee (R.S.); martin.thalfeldt@taltech.ee (M.T.); jarek.kurnitski@taltech.ee (J.K.)

[2]  School of Engineering, Aalto University, Otakaari 4, 02150 Espoo, Finland

*  Correspondence: martin.kiil@taltech.ee

Received: 22 December 2019; Accepted: 19 January 2020; Published: 23 January 2020

**Abstract:** Local thermal comfort and draught rate has been studied widely. There has been more meaningful research performed in controlled boundary condition situations than in actual work environments involving occupants. Thermal comfort conditions in office buildings in Estonia have been barely investigated in the past. In this paper, the results of thermal comfort and draught rate assessment in five office buildings in Tallinn are presented and discussed. Studied office landscapes vary in heating, ventilation and cooling system parameters, room units, and elements. All sample buildings were less than six years old, equipped with dedicated outdoor air ventilation system and room conditioning units. The on-site measurements consisted of thermal comfort and draught rate assessment with indoor climate questionnaire. The purpose of the survey is to assess the correspondence between heating, ventilation and cooling system design, and the actual situation. Results show, whether and in what extent the standard-based criteria for thermal comfort is suitable for actual usage of the occupants. Preferring one room conditioning unit type or system may not guarantee better thermal environment without draught. Although some heating, ventilation and cooling systems observed in this study should create the prerequisites for ensuring more comfort, results show that this is not the case for all buildings in this study.

**Keywords:** thermal comfort; draught; cooling period; open office

---

## 1. Introduction

Modern low energy office buildings require energy efficient heating, ventilation, and air conditioning (HVAC) systems which can provide comfortable and healthy indoor environment. In temperate climate countries, mechanical ventilation and active cooling systems are common practice in such buildings. However, mechanical HVAC systems do not always provide satisfactory thermal conditions [1]. It is important to properly apply control strategies, design and install room cooling units and ventilation supply air elements, as well as to operate and maintain the systems to provide comfortable indoor climate without temperature fluctuations and draught risk in the cooling season [2–8]. Office plans, in terms of occupant positions and density, can be very different from initial design and vary significantly, resulting in changing conditions and dynamic settings which makes it difficult to design the systems adequately to ensure stable thermal environment. Open office layout design is used commonly in most office buildings mainly to allow flexibility in workspaces allocation [9]. This creates a difficult task for HVAC systems design, requiring careful planning to assure adequate conditions in the occupied zone in different layout cases.

As occupant satisfaction with thermal environment is dependent on many factors, such as gender, age, health, activity, mood, and other physiological and psychological factors, assessing thermal comfort

(TC) based on temperature and air movement measurements is usually not sufficient for adequate estimation [6,10–13]. Thus, evaluation by questioning the occupants is usually also needed to specify the problems and get a comprehensive overview of the TC situation. Studies on office workers thermal sensation have shown that the predicted TC and actual sensation can differ significantly [12,14,15]. For example, gender specific analysis indicates higher dissatisfaction rates for female occupants [11,16–19]. Recent research has widely focused on individual perception of TC [20–22], developed methods to analyze the preferences for TC using machine learning algorithms [23,24] and adapt systems to provide preferable personal comfort by implementing Personalized Comfort Systems [20–22,25]. Utilization of such systems in buildings requires paradigm shifts in occupant interaction with HVAC systems as well as system design practices, integration of advanced controls and information technologies solutions [26,27].

In addition to the individual preferences and system specific aspects influencing thermal comfort (TC), there are many building related design factors that can affect the performance of HVAC systems and in turn influence the thermal environment. Of these factors, façade design, namely window sizes, layout, and glazing parameters, can have large impact on cooling load as well as radiant temperature asymmetry and thus major influence on the overall thermal conditions in the office [28,29]. Thalfeldt et al. [28] showed the importance of façade design by analyzing the effect on office buildings energy efficiency and cooling load in cold climate countries. Window-to-wall ratio (*WWR*) of 0.25 was found optimal for triple glazing window solutions. Larger glazing results in higher cooling loads and increase the need for larger room cooling units, higher cooled airflow rates or lower supply air temperatures to maintain the room temperature. The latter factors also increase the risk of draught in occupied spaces. In several studies, draught rate (*DR*) has been identified as the main cause of discomfort even if other thermal environment factors are at satisfactory levels [6,15,29,30].

Depending mainly on the cooling load, cooling plant solution, and interior design, different water based room cooling solutions are used in offices, which can be classified by supply water temperature as low temperature room cooling units e.g., fan coil units and high temperature units, such as thermally active building systems (TABS), passive cooling beams, or active cooling beams, combined with ventilation supply air terminals [31,32]. In low energy buildings, high temperature cooling is usually preferred to achieve higher energy efficiency for cooled water production by cooling plants [32]. The performance of these systems is extensively analyzed in various recent studies. Most of the research is based on either computer simulations, mainly computational fluid dynamics (CFD) studies or studies conducted in controlled laboratory environments [33–47]. The research in real office settings is mainly focused on buildings located in warm and hot climate countries, dominated by cooling need [48–51]. To the knowledge of the authors, only few extensive studies have been carried out in cold and temperate climates and in low energy buildings. In Germany, Pfafferott, Herkel, Kalz, and Zeuschner [14] have conducted research on summertime TC in 12 low energy office buildings which are passively cooled with local heat sink based TABS. Results showed, that 41% of occupants were dissatisfied with thermal environment in summer, but assessment, according to the standard CEN EN 15251 [52], showed measured indoor temperature-based classification relative to the indoor climate category I (highest) and II, indicating a gap between perceived and assessed TC conditions and the need for more detailed comfort assessment. Hens [15] investigated TC in two office buildings in Belgium cooled with active chilled beams and air-cooling systems. He found that the Fanger [53] predicted mean vote (*PMV*)/predicted percentage dissatisfied (*PPD*) curve underestimated the actual number of dissatisfied occupants and that standards should not be considered as absolute references. It was also concluded that one should be very careful when interpreting the results of TC studies.

The theoretical knowledge involved or access to expert engineers during office building design is feasible and implemented in practice in Estonia. However, the volatile quality of different parts of the design, building phase simplifications with budget cuts and the contradiction between the initial task and the actual situation leads to a risk of an outcome failure regarding TC for the occupant. In the Estonian construction market, great emphasis is given on diplomas, professional certificates, software for both building information modeling (BIM), and product selection programs integrating BIM solutions in the building process. In reality, during the construction process the HVAC designer and after the warranty period, the constructor retreat. Therefore, in a short period of time during the design a huge effort is invested in the project definition, while after the realization phase much of the expert advice is ignored. Otherwise, complaints regarding draught or room temperature were not topical issues. In Estonia as well, in-depth research on cooling season TC and occupant satisfaction is practically non-existent, a few studies in office buildings have been conducted with the main focus on heating season performance and mostly aimed towards energy efficiency analysis. The conducted studies indicate problems and dissatisfaction with thermal environment but lack the detail to specify the causes and details of occupants' thermal conditions and HVAC systems performance in terms of room equipment.

Regardless of the design and performance of the HVAC systems installed in actual open offices, the hypothesis of this study proposes that high proportion of occupants are dissatisfied with the TC conditions. The goal of this study is to determine, whether the thermal sensation dissatisfaction of the occupants in modern office spaces is verifiable and in accordance with the valid standard criteria. This paper aims to fill the gap of summer TC assessment by extensive field studies and thorough occupant survey in modern office buildings in Estonia, a temperate climate country. We have investigated four recently constructed and one reconstructed office buildings with open plan office layouts designed with different ventilation and cooling solutions, including mixing and displacement ventilation, TABS, radiant cooling panels, fan coil units, and active cooling beams. The on-site measurements conducted in the offices consist of high resolution and accuracy temperature and air velocity measurements with *DR* and *TC* calculations, which are described in the following chapter.

## 2. Methods

The flow chart of research methodology is shown in Figure 1. Section of methods is divided between description of reference objects, measurement set-up and equipment specifications, data analysis, and indoor climate questionnaire (ICQ). We used standard-based [52,54,55] methods in this study to measure and calculate TC parameters and to perform an online ICQ survey. The TC measuring probe and tripod mobile and flexible kit set [56] we used was designed for research and development purposes.

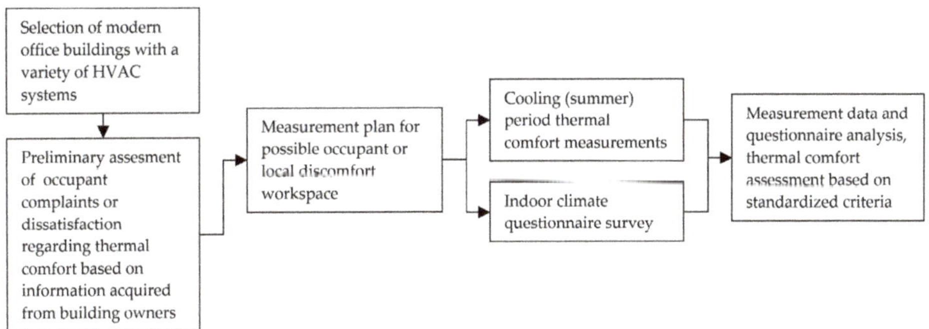


**Figure 1.** Flow chart of research methodology.

### 2.1. Reference Objects

General information regarding reference objects are provided in Table 1. The scope and range of measurement points with main building envelope characteristics, such as thermal transmittance for main surfaces, such as walls, windows, floors on ground and roofs are listed with specific heat loss of external envelopes and window-to-wall ratios.

**Table 1.** General building information of reference objects.

| Bldg | Year of Constr. | Net Floor Area (m$^2$)/appr. Total Measured Area (%) | No of Floors/ No of Measured Floors | Thermal Transmittance W/(m$^2$ × K) | Specific Heat Loss of External Envelopes W/(m$^2$ × K)/Window-to-Wall Ratio/Glazing $g$ Value |
|---|---|---|---|---|---|
| A | 2015 | 10,800/30 | 13/4 | $U_{window}$ 0.80/$U_{wall}$ 0.18 $U_{roof}$ 0.09/$U_{floor}$ 0.14 | H/A 0.50 WWR 0.69 $g$ 0.25 |
| B | 2018 | 7000/20 | 5/3 | $U_{window}$ 0.83/$U_{wall}$ 0.12 $U_{roof}$ 0.09/$U_{floor}$ 0.13* (*above ambient air) | H/A 0.31 WWR 0.59 $g$ 0.24 |
| C | 2017 | 18,900/10 | 14/2 | $U_{window}$ 0.65/$U_{wall}$ 0.10 $U_{roof}$ 0.10/$U_{floor}$ 0.15 | H/A 0.30 WWR 0.38 $g$ 0.30 |
| D | 2018 | 13,900/100 (available office space) | 2/2 | $U_{window}$ 1.0/$U_{wall}$ 0.15 $U_{roof}$ 0.14/$U_{floor}$ < 0.15 | H/A < 0.20 WWR < 0.25 $g$ 0.30 |
| E | Reconstr, 2014 (1982) | 5300/20 | 6/1 | $U_{window}$ > 1.2/$U_{wall}$ > 0.25 $U_{roof}$ N/A/$U_{floor}$ N/A | H/A > 0.50 WWR 0.90 $g$ 0.40 |

In Buildings A, B, C, and D, measurements were also taken on the highest floor and in Buildings B and D on the lowest floor. The temperature of slabs was considered to be close to $t_i$ and therefore the impact on operative temperature was not accounted for, as heat transmission through the building envelope in such low energy buildings is negligible compared to the heat gains through glazed surfaces and have little effect on TC. A variety of HVAC systems was involved in measurements zones (Table 2) including new and innovative solutions in the Estonian construction market.

**Table 2.** Heating, ventilation, and air conditioning room design solutions of reference objects.

| Bldg. | Heating | Ventilation | Cooling |
|---|---|---|---|
| A | Water-based convectors (height 300 mm, length 700–1800 mm) below the windowsill. Installed room unit heating power 18 W/m$^2$. | Mixing ventilation 1.4 l/(s × m$^2$) using active exposed chilled beams (effective length 2700–3300 mm) mounted in the open ceiling (height 2.75 m) for supply and circular valves (Ø 125 mm) for extract air (height 2.7 m). | Active exposed chilled beams (effective length 2700–3300 mm) mounted in the open ceiling (height 2.75 m). Installed room unit sensible cooling power 52 W/m$^2$. |
| B | Thermally active building system (slab, room height 3.0 m). Installed heating power 43 W/m$^2$. | Displacement ventilation 1.4 l/(s × m$^2$) including duct diffusers (Ø 160–315 mm, nozzle angle 120–180 °C) for supply (height 2.7–2.8 m), mounted in the open ceiling to the perimeter of rooms. Wall and ceiling grilles with plenum box serving extract air (height 2.8 m on cornice, 2.6 m for ribbed suspended ceiling). | Thermally active building system (slab, room height 3.0 m). Installed sensible cooling power 41 W/m$^2$. |
| C | 4-pipe active ceiling integrated chilled beams (effective length 900–1500 mm) mounted in suspended ceiling (height 2.7 m). Installed room unit heating power 17 W/m$^2$. | Mixing ventilation 1.7 l/(s × m$^2$) using 4-pipe ceiling integrated chilled beams (effective length 900–1500 mm) for supply air and circular valves (Ø 100 mm) for extract air (height 2.7 m). | 4-pipe active ceiling integrated chilled beams (effective length 900–1500 mm) mounted in suspended ceiling (height 2.7 m). Installed room unit sensible cooling power 46 W/m$^2$. |
| D | 4-pipe radiant panels mounted in the open ceiling on the height of 2.4 m. Installed room unit heating power 24 W/m$^2$. | Mixing ventilation 2.1 l/(s × m$^2$). Rectangular diffusers including directionally adjustable nozzles (plates 160 × 160 / 200 × 200 mm) mounted on plenum box for supply air and circular plate (Ø 200–250 mm) combined with plenum box for extract air in the open ceiling (height 2.7 m). | 4-pipe radiant panels mounted in the open ceiling on the height of 2.4 m. Installed room unit sensible cooling power 10 W/m$^2$. |
| E | Electrical convectors (height 200 mm, length 1500 mm) in front of windows. Installed room unit heating power 60 W/m$^2$. | Mixing ventilation 1.3 l/(s × m$^2$) with circular supply and extract air valves (Ø 160–250 mm) mounted in the suspended ceiling (height 2.5–2.7 m). | Multi-split fan coil units (without heating function) mounted in the suspended ceiling (height 2.7 m). Installed total cooling power 78 W/m$^2$. (Ventilation supply air is not chilled) |

Ø: diameter of air diffuser connection duct.

The buildings involved in this study were chosen from a range of modern office spaces in Tallinn. First criteria for reference objects was the correspondence with the Estonian energy efficiency regulations, which were first set in 2007 [57]. This created the prerequisites for new buildings and HVAC systems criteria, such as envelope related parameters, such as air tightness, external wall insulation thickness, window glazing solutions, and HVAC system parameters, e.g., effectiveness of room units and energy sources, heat recovery effectiveness, specific fan power of ventilation units etc.

Buildings A and B have high temperature heating systems and district heating. Building B is using low temperature heating and a ground source heat pump, Building D has high temperature heating water produced and a gas boiler and electrical heating convectors are installed in Building E. All of the studied buildings are equipped with dedicated outdoor air ventilation systems with heat recovery. Ventilation air distribution methods were classified as mixing ventilation, except for Building B, where supply air systems were built in a way to support displacement ventilation method. Buildings A and C were using active chilled beams for supply air distribution. Buildings A, C, and D are built with chillers to supply the cooling system. In all Buildings, except for E, high temperature cooling is used in room conditioning units as supply air is dehumidified in the air handling units. Multi-split fan coil units with refrigerant without the option of heating function were in operation in Building E. Room conditioning units in Buildings C and D, including the Building B with thermally active buildings system, operated both for heating and cooling purposes.

## 2.2. Measurement Equipment

Experimental measurements in this study were carried out with a TC measurement system Dantec Dynamics ComfortSense [56]. The system is designed for high quality multi-point measurements of $v_a$, $t_i$, $RH$, and $t_o$. The set is equipped with software, what allows easy setup for measuring sequence and positions giving researchers a comprehensive overview if the measured data. Measurement equipment probe data is described in Table 3.

**Table 3.** Specifications of measuring equipment [56].

| | 54T33 Draft Probe | 54T37 Relative Humidity Probe | 54T38 Operative Temperature Probe |
|---|---|---|---|
| Image | | | |
| Range | 0.05–5 m/s<br>−20 °C to +80 °C | 0–100% | 0 to +45 °C |
| Accuracy | ±0.02 m/s<br>±0.2 K | +1.5% | ±0.2 K |

The set is mounted on a tripod including five draft probes, one humidity and one $t_o$ probe. For a sitting position, ISO standard [55] recommends measuring heights for ankle level 0.1 m, abdomen level 0.6 m, and head level 1.1 m. Conformably to Fanger and Christensen [6], mean $v_a$ and standard deviation at three heights around the sitting occupant body were measured according to heights shown on Figure 2. RH probe was set at 1.0 m as a fixed height for measuring has not been fixed for measurements. The $t_o$ probe was mounted with the angle of 30° at the height of 0.6 m as the abdomen level of a sitting person [55].

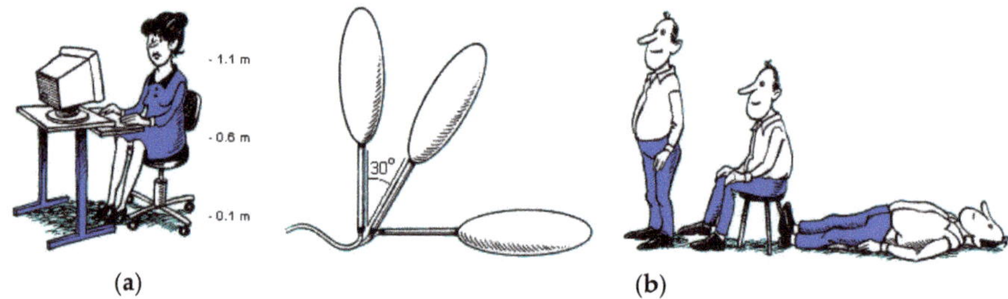



**Figure 2.** Recommended air velocity probe heights behind the feet, elbow, and neck for a sitting person (**a**); Recommended operative temperature probe person's angle factor to their surroundings (**b**) [58].

Probes were connected with 54N90 ComfortSense main frame [56], using 7 channels of 16. Main frame was in turn connected with laptop computer where the measurement data was stored using ComfortSense software version 4, (Dantec Dynamics A/S, Skovlunde, Denmark) [56]. Measurement period of 180 seconds as the least time recommended [59] was used.

*2.3. Data Analysis*

Measurement data, including $t_i$, $v_a$, *RH*, and $t_o$ was recorded with the sampling rate 20 Hz with ComfortSense [56] and processed in Microsoft Excel. TC parameters are calculated for each measurement positions with equations for *Tu*, *DR*, *PMV*, and *PPD* followed. To assess *DR*, the fluctuation rate of $v_a$ is described as *Tu*, which is calculated by [59]

$$Tu = \frac{SD}{v_a} \times 100 \ (\%),\tag{1}$$

where *SD* is standard deviation of measured local mean $v_a$ (m/s) for one measurement. With $t_i$, $v_a$, and *Tu*, the percentage of people predicted to be dissatisfied because of draught may be calculated as [60]

$$DR = (34 - t_i) \times (v_a - 0.05)^{0.62} \times (0.37 \times v_a \times Tu + 3.14) \ (\%)\tag{2}$$

To predict the mean value of the subjective ratings of a group of people in a given environment, *PMV* index is used. Consisting of a set of parameters with sub-formulas, the *PMV* equation is given by [60]

$$PMV = [0.303 \times \exp(-0.036 \times M) + 0.028] \times [(M - W) - H_d - E_c - C_{res} - E_{res}]\tag{3}$$

The *PMV* index in Equation (3) was calculated using Equations (4)–(11). In the equations provided, $M$ (W/m$^2$) is metabolic rate and $W$ (W/m$^2$) is the effective mechanical power. Assumption of metabolic rate 1.2 met for sedentary activity for summer season provided in EN 16798-1 was used. Sedentary activity does not suppose producing effective mechanical power, therefore 0 (W/m$^2$) was used in analysis. The next symbol $H_d$ in Equation (3) represents dry heat loss, which is found as

$$H_d = \frac{(mt_{sk} - t_{cl})}{I_{cl}}\left(\text{W}/\text{m}^2\right),\tag{4}$$

where $mt_{sk}$ is mean skin temperature [°C] and in Equation (4), $t_{cl}$ is expressed using $t_o$ and calculated through iterative process, by

$$t_{cl} = 35.7 - 0.028 \times (M - W) - I_{cl} \times \{3.96 \times 10^{-8} \times f_{cl} \times [(t_{cl} + 273)^4 - (t_o + 273)^4] + I_{cl} \times f_{cl} \times h_c \times (t_{cl} - t_o)\} \ (°C)\tag{5}$$

In Equation (5), $I_{cl}$ [(m$^2$ × K)/W] is the clothing insulation, $f_{cl}$ is the clothing surface area factor, $v_{ar}$ (m/s) is the relative air velocity, $h_c$ [W/(m$^2$ × K)] is the convective heat transfer coefficient, and $t_{cl}$ (°C) is the clothing surface temperature. Clothing unit 0.5 clo for summer season provided in EN 16798-1 was used in calculations. Equation (5) includes $h_c$, which is given as

$$h_c = 2.38 \times |t_{cl} - t_i|^{0.25} \quad \text{for} \quad 2.38 \times |t_{cl} - t_i|^{0.25} > 12.1 \times \sqrt{v_{ar}} \quad \text{and}$$
$$12.1 \times \sqrt{v_{ar}} \quad \text{for} \quad 2.38 \times |t_{cl} - t_i|^{0.25} < 12.1 \times \sqrt{v_{ar}} \left[ W/\left(m^2 \times K\right)\right], \tag{6}$$

where $v_{ar}$ was set equal to the $v_a$ as occupants were intended to be stationary sensing draught. Equation (5) includes $f_{cl}$, which is calculated by

$$f_{cl} = 1.00 + 1.290 \times I_{cl} \text{ for } I_{cl} \leq 0.078 \text{ and}$$
$$1.05 + 0.645 \times I_{cl} \text{ for } I_{cl} > 0.078 \tag{7}$$

Continuing the *PMV* index calculation, in Equation (3), evaporative heat exchange at the skin, when the person experiences a sensation of thermal neutrality $E_c$ given as

$$E_c = 3.05 \times 10^{-3} \times [5733 - 6.99 \times (M - W) - p_a] + 0.42 \times (M - W - 58.15) \, (W/m^2), \tag{8}$$

where $p_a$ is the water vapor partial pressure [Pa], calculated using measured *RH* by

$$p_a = \frac{RH}{100} \times 479 + (11.52 + 1.62 \times t_i)^2 (\text{Pa}), \tag{9}$$

In addition, Equation (3) for *PMV* includes respiratory convective heat exchange $C_{res}$, calculated as

$$C_{res} = 0.0014 \times M \times (34 - t_i) \, (W/m^2), \tag{10}$$

and Equation (3) includes also respiratory evaporative heat exchange $E_{res}$, given as

$$E_{res} = 1.72 \times 10^{-5} \times M \times (5867 - p_a) \, (W/m^2), \tag{11}$$

Finally, to predict the rate of people dissatisfied in a thermal environment, the *PPD* index is used. Knowing *PMV*, *PPD* can be calculated as [60]

$$PPD = 100 - 95 \times \exp(-0.03353 \times PMV^4 - 0.2179 \times PMV^2) \tag{12}$$

Measured values are shown in Building result figures in the results chapter and used in Equations (1) and (2) for calculating *Tu* and *DR*, and in Equations (3) and (12) to calculate *PMV* and *PPD*.

### 2.4. Indoor Climate Questionnaire

To study occupant satisfaction we provided online questionnaires to the employees of the measured office spaces. As some organizations involved in this study are moving towards policy of a paperless work management, we used Google Forms [61] application. In addition to standard CEN EN 15251 [52] suggestions, we added also questions about age, gender, amount of time behind the desk during workday, and the working environment regarding cabinet or open office plan. The ICQ is presented in Appendix A.

### 2.5. On-Site Measurements

This section provides an overview of the TC measurement time and weather information (Table 4), followed by measurement results with calculated TC indicative parameters *Tu*, *DR*, *PMV*, and *PPD*. ICQ survey results are summarized at the end of the results sections.

**Table 4.** Time of measurements and weather information from the Estonian Weather Service [62].

| Building | Time of Measurements | Weather Conditions | Maximum Outdoor Temp. °C | Mean Outdoor Temp. °C |
|---|---|---|---|---|
| A | 06.08.2019 before midday | cloudy skies showers | +20.9 | +15.2 |
| B | 14.08.2019 after midday | cloudy skies no precipitation | +19.7 | +13.8 |
| C | 12.08.2019 after midday | cloudy skies light showers | +22.0 | +17.3 |
| D | 29.08.2019 after midday | sunny skies no precipitation | +26.5 | +20.6 |
| E | 05.08.2019 after midday | cloudy skies no precipitation | +19.7 | +13.8 |

The experiments were carried out on regular workdays during August. Measurements were taken by two persons, by the main author of this article assisted by graduate students in different buildings. HVAC systems were in regular performance mode without disfunctions or failures recorded. Internal gains by occupants, office equipment, and lighting were in use by default as some desks were empty by unused space, duties, or vacation. No serious defects in HVAC design or construction were observed. Although, some air flow and velocity aspects were noticeable. As in Buildings A and C, active beams were in use, occupants were not always placed sitting according to rule of thumbs, according to the architectural layout, or number of persons. Possible air flow obstacles by lighting fixture (Figure 3a) were noticed with open ceiling in Building A. *DR* risk was also predictable in building E (Figure 3b) where some vanes were taped to closed position. *DR* risk was more carefully considered in Buildings B and D.

(a)  (b)

**Figure 3.** Possible air flow obstacles with open active beam solution (**a**); modified airflow distribution with fan coil unit (**b**).

## 3. Results

Based on on-site measurements, the summary of $v_a$ in each measurement position are shown below for each Building. According to three heights provided in Figure 2, $v_a$ values during measurement period are shown with box and whiskers plot. Minimum and maximum are at the end of the whiskers, the lower and the upper line of the box are first and third quartiles, the line between is median and the cross shows mean $v_a$ value of the measurement in one position.

On the box and whiskers plot, the category of the indoor climate category is colored according to the lowest criteria achieved during measurements, meaning if one of the three height is in III category, the measurement point is placed in the least, III category. In the table part on the result figures below the box and whiskers plot, measured values and calculated parameters are colored according to the category reached to be more easily distinguishable. *Tu* and *RH* measured values are not colored as being not categorized. Nonetheless, measurement point indoor climate category is defined by the inferior measured value or calculated parameter reached altogether.

### 3.1. Building A Results

The $v_a$ results and TC parameters in Building A equipped with open ceiling active chilled beams are provided below in Figure 4. In Building A, in 2/3 of the measured positions the $v_a$ was below the first indoor climate category threshold. Five positions met the II category requirement and in one position the $v_a$ was above the category II threshold. Measurement No 14 was taken in an office space with unusually high internal gains, where also multi-split fan coil units were additionally added to the environment due to the specifics of the lessee. The results of $t_i$ and $v_a$ including *DR*, *PMV*, and *PPD* are placed in the first category mainly.

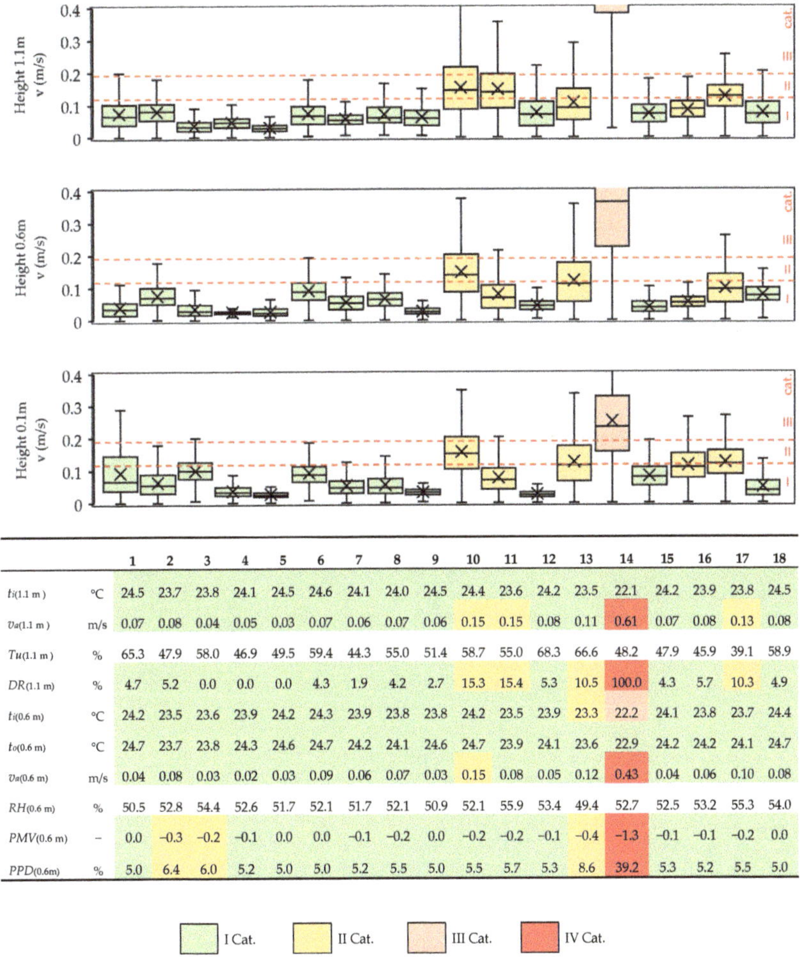

| | | 1 | 2 | 3 | 4 | 5 | 6 | 7 | 8 | 9 | 10 | 11 | 12 | 13 | 14 | 15 | 16 | 17 | 18 |
|---|---|---|---|---|---|---|---|---|---|---|---|---|---|---|---|---|---|---|---|
| $t_{i(1.1\,m)}$ | °C | 24.5 | 23.7 | 23.8 | 24.1 | 24.5 | 24.6 | 24.1 | 24.0 | 24.5 | 24.4 | 23.6 | 24.2 | 23.5 | 22.1 | 24.2 | 23.9 | 23.8 | 24.5 |
| $v_{a(1.1\,m)}$ | m/s | 0.07 | 0.08 | 0.04 | 0.05 | 0.03 | 0.07 | 0.06 | 0.07 | 0.06 | 0.15 | 0.15 | 0.08 | 0.11 | 0.61 | 0.07 | 0.08 | 0.13 | 0.08 |
| $Tu_{(1.1\,m)}$ | % | 65.3 | 47.9 | 58.0 | 46.9 | 49.5 | 59.4 | 44.3 | 55.0 | 51.4 | 58.7 | 55.0 | 68.3 | 66.6 | 48.2 | 47.9 | 45.9 | 39.1 | 58.9 |
| $DR_{(1.1\,m)}$ | % | 4.7 | 5.2 | 0.0 | 0.0 | 0.0 | 4.3 | 1.9 | 4.2 | 2.7 | 15.3 | 15.4 | 5.3 | 10.5 | 100.0 | 4.3 | 5.7 | 10.3 | 4.9 |
| $t_{i(0.6\,m)}$ | °C | 24.2 | 23.5 | 23.6 | 23.9 | 24.2 | 24.3 | 23.9 | 23.8 | 23.8 | 24.2 | 23.5 | 23.9 | 23.3 | 22.2 | 24.1 | 23.8 | 23.7 | 24.4 |
| $t_{o(0.6\,m)}$ | °C | 24.7 | 23.7 | 23.8 | 24.3 | 24.6 | 24.7 | 24.2 | 24.1 | 24.6 | 24.7 | 23.9 | 24.1 | 23.6 | 22.9 | 24.2 | 24.2 | 24.1 | 24.7 |
| $v_{a(0.6\,m)}$ | m/s | 0.04 | 0.08 | 0.03 | 0.02 | 0.03 | 0.09 | 0.06 | 0.07 | 0.03 | 0.15 | 0.08 | 0.05 | 0.12 | 0.43 | 0.04 | 0.06 | 0.10 | 0.08 |
| $RH_{(0.6\,m)}$ | % | 50.5 | 52.8 | 54.4 | 52.6 | 51.7 | 52.1 | 51.7 | 52.1 | 50.9 | 52.1 | 55.9 | 53.4 | 49.4 | 52.7 | 52.5 | 53.2 | 55.3 | 54.0 |
| $PMV_{(0.6\,m)}$ | – | 0.0 | −0.3 | −0.2 | −0.1 | 0.0 | 0.0 | −0.1 | −0.2 | 0.0 | −0.2 | −0.2 | −0.1 | −0.4 | −1.3 | −0.1 | −0.1 | −0.2 | 0.0 |
| $PPD_{(0.6\,m)}$ | % | 5.0 | 6.4 | 6.0 | 5.2 | 5.0 | 5.0 | 5.2 | 5.5 | 5.0 | 5.5 | 5.7 | 5.3 | 8.6 | 39.2 | 5.3 | 5.2 | 5.5 | 5.0 |



**Figure 4.** Building A air velocity results in measurement points 1–18 and the thermal comfort parameters.

### 3.2. Building B Results

The $v_a$ results and TC parameters of Building B with slab-based TABS system are given below in Figure 5. Building B had more measured points in the second category by *PMV* and *PPD* compared to Building A. *DR* met the II category in four measurement positions. Positions 4–8 were in an office, where the ventilation rate had been doubled by the request of the lessee. These four measurements stand out above the others. Regarding the other four buildings observed, displacement ventilation effect can be seen, as $v_a$ fluctuates more near the floor.

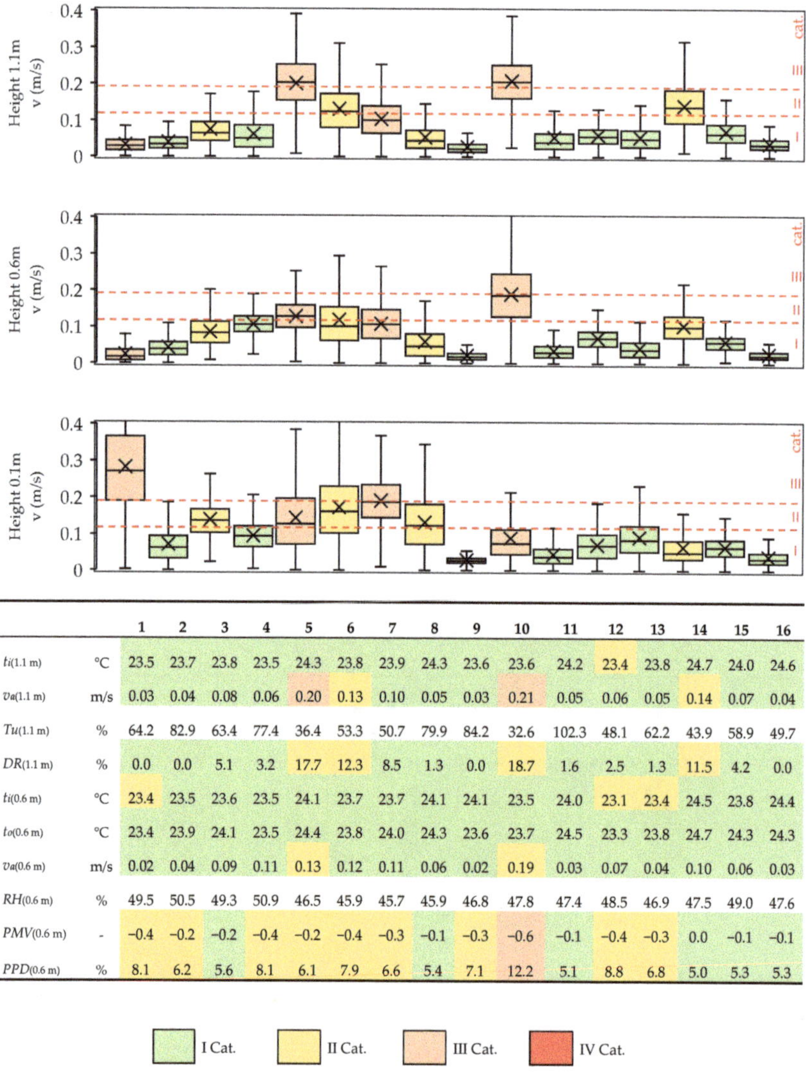

| | | 1 | 2 | 3 | 4 | 5 | 6 | 7 | 8 | 9 | 10 | 11 | 12 | 13 | 14 | 15 | 16 |
|---|---|---|---|---|---|---|---|---|---|---|---|---|---|---|---|---|---|
| $t_{i(1.1\,m)}$ | °C | 23.5 | 23.7 | 23.8 | 23.5 | 24.3 | 23.8 | 23.9 | 24.3 | 23.6 | 23.6 | 24.2 | 23.4 | 23.8 | 24.7 | 24.0 | 24.6 |
| $v_{a(1.1\,m)}$ | m/s | 0.03 | 0.04 | 0.08 | 0.06 | 0.20 | 0.13 | 0.10 | 0.05 | 0.03 | 0.21 | 0.05 | 0.06 | 0.05 | 0.14 | 0.07 | 0.04 |
| $Tu_{(1.1\,m)}$ | % | 64.2 | 82.9 | 63.4 | 77.4 | 36.4 | 53.3 | 50.7 | 79.9 | 84.2 | 32.6 | 102.3 | 48.1 | 62.2 | 43.9 | 58.9 | 49.7 |
| $DR_{(1.1\,m)}$ | % | 0.0 | 0.0 | 5.1 | 3.2 | 17.7 | 12.3 | 8.5 | 1.3 | 0.0 | 18.7 | 1.6 | 2.5 | 1.3 | 11.5 | 4.2 | 0.0 |
| $t_{i(0.6\,m)}$ | °C | 23.4 | 23.5 | 23.6 | 23.5 | 24.1 | 23.7 | 23.7 | 24.1 | 24.1 | 23.5 | 24.0 | 23.1 | 23.4 | 24.5 | 23.8 | 24.4 |
| $t_{o(0.6\,m)}$ | °C | 23.4 | 23.9 | 24.1 | 23.5 | 24.4 | 23.8 | 24.0 | 24.3 | 23.6 | 23.7 | 24.5 | 23.3 | 23.8 | 24.7 | 24.3 | 24.3 |
| $v_{a(0.6\,m)}$ | m/s | 0.02 | 0.04 | 0.09 | 0.11 | 0.13 | 0.12 | 0.11 | 0.06 | 0.02 | 0.19 | 0.03 | 0.07 | 0.04 | 0.10 | 0.06 | 0.03 |
| $RH_{(0.6\,m)}$ | % | 49.5 | 50.5 | 49.3 | 50.9 | 46.5 | 45.9 | 45.7 | 45.9 | 46.8 | 47.8 | 47.4 | 48.5 | 46.9 | 47.5 | 49.0 | 47.6 |
| $PMV_{(0.6\,m)}$ | - | −0.4 | −0.2 | −0.2 | −0.4 | −0.2 | −0.4 | −0.3 | −0.1 | −0.3 | −0.6 | −0.1 | −0.4 | −0.3 | 0.0 | −0.1 | −0.1 |
| $PPD_{(0.6\,m)}$ | % | 8.1 | 6.2 | 5.6 | 8.1 | 6.1 | 7.9 | 6.6 | 5.4 | 7.1 | 12.2 | 5.1 | 8.8 | 6.8 | 5.0 | 5.3 | 5.3 |



**Figure 5.** Building B air velocity results in measurement points 1–16 and the thermal comfort parameters.

### 3.3. Building C Results

Building C was equipped with suspended ceiling active chilled beams and the results of $v_a$ and parameters of TC are presented below in Figure 6. *PMV*, *PPD*, and $t_i$ were similar to Buildings A and

B, at the same time $v_a$ and *DR* were measured at two positions in the II category and three times in the III category. The $v_a$ is more fluctuating on the height of the sitting person neck.

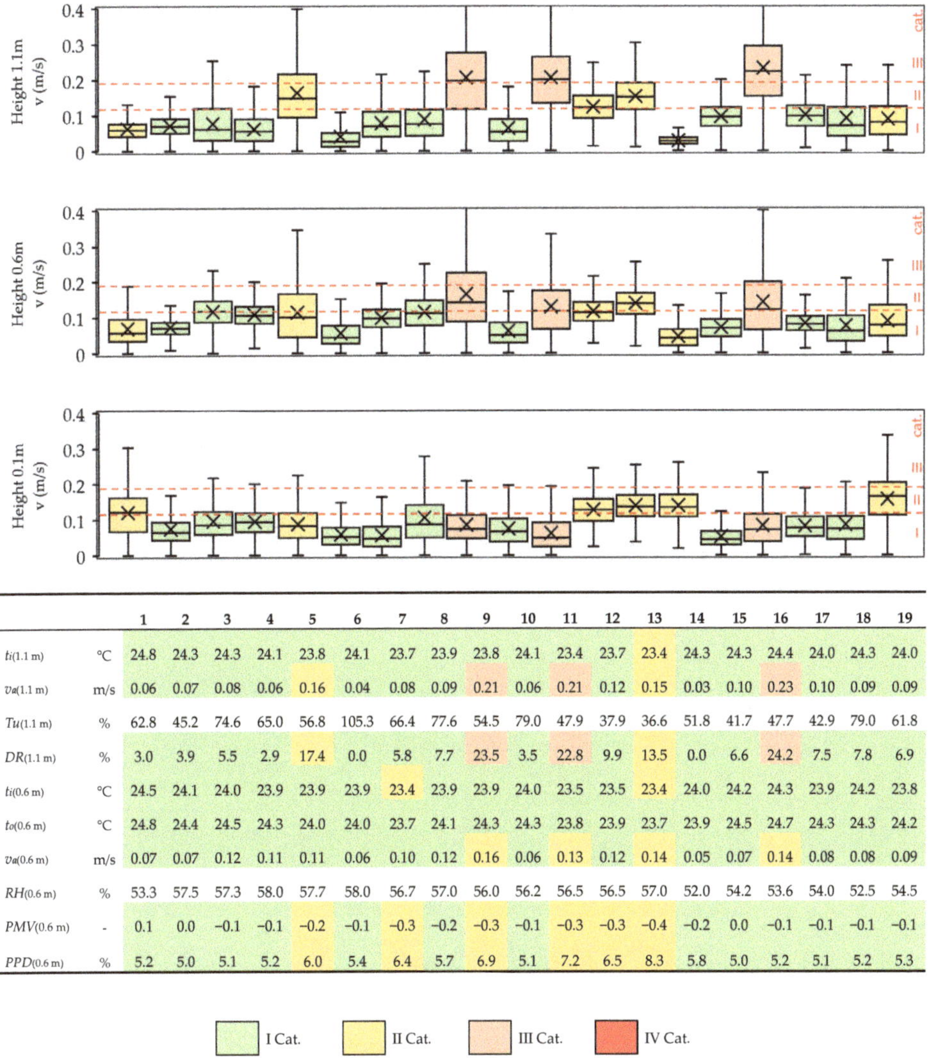

| | | 1 | 2 | 3 | 4 | 5 | 6 | 7 | 8 | 9 | 10 | 11 | 12 | 13 | 14 | 15 | 16 | 17 | 18 | 19 |
|---|---|---|---|---|---|---|---|---|---|---|---|---|---|---|---|---|---|---|---|---|
| $ti_{(1.1\ m)}$ | °C | 24.8 | 24.3 | 24.3 | 24.1 | 23.8 | 24.1 | 23.7 | 23.9 | 23.8 | 24.1 | 23.4 | 23.7 | 23.4 | 24.3 | 24.3 | 24.4 | 24.0 | 24.3 | 24.0 |
| $va_{(1.1\ m)}$ | m/s | 0.06 | 0.07 | 0.08 | 0.06 | 0.16 | 0.04 | 0.08 | 0.09 | 0.21 | 0.06 | 0.21 | 0.12 | 0.15 | 0.03 | 0.10 | 0.23 | 0.10 | 0.09 | 0.09 |
| $Tu_{(1.1\ m)}$ | % | 62.8 | 45.2 | 74.6 | 65.0 | 56.8 | 105.3 | 66.4 | 77.6 | 54.5 | 79.0 | 47.9 | 37.9 | 36.6 | 51.8 | 41.7 | 47.7 | 42.9 | 79.0 | 61.8 |
| $DR_{(1.1\ m)}$ | % | 3.0 | 3.9 | 5.5 | 2.9 | 17.4 | 0.0 | 5.8 | 7.7 | 23.5 | 3.5 | 22.8 | 9.9 | 13.5 | 0.0 | 6.6 | 24.2 | 7.5 | 7.8 | 6.9 |
| $ti_{(0.6\ m)}$ | °C | 24.5 | 24.1 | 24.0 | 23.9 | 23.9 | 23.9 | 23.4 | 23.9 | 23.9 | 24.0 | 23.5 | 23.5 | 23.4 | 24.0 | 24.2 | 24.3 | 23.9 | 24.2 | 23.8 |
| $to_{(0.6\ m)}$ | °C | 24.8 | 24.4 | 24.5 | 24.3 | 24.0 | 24.0 | 23.7 | 24.1 | 24.3 | 24.3 | 23.8 | 23.9 | 23.7 | 23.9 | 24.5 | 24.7 | 24.3 | 24.3 | 24.2 |
| $va_{(0.6\ m)}$ | m/s | 0.07 | 0.07 | 0.12 | 0.11 | 0.11 | 0.06 | 0.10 | 0.12 | 0.16 | 0.06 | 0.13 | 0.12 | 0.14 | 0.05 | 0.07 | 0.14 | 0.08 | 0.08 | 0.09 |
| $RH_{(0.6\ m)}$ | % | 53.3 | 57.5 | 57.3 | 58.0 | 57.7 | 58.0 | 56.7 | 57.0 | 56.0 | 56.2 | 56.5 | 56.5 | 57.0 | 52.0 | 54.2 | 53.6 | 54.0 | 52.5 | 54.5 |
| $PMV_{(0.6\ m)}$ | - | 0.1 | 0.0 | −0.1 | −0.1 | −0.2 | −0.1 | −0.3 | −0.2 | −0.3 | −0.1 | −0.3 | −0.3 | −0.4 | −0.2 | 0.0 | −0.1 | −0.1 | −0.1 | −0.1 |
| $PPD_{(0.6\ m)}$ | % | 5.2 | 5.0 | 5.1 | 5.2 | 6.0 | 5.4 | 6.4 | 5.7 | 6.9 | 5.1 | 7.2 | 6.5 | 8.3 | 5.8 | 5.0 | 5.2 | 5.1 | 5.2 | 5.3 |



**Figure 6.** Building C air velocity results in measurement points 1–19 and the thermal comfort parameters.

## 3.4. Building D Results

Equipped with radiant cooling panels, results of $v_a$ and parameters of TC in Building D are showed below in Figure 7. Compared to other buildings, Building D with the least number of positions had the best results on all analyzed parameters. In all cases, I category *DR* was achieved. At all times, mean $v_a$ remained below 0.10 m/s being more fluctuating near the floor.

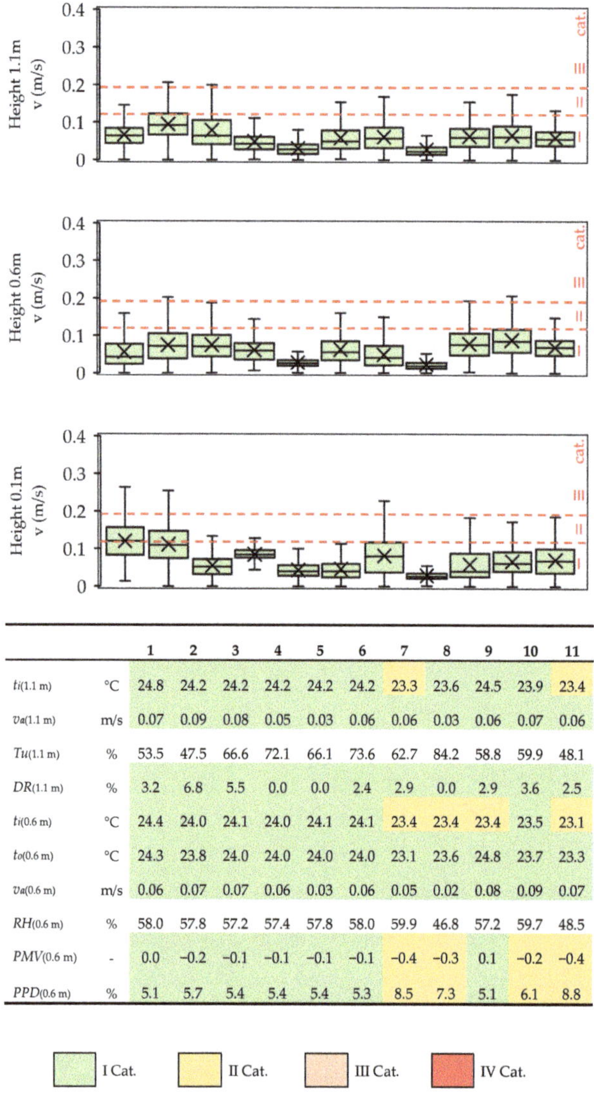

| | | 1 | 2 | 3 | 4 | 5 | 6 | 7 | 8 | 9 | 10 | 11 |
|---|---|---|---|---|---|---|---|---|---|---|---|---|
| $t_{i(1.1\ m)}$ | °C | 24.8 | 24.2 | 24.2 | 24.2 | 24.2 | 24.2 | 23.3 | 23.6 | 24.5 | 23.9 | 23.4 |
| $v_{a(1.1\ m)}$ | m/s | 0.07 | 0.09 | 0.08 | 0.05 | 0.03 | 0.06 | 0.06 | 0.03 | 0.06 | 0.07 | 0.06 |
| $Tu_{(1.1\ m)}$ | % | 53.5 | 47.5 | 66.6 | 72.1 | 66.1 | 73.6 | 62.7 | 84.2 | 58.8 | 59.9 | 48.1 |
| $DR_{(1.1\ m)}$ | % | 3.2 | 6.8 | 5.5 | 0.0 | 0.0 | 2.4 | 2.9 | 0.0 | 2.9 | 3.6 | 2.5 |
| $t_{i(0.6\ m)}$ | °C | 24.4 | 24.0 | 24.1 | 24.0 | 24.1 | 24.1 | 23.4 | 23.4 | 23.4 | 23.5 | 23.1 |
| $t_{o(0.6\ m)}$ | °C | 24.3 | 23.8 | 24.0 | 24.0 | 24.0 | 24.0 | 23.1 | 23.6 | 24.8 | 23.7 | 23.3 |
| $v_{a(0.6\ m)}$ | m/s | 0.06 | 0.07 | 0.07 | 0.06 | 0.03 | 0.06 | 0.05 | 0.02 | 0.08 | 0.09 | 0.07 |
| $RH_{(0.6\ m)}$ | % | 58.0 | 57.8 | 57.2 | 57.4 | 57.8 | 58.0 | 59.9 | 46.8 | 57.2 | 59.7 | 48.5 |
| $PMV_{(0.6\ m)}$ | - | 0.0 | −0.2 | −0.1 | −0.1 | −0.1 | −0.1 | −0.4 | −0.3 | 0.1 | −0.2 | −0.4 |
| $PPD_{(0.6\ m)}$ | % | 5.1 | 5.7 | 5.4 | 5.4 | 5.4 | 5.3 | 8.5 | 7.3 | 5.1 | 6.1 | 8.8 |



**Figure 7.** Building D air velocity results in measurement points 1–11 and the thermal comfort parameters.

### 3.5. Building E Results

According to the results, Building E achieved the worst TC values by categories. $DR$ was in the II category in 4 positions of 14, $t_i$ was in III category four times. $PMV$ and $PPD$ second category was not reached 5 times. Fluctuations of $v_a$ were random depending on the height. The $v_a$ results and TC parameters in Building E, with fan coil units mounted in the suspended ceiling, are compared below in Figure 8.

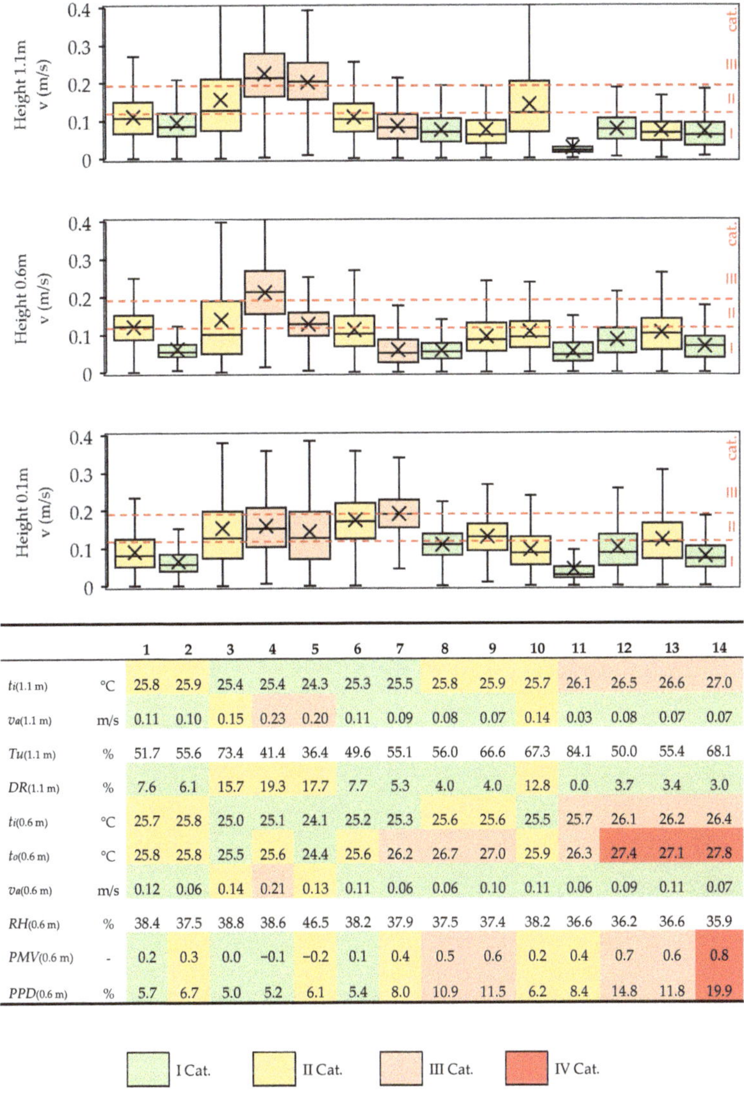

| | | 1 | 2 | 3 | 4 | 5 | 6 | 7 | 8 | 9 | 10 | 11 | 12 | 13 | 14 |
|---|---|---|---|---|---|---|---|---|---|---|---|---|---|---|---|
| $t_{i(1.1\,m)}$ | °C | 25.8 | 25.9 | 25.4 | 25.4 | 24.3 | 25.3 | 25.5 | 25.8 | 25.9 | 25.7 | 26.1 | 26.5 | 26.6 | 27.0 |
| $v_{a(1.1\,m)}$ | m/s | 0.11 | 0.10 | 0.15 | 0.23 | 0.20 | 0.11 | 0.09 | 0.08 | 0.07 | 0.14 | 0.03 | 0.08 | 0.07 | 0.07 |
| $Tu_{(1.1\,m)}$ | % | 51.7 | 55.6 | 73.4 | 41.4 | 36.4 | 49.6 | 55.1 | 56.0 | 66.6 | 67.3 | 84.1 | 50.0 | 55.4 | 68.1 |
| $DR_{(1.1\,m)}$ | % | 7.6 | 6.1 | 15.7 | 19.3 | 17.7 | 7.7 | 5.3 | 4.0 | 4.0 | 12.8 | 0.0 | 3.7 | 3.4 | 3.0 |
| $t_{i(0.6\,m)}$ | °C | 25.7 | 25.8 | 25.0 | 25.1 | 24.1 | 25.2 | 25.3 | 25.6 | 25.6 | 25.5 | 25.7 | 26.1 | 26.2 | 26.4 |
| $t_{o(0.6\,m)}$ | °C | 25.8 | 25.8 | 25.5 | 25.6 | 24.4 | 25.6 | 26.2 | 26.7 | 27.0 | 25.9 | 26.3 | 27.4 | 27.1 | 27.8 |
| $v_{a(0.6\,m)}$ | m/s | 0.12 | 0.06 | 0.14 | 0.21 | 0.13 | 0.11 | 0.06 | 0.06 | 0.10 | 0.11 | 0.06 | 0.09 | 0.11 | 0.07 |
| $RH_{(0.6\,m)}$ | % | 38.4 | 37.5 | 38.8 | 38.6 | 46.5 | 38.2 | 37.9 | 37.5 | 37.4 | 38.2 | 36.6 | 36.2 | 36.6 | 35.9 |
| $PMV_{(0.6\,m)}$ | - | 0.2 | 0.3 | 0.0 | −0.1 | −0.2 | 0.1 | 0.4 | 0.5 | 0.6 | 0.2 | 0.4 | 0.7 | 0.6 | 0.8 |
| $PPD_{(0.6\,m)}$ | % | 5.7 | 6.7 | 5.0 | 5.2 | 6.1 | 5.4 | 8.0 | 10.9 | 11.5 | 6.2 | 8.4 | 14.8 | 11.8 | 19.9 |



**Figure 8.** Building E air velocity results in measurement points 1–14 and the thermal comfort parameters.

## 3.6. Results of the Indoor Climate Questionnaire

Based on ICQ survey, summary of the results for thermal environment are shown below in Figure 9, the ICQ results for *PMV* and *PPD* are presented below in Figure 10. The highest number of answers were in the Building A with 36 responses divided between all age groups equally between men and women. A total of 83% were working in open office layout and 86% were spending most of the day at their workplace. For 83% of the respondents, $t_i$ was described as suitable. Meanwhile, 6 occupants found it to be warm and 7 slightly cooler. A total of 89% had not or had perceived slight odor, 72% did not find lighting fixtures or sunlight to be disturbing, and 81% found ICQ to be suitable or better. A total of 61% perceived overall acoustics and 36% perceived other noises to be disturbing. Roughly half of the respondents rarely felt eye problems, headaches, or concentration matters and

64% rarely felt nasal or throat irritation. Extra comments mentioned occasional lack of ventilation and air dryness.

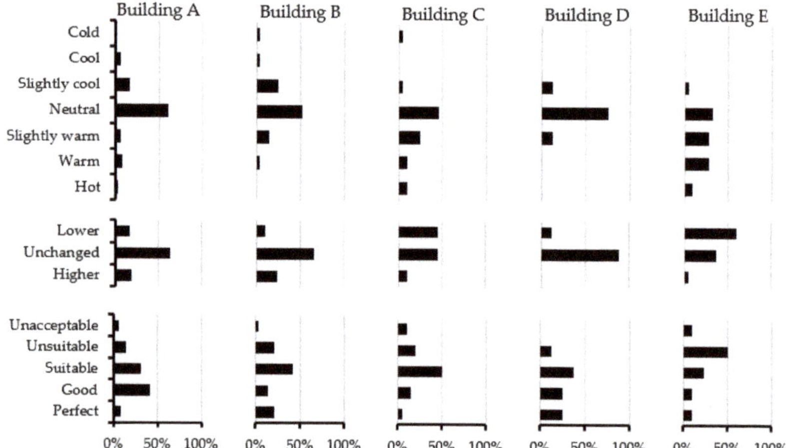

**Figure 9.** Indoor climate questionnaire results for indoor air temperature. The descriptions of y-axis are the room air temperature sensation question (upper) and verification questions (middle and lower) from the indoor climate questionnaire (see Appendix A).

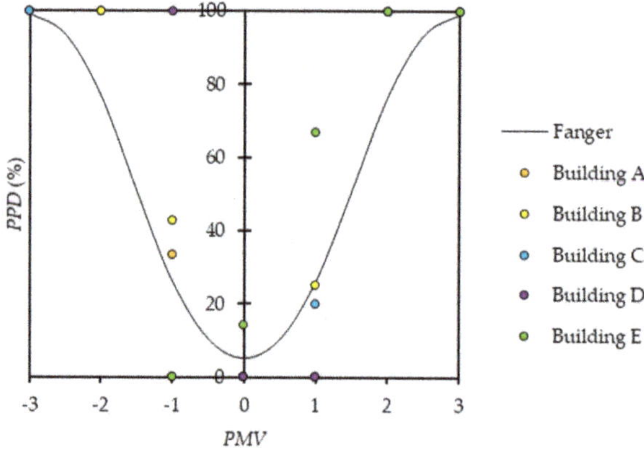

**Figure 10.** Indoor climate questionnaire results for predicted percentage of dissatisfied and predicted mean vote.

Respondents in the Building B were 38% females, 2/3 aged between 26–35 or 36–45 and 1/2 spending half of the workday behind the desk. A total of 72% of them working in open office environment. Ninety percent found $t_i$ to be suitable. A total of 13 of the 29 respondents did not perceive odor. Lighting was disturbing for 21% and sunlight for 14%, meanwhile 14% were dissatisfied with ICQ. Seven percent did not find room acoustics and 17% general noise in office to be disturbing. Half of the respondents had rarely felt eye dryness or irritation, occur headaches or fatigue, and felt nasal problem or dry throat. A total of 62% had rarely felt concentration problems.

Seventy percent of the 20 ICQ respondents in Building C were women. Answers were divided between the age of 26 to 65 with the majority of them working in open office landscape, 2/3 working behind their desk most of the day. Perceived as too warm by 20%, $t_i$ was suitable by 75% of the occupants. Ninety percent had not perceived or had perceived slight odor. A total of 70% did not find lighting equipment to be disturbing and 75% was not disturbed by the sunlight. Forty percent of the respondents found air quality to be not suitable or unacceptable. A total of 85% perceived colleagues' speech and overall room acoustic to be somewhat disturbing, while 65% claimed other noises to be distracting. A total of 1/3 had rarely felt eye problems, occurred headaches, or tiredness. A total of 45% had rarely felt nasal or throat irritation and 20% had rarely had concentration issues. Extra comments mentioned lower fresh air rate in the end of the day.

Building D had only 8 responses for the online ICQ all of them working in the open office. For the majority of the answers, $t_i$ was suitable. Odor was rarely noticed, lighting or sunlight was not disturbing. ICQ was suitable or better, while room acoustics was more disturbing than other noises. Nasal issues were more often to occur compared to eye dryness or headaches and concentration issues. Extra comments noted that open office may be cheaper option for the employer being unsuitable for the employees.

A total of 2/3 of the 22 respondents in Building E were in the second age group between 26–35 years and 36% in overall were females. A total of 77% of the tenants were working in an open office environment, while 2/3 of them were spending most of their day behind the desk. One-third found $t_i$ to be suitable and 2/3 claimed the $t_i$ to be slightly warm, warm, or hot. Fifty percent perceived weak or moderate odor. Room lighting equipment did not disturb 82% and the sun did not disturb 60% of the respondents. A total of 2/3 marked ICQ suitable, good, or very good. Room acoustic level was not claimed to be disturbing for 40% and other noise for 23% of the respondents. Fifty percent had rarely felt eye dryness or irritation, 64% had rarely occur headaches or fatigue, 82% had rarely felt nasal problems or dry throat, and 50% mentioned concentration issues sometimes, often, or all the time. Extra comments noted that air quality decreases in the second phase of the day and the missing option for opening windows was also described as a disadvantage.

Number of respondents of the ICQ is below the least recommended sample size [63], therefore the results of the ICQ include higher uncertainty (Figure 10). Thermal sensation voted by occupants covers significantly wider range than *PMV* calculated from measurements. Majority of the respondents were working in open office. The most unsatisfying $t_i$ was in the Building E and the most suitable $t_i$ was in the Building D. In general, unsuitable $t_i$ was perceived more as warmer than cooler. In Buildings A, B, and D the $t_i$ was perceived suitable for over 80% of the employees, while it was 67% in the Building E and the 60% in the Building C.

## 4. Discussion

The on-site measurement results showed, that the during cooling summertime *DR* risk can be stated in all observed buildings. Preconception of avoiding fan coil units for cooling does not immediately guarantee a superior thermal environment without draught. However, draught risk was the lowest in Building D with radiant cooling panels as room conditioning units.

Possible causes, $v_a$ and *DR* was not significantly higher in the case of fan coil units in Building F. was the taping of air distribution vanes (Figure 3b) and also positioning of the working stations was carried out avoiding direct draught from the fan coil units. This could explain the higher thermal environment temperatures. The induced airflow rate is manually adjustable for open ceiling active chilled beams in Building A and was adjusted into different positions for avoiding possible draught between two beams in various places. In Building C, few suspended ceiling active chilled beams had paper covers blocking air flow from the nozzles. These modifications were made due to the complaints, decrease in productivity or spatial plan and the layout of the workspaces. Described modifications in Buildings A, C, and E refer to possible ineffective floor space areas. Therefore, whether the design or construction may have been inaccurate or user-based thermal environment setpoints do not meet the requirements for $v_a$ and *DR*.

The $v_a$ limit values in EN 16798–1:2019 [54] have been calculated assuming $t_0$ +23 °C and *Tu* 40%. Figure 11a illustrates that the *Tu* is considerably higher than the default value, which increases the unsatisfaction with local TC. However, the measured $t_o$ was higher than the default value in most of the measured positions in all buildings, which decreases the number of dissatisfied. Figure 11b shows that, in general, the *DR* calculated based on measured $t_o$ and *Tu* is in the same scale with the one calculated with the default values.

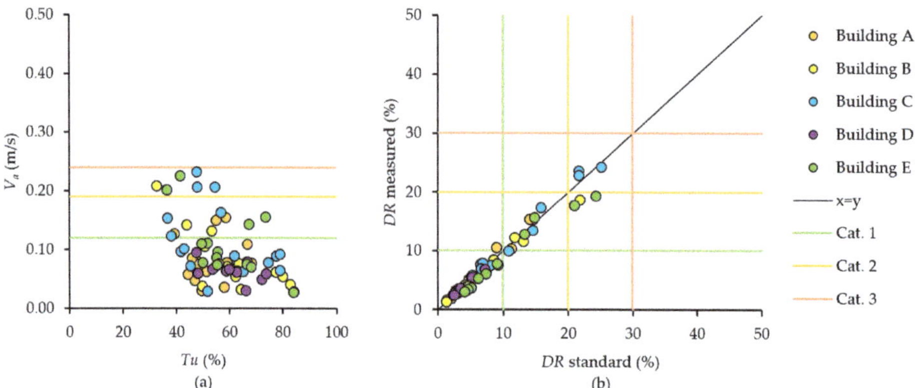

**Figure 11.** Air velocity and turbulence intensity results according to maximum air velocity categories I–III in summer (**a**); Draught rate correlation in measured and standard-based [54] conditions according to draught rate categories I–III (**b**).

In further analysis of this study $t_i$ will be more deeply discussed, foreseeing to include transitional period and heating period measurements, façade inspection and $t_i$ periodical data analysis in the reference buildings. Therefore, the performance of the cooling units according to $t_i$ could be more clearly presented by period or duration curve. Periodical data analysis on $t_i$ is mandatory as $t_i$ presented in this study reflects only a fragment of the thermal environment. Positioning TC measurement values on periodical $t_i$ data can indicate TC measurement accuracy and dispersion. IQC survey number of respondents also needs additional attention, how to achieve a higher response rate.

There are several limitations to this work. Authors had no control over the boundary conditions during measurements. This study only focuses on a few office spaces in five different building in Tallinn. More further studies of actual work environment need to be performed in order to be able to draw general conclusions about studied room conditioning solutions air distribution performance.

## 5. Conclusions

This study was based on TC measurements in open office environments in Tallinn. First or second category measured general thermal comfort in four buildings out of five were still inconvenient for significant number of occupants because of local thermal discomfort caused by draught and by some additional dissatisfaction indicated by questionnaires. Questionnaire survey showed deviation from predicted *PPD* in both directions. Some small occupant groups were either more satisfied or less satisfied at slightly cool or slightly warm thermal sensation, but at neutral sensation the results were more consistent. Less satisfied occupant groups exposed to higher air velocities has likely affected their satisfaction reported in thermal sensation questions because there was no specific draught question available.

Temperature measurements showed that air and operative temperature was the worst in Building E which was close to drop out from category III, while measurement results in Buildings A–D remained in between I and II category. According to the questionnaire over 80% of the employees in Buildings A, B, and C, and 75% in Building D were thermally satisfied. In the Building E, 59% of occupants found the thermal environment unsuitable or unacceptable. Generally, the average thermal satisfaction of occupants was well in line with the measurements.

Air velocity and draught rate measurements showed that modern offices do not necessarily reach to generally expected good indoor climate category II air velocity and draught rate values. A room conditioning solution with suspended ceiling active chilled beams in Building C, displacement ventilation in Building B with TABS and fan coil units in Building E showed category III performance only. Open ceiling active chilled beams in Building A corresponded to category II requirements and ceiling panels for room conditioning in Building D showed superior Category I performance. Category II and III results with active chilled beams indicate that dedicated air distribution solution together with proper design and sizing is needed to reach category II.

We found that existing standards do not provide enough detailed questionnaire for the assessment of occupant dissatisfaction. Our results suggest that questionnaire could be an easier compliance assessment method compared to measurements, which need an expensive equipment and carefully selected measurement days. For the compliance assessment with the measurement, there is more guidance needed especially how to select relevant measurement conditions and locations for draught rate measurement. Future office buildings with open-plan layouts revealed to be demanding environments where careful air distribution design is needed in order to meet comfort requirements.

**Author Contributions:** J.K. conceived and designed the experiments. M.K. prepared agreements with the building owners, performed the measurements and analyzed the data. M.T. and J.K. helped to perform the data analysis. M.K., R.S., M.T., and J.K. wrote this paper. All authors have read and agreed to the published version of the manuscript.

**Funding:** This research was supported by the Estonian Centre of Excellence in Zero Energy and Resource Efficient Smart Buildings and Districts, ZEBE (grant 2014-2020.4.01.15-0016) funded by the European Regional Development Fund, by the programme Mobilitas Pluss (Grant No—2014-2020.4.01.16-0024, MOBTP88), by the European Commission through the H2020 project Finest Twins (grant No. 856602) and the Estonian Research Council grant (PSG409).

**Acknowledgments:** The authors are grateful for the provided cooperation of the building owners, questionnaire respondents for their time and the valuable help from Tallinn University of Technology graduate students.

**Conflicts of Interest:** The authors declare no conflict of interest.

## Nomenclature

| | |
|---|---|
| $A$ | net floor area (m$^2$) |
| $C_{res}$ | respiratory convective heat exchange (W/m$^2$) |
| $DR$ | draught rate (%) |
| $E_c$ | evaporative heat exchange at the skin, when the person experiences a sensation of thermal neutrality (W/m$^2$) |
| $E_{res}$ | respiratory evaporative heat exchange (W/m$^2$) |
| $f_{cl}$ | clothing surface area factor |
| $g$ | solar radiation transmittance through window glass |
| $H$ | heat loss (W/K) |
| $H_d$ | dry heat loss (W/m$^2$) |
| $h_c$ | convective heat transfer coefficient [W/(m$^2 \times$ K)] |
| HVAC | heating, ventilation, and air conditioning |
| $I_{cl}$ | clothing insulation [(m$^2 \times$ K)/W] |
| ICQ | indoor climate questionnaire |
| $M$ | metabolic rate (W/m$^2$) |
| $mt_{sk}$ | mean skin temperature (°C) |
| $p_a$ | water vapor partial pressure (Pa) |
| $PMV$ | predicted mean vote |
| $PPD$ | predicted percentage dissatisfied (%) |
| $RH$ | relative humidity (%) |
| $SD$ | standard deviation |
| TABS | thermally active building systems |

| | |
|---|---|
| TC | thermal comfort |
| $t_{cl}$ | clothing surface temperature (°C) |
| $t_i$ | indoor air temperature (°C) |
| $t_o$ | operative temperature (°C) |
| $Tu$ | turbulence intensity |
| $U$ | thermal transmittance [W/(m$^2$×K)] |
| $v_a$ | air velocity (m/s) |
| $v_{ar}$ | relative air velocity (m/s) |
| $W$ | effective mechanical power (W/m$^2$) |
| $WWR$ | window-to-wall ratio |

## Appendix A

The ICQ form is for online survey is provided below.

1. Gender—( ) Female, ( ) Male
2. Age—( ) 18–25, ( ) 26–35, ( ) 36–45, ( ) 46–55, ( ) 56–65, ( ) 66+
3. Workstation—( ) Private office (max 3 people), ( ) Open office
4. Which amount of the workday you spend at your desk—( ) Whole workday, ( ) Half of the workday (up to 4–5 h), ( ) Few hours (max 1–2 h)
5. In which zone do you spend the most of your workday (1–n in picture)—( ) 1–n
6. How do you rate you thermal sensation (choose neutral if you do not want a change in temperature)—( ) Hot, ( ) Warm, ( ) Slightly warm, ( ) Comfortable, ( ) Slightly cool, ( ) Cool, ( ) Cold
7. How do you perceive odor intensity—( ) No odor, ( ) Weak, ( ) Moderate, ( ) Strong, ( ) Very strong, ( ) Unbearable
8. Would you prefer the room temperature to be—( ) Higher, ( ) Unchanged, ( ) Lower
9. Does the room lighting disturb working—( ) Yes, ( ) No
10. Does the sunlight disturb working—( ) Yes, ( ) No
11. Please rate (room temperature is)—( ) Perfect, ( ) Good, ( ) Suitable, ( ) Unsuitable, ( ) Unbearable
12. Please rate (air quality is)—( ) Perfect, ( ) Good, ( ) Suitable, ( ) Unsuitable, ( ) Unbearable
13. How do you perceive acoustic level (colleagues' speech and overall room acoustics)—( ) Does not disturb at all, ( ) Rarely disturbs, ( ) Sometimes disturbs, ( ) Often disturbs, ( ) Disturbs all the time
14. How do you perceive other noise in your workplace—( ) Does not disturb at all, ( ) Rarely disturbs, ( ) Sometimes disturbs, ( ) Often disturbs, ( ) Disturbs all the time
15. Whether and how often have you experienced the following symptoms (eye dryness or irritation)—( ) Never, ( ) Rarely, ( ) Sometimes, ( ) Often, ( ) All the time
16. Whether and how often have you experienced the following symptoms (headache or fatigue)—( ) Never, ( ) Rarely, ( ) Sometimes, ( ) Often, ( ) All the time
17. Whether and how often have you experienced the following symptoms (nasal or throat dryness or irritation)—( ) Never, ( ) Rarely, ( ) Sometimes, ( ) Often, ( ) All the time
18. Whether and how often have you experienced the following symptoms (concentration problems)—( ) Never, ( ) Rarely, ( ) Sometimes, ( ) Often, ( ) All the time

## References


1. Seppanen, O. Ventilation Strategies for Good Indoor Air Quality and Energy Efficiency. *Int. J. Vent.* **2008**, *6*, 297–306.
2. Yang, Z.; Ghahramani, A.; Becerik-Gerber, B. Building occupancy diversity and HVAC (heating, ventilation, and air conditioning) system energy efficiency. *Energy* **2016**, *109*, 641–649. [CrossRef]
3. Mathews, E.H.; Botha, C.P.; Arndt, D.C.; Malan, A.G. HVAC control strategies to enhance comfort and minimise energy usage. *Energy Build.* **2001**, *33*, 853–863. [CrossRef]
4. Simmonds, P. The Utilization of Optimal-Design and Operation Strategies in Lowering the Energy-Consumption in Office Buildings. *Renew. Energy* **1994**, *5*, 1193–1201. [CrossRef]
5. Guo, W.; Zhou, M. Technologies toward thermal comfort-based and energy-efficient HVAC systems: A review. In Proceedings of the 2009 IEEE International Conference on Systems, Man and Cybernetics, San Antonio, TX, USA, 11–14 October 2009; pp. 3883–3888.
6. Fanger, P.O.; Christensen, N.K. Perception of draught in ventilated spaces. *Ergonomics* **1986**, *29*, 215–235. [CrossRef] [PubMed]

7.   Shahrestani, M.; Yao, R.M.; Cook, G.K. Decision Making for HVAC&R System Selection for a Typical Office Building in the UK. *Ashrae Trans.* **2012**, *118*, 222–229.

8.   Nemethova, E.; Stutterecker, W.; Schoberer, T. Thermal Comfort and HVAC Systems Operation Challenges in a Modern Office Building—Case Study. *Sel. Sci. Pap. J. Civ. Eng.* **2016**, *11*, 103–114. [CrossRef]

9.   Shahzad, S.S.; Brennan, J.; Theodossopoulos, D.; Hughes, B.; Calautit, J.K. Energy Efficiency and User Comfort in the Workplace: Norwegian Cellular vs. British Open Plan Workplaces. *Energy Procedia* **2015**, *75*, 807–812. [CrossRef]

10.   Choi, J.H.; Loftness, V.; Aziz, A. Post-occupancy evaluation of 20 office buildings as basis for future IEQ standards and guidelines. *Energy Build.* **2012**, *46*, 167–175. [CrossRef]

11.   Karjalainen, S. Thermal comfort and gender: A literature review. *Indoor Air* **2012**, *22*, 96–109. [CrossRef]

12.   Schellen, L.; Loomans, M.G.L.C.; de Wit, M.H.; Olesen, B.W.; Lichtenbelt, W.D.V. The influence of local effects on thermal sensation under non-uniform environmental conditions-Gender differences in thermophysiology, thermal comfort and productivity during convective and radiant cooling. *Physiol. Behav.* **2012**, *107*, 252–261. [CrossRef] [PubMed]

13.   Rupp, R.F.; Vasquez, N.G.; Lamberts, R. A review of human thermal comfort in the built environment. *Energy Build.* **2015**, *105*, 178–205. [CrossRef]

14.   Pfafferott, J.U.; Herkel, S.; Kalz, D.E.; Zeuschner, A. Comparison of low-energy office buildings in summer using different thermal comfort criteria. *Energy Build.* **2007**, *39*, 750–757. [CrossRef]

15.   Hens, H.S.L.C. Thermal comfort in office buildings: Two case studies commented. *Build. Environ.* **2009**, *44*, 1399–1408. [CrossRef]

16.   Kolarik, J.; Toftum, J.; Olesen, B.W. Operative temperature drifts and occupant satisfaction with thermal environment in three office buildings using radiant heating/ cooling system. In Proceedings of the Healthy Buildings Europe 2015, Eindhoven, The Netherlands, 18–20 May 2015.

17.   Griefahn, B.; Kunemund, C. The effects of gender, age, and fatigue on susceptibility to draft discomfort. *J. Therm. Biol.* **2001**, *26*, 395–400. [CrossRef]

18.   Maykot, J.K.; Rupp, R.F.; Ghisi, E. A field study about gender and thermal comfort temperatures in office buildings. *Energy Build.* **2018**, *178*, 254–264. [CrossRef]

19.   Maula, H.; Hongisto, V.; Ostman, L.; Haapakangas, A.; Koskela, H.; Hyona, J. The effect of slightly warm temperature on work performance and comfort in open-plan offices - a laboratory study. *Indoor Air* **2016**, *26*, 286–297. [CrossRef]

20.   Wang, Z.; de Dear, R.; Luo, M.H.; Lin, B.R.; He, Y.D.; Ghahramani, A.; Zhu, Y.X. Individual difference in thermal comfort: A literature review. *Build. Environ.* **2018**, *138*, 181–193. [CrossRef]

21.   Kim, J.; Zhou, Y.X.; Schiavon, S.; Raftery, P.; Brager, G. Personal comfort models: Predicting individuals' thermal preference using occupant heating and cooling behavior and machine learning. *Build. Environ.* **2018**, *129*, 96–106. [CrossRef]

22.   Pazhoohesh, M.; Zhang, C. A satisfaction-range approach for achieving thermal comfort level in a shared office. *Build. Environ.* **2018**, *142*, 312–326. [CrossRef]

23.   Enescu, D. A review of thermal comfort models and indicators for indoor environments. *Renew. Sustain. Energy Rev.* **2017**, *79*, 1353–1379. [CrossRef]

24.   Laftchiev, E.; Nikovski, D. An IoT system to estimate personal thermal comfort. In Proceedings of the 2016 IEEE 3rd World Forum on Internet of Things (WF-IoT), Reston, VA, USA, 12–14 December 2016; pp. 672–677.

25.   Ghahramani, A.; Castro, G.; Karvigh, S.A.; Becerik-Gerber, B. Towards unsupervised learning of thermal comfort using infrared thermography. *Appl. Energy* **2018**, *211*, 41–49. [CrossRef]

26.   Jung, W.; Jazizadeh, F. Human-in-the-loop HVAC operations: A quantitative review on occupancy, comfort, and energy-efficiency dimensions. *Appl. Energy* **2019**, *239*, 1471–1508. [CrossRef]

27.   Shi, J.; Yu, N.P.; Yao, W.X. Energy efficient building HVAC control algorithm with real-time occupancy prediction. In Proceedings of the 8th International Conference on Sustainability in Energy and Buildings, Turin, Italy, 11–13 September 2017.

28.   Thalfeldt, M.; Pikas, E.; Kurnitski, J.; Voll, H. Facade design principles for nearly zero energy buildings in a cold climate. *Energy Build.* **2013**, *67*, 309–321. [CrossRef]

29.   Kähkönen, E. Draught, Radiant Temperature Asymmetry and Air Temperature – a Comparison between Measured and Estimated Thermal Parameters. *Indoor Air* **1991**, *1*, 439–447. [CrossRef]

30. Kiil, M.; Mikola, A.; Thalfeldt, M.; Kurnitski, J. Thermal comfort and draught assessment in a modern open office building in Tallinn. *E3S Web Conf.* **2019**, *111*, 02013. [CrossRef]

31. Rhee, K.N.; Olesen, B.W.; Kim, K.W. Ten questions about radiant heating and cooling systems. *Build. Environ.* **2017**, *112*, 367–381. [CrossRef]

32. Saber, E.M.; Tham, K.W.; Leibundgut, H. A review of high temperature cooling systems in tropical buildings. *Build. Environ.* **2016**, *96*, 237–249. [CrossRef]

33. Schellen, L.; Loomans, M.G.L.C.; de Wit, M.H.; Olesen, B.W.; Lichtenbelt, W.D.V.M. Effects of different cooling principles on thermal sensation and physiological responses. *Energy Build.* **2013**, *62*, 116–125. [CrossRef]

34. Maula, H.; Hongisto, V.; Koskela, H.; Haapakangas, A. The effect of cooling jet on work performance and comfort in warm office environment. *Build. Environ.* **2016**, *104*, 13–20. [CrossRef]

35. Gao, S.; Wang, Y.A.; Zhang, S.M.; Zhao, M.; Meng, X.Z.; Zhang, L.Y.; Yang, C.; Jin, L.W. Numerical investigation on the relationship between human thermal comfort and thermal balance under radiant cooling system. *Energy Procedia* **2017**, *105*, 2879–2884. [CrossRef]

36. Cen, C.; Jia, Y.H.; Liu, K.X.; Geng, R.X. Experimental comparison of thermal comfort during cooling with a fan coil system and radiant floor system at varying space heights. *Build. Environ.* **2018**, *141*, 71–79. [CrossRef]

37. Kolarik, J.; Toftum, J.; Olesen, B.W.; Jensen, K.L. Simulation of energy use, human thermal comfort and office work performance in buildings with moderately drifting operative temperatures. *Energy Build.* **2011**, *43*, 2988–2997. [CrossRef]

38. Fonseca, N. Experimental study of thermal condition in a room with hydronic cooling radiant surfaces. *Int. J. Refrig.* **2011**, *34*, 686–695. [CrossRef]

39. Li, R.L.; Yoshidomi, T.; Ooka, R.; Olesen, B.W. Field evaluation of performance of radiant heating/cooling ceiling panel system. *Energy Build.* **2015**, *86*, 58–65. [CrossRef]

40. Saber, E.M.; Iyengar, R.; Mast, M.; Meggers, F.; Tham, K.W.; Leibundgut, H. Thermal comfort and IAQ analysis of a decentralized DOAS system coupled with radiant cooling for the tropics. *Build. Environ.* **2014**, *82*, 361–370. [CrossRef]

41. Chiang, W.H.; Wang, C.Y.; Huang, J.S. Evaluation of cooling ceiling and mechanical ventilation systems on thermal comfort using CFD study in an office for subtropical region. *Build. Environ.* **2012**, *48*, 113–127. [CrossRef]

42. Mustakallio, P.; Bolashikov, Z.; Kostov, K.; Melikov, A.; Kosonen, R. Thermal environment in simulated offices with convective and radiant cooling systems under cooling (summer) mode of operation. *Build. Environ.* **2016**, *100*, 82–91. [CrossRef]

43. Cehlin, M.; Karimipanah, T.; Larsson, U.; Ameen, A. Comparing thermal comfort and air quality performance of two active chilled beam systems in an open-plan office. *J. Build. Eng.* **2019**, *22*, 56–65. [CrossRef]

44. Kim, T.; Kato, S.; Murakami, S.; Rho, J. Study on indoor thermal environment of office space controlled by cooling panel system using field measurement and the numerical simulation. *Build. Environ.* **2005**, *40*, 301–310. [CrossRef]

45. Fredriksson, J.; Sandberg, M.; Moshfegh, B. Experimental investigation of the velocity field and airflow pattern generated by cooling ceiling beams. *Build. Environ.* **2001**, *36*, 891–899. [CrossRef]

46. Rhee, K.N.; Shin, M.S.; Choi, S.H. Thermal uniformity in an open plan room with an active chilled beam system and conventional air distribution systems. *Energy Build.* **2015**, *93*, 236–248. [CrossRef]

47. Koskela, H.; Haggblom, H.; Kosonen, R.; Ruponen, M. Air distribution in office environment with asymmetric workstation layout using chilled beams. *Build. Environ.* **2010**, *45*, 1923–1931. [CrossRef]

48. Indraganti, M.; Ooka, R.; Rijal, H.B. Thermal comfort in offices in summer: Findings from a field study under the 'setsuden' conditions in Tokyo, Japan. *Build. Environ.* **2013**, *61*, 114–132. [CrossRef]

49. De Vecchi, R.; Candido, C.; de Dear, R.; Lamberts, R. Thermal comfort in office buildings: Findings from a field study in mixed-mode and fully-air conditioning environments under humid subtropical conditions. *Build. Environ.* **2017**, *123*, 672–683. [CrossRef]

50. Azad, A.S.; Rakshit, D.; Wan, M.P.; Babu, S.; Sarvaiya, J.N.; Kumar, D.E.V.S.K.; Zhang, Z.; Lamano, A.S.; Krishnasayee, K.; Gao, C.P.; et al. Evaluation of thermal comfort criteria of an active chilled beam system in tropical climate: A comparative study. *Build. Environ.* **2018**, *145*, 196–212. [CrossRef]

51. He, Y.D.; Li, N.P.; Huang, Q. A field study on thermal environment and occupant local thermal sensation in offices with cooling ceiling in Zhuhai, China. *Energy Build.* **2015**, *102*, 277–283. [CrossRef]

52. CEN EN 15251:2007. *European Committee for Standardization, Indoor Environmental Input Parameters for Design and Assessment of Energy Performance of Buildings Addressing Indoor Air Quality, Thermal Environment, Lighting and Acoustics*; European Committee for Standardization: Brussels, Belgium, 2007.

53. Fanger, P.O. *Thermal Comfort, Analysis and Applications in Environmental Engineering*; Danish Technical Press: Manhattan, KS, USA, 1970.

54. CEN EN 16798-1:2019. *Energy Performance of Buildings—Ventilation for Buildings—Part 1: Indoor Environmental Input Parameters for Design and Assessment of Energy Performance of Buildings Addressing Indoor Air Quality, Thermal Environment, Lighting and Acoustics—Module M1-6*; European Committee for Standardization: Brussels, Belgium, 2019.

55. ISO 7726:1998. *Ergonomics of the Thermal Environment—Instruments for Measuring Physical Quantities*; International Organization for Standardization: Geneva, Switzerland, 1998.

56. Dantec Dynamics. *ComfortSense specification*; Dantec Dynamics A/S, A Nova Instruments Company: Denmark, Skovlunde, 2019; Available online: http://www.dantecdynamics.jp/docs/products-and-services/thermal-comfort/PI264_ComfortSense.pdf (accessed on 5 June 2019).

57. Ministry of Economic Affairs and Communications. Estonian Regulation No 258: Minimum Requirements for Energy Performance. *Riigi Teataja* **2007**, *72*, 445.

58. Thermal Comfort. Innova AirTech Instruments A/S. 2002. Available online: http://www.labeee.ufsc.br/sites/default/files/disciplinas/Thermal%20Booklet.pdf (accessed on 10 July 2019).

59. CEN EN 15726:2011. *Ventilation for Buildings—Air Diffusion—Measurements in the Occupied Zone of Air-Conditioned/Ventilated Rooms to Evaluate Thermal and Acoustic Conditions*; European Committee for Standardization: Brussels, Belgium, 2011.

60. ISO 7730:2005. *Ergonomics of the Thermal Environment—Analytical Determination and Interpretation of Thermal Comfort Using Calculation of the PMV and PPD Indices and Local Thermal Comfort Criteria*; International Organization for Standardization: Geneva, Switzerland, 2005.

61. Google. Google Forms. Google Inc., Mountain View (CA), USA. 2019. Available online: https://www.google.com/forms/about/ (accessed on 1 July 2019).

62. EMHI. Observation Data. Estonian Weather Service: Tallinn, Estonia. 2019. Available online: https://www.ilmateenistus.ee/ilm/ilmavaatlused/vaatlusandmed/tunniandmed/?lang=en (accessed on 15 September 2019).

63. Wang, J.Y.; Wang, Z.; de Dear, R.; Luo, M.H.; Ghahramani, A.; Lin, B.R. The uncertainty of subjective thermal comfort measurement. *Energy Build.* **2018**, *181*, 38–49. [CrossRef]

*Review*

# Determination of Thermal Comfort in Indoor Sport Facilities Located in Moderate Environments: An Overview

**Fabio Fantozzi and Giulia Lamberti ***

DESTEC, Dept. of Energy, Systems, Territory and Constructions Engineering, School of Engineering, University of Pisa, Largo Lucio Lazzarino, 56122 Pisa, Italy; f.fantozzi@ing.unipi.it
* Correspondence: giulia.lamberti@phd.unipi.it

Received: 31 October 2019; Accepted: 29 November 2019; Published: 3 December 2019

**Abstract:** In previous years, providing comfort in indoor environments has become a major question for researchers. Thus, indoor environmental quality (IEQ)—concerning the aspects of air quality, thermal comfort, visual and acoustical quality—assumed a crucial role. Considering sport facilities, the evaluation of the thermal environment is one of the main issues that should be faced, as it may interfere with athletes' performance and health. Thus, the necessity of a review comprehending the existing knowledge regarding the evaluation of the thermal environment and its application to sport facilities becomes increasingly relevant. This paper has the purpose to consolidate the aspects related to thermal comfort and their application to sport practice, through a deep study concerning the engineering, physiological, and psychological approaches to thermal comfort, a review of the main standards on the topic and an analysis of the methodologies and the models used by researchers to determine the thermal sensation of sport facilities' occupants. Therefore, this review provides the basis for future research on the determination of thermal comfort in indoor sport facilities located in moderate environments.

**Keywords:** thermal comfort models; thermal comfort assessment; Fanger's models; moderate environments; sport facilities

---

## 1. Introduction

In recent years, ensuring comfort in indoor environments has become a real challenge involving different disciplines. However, in the past, the parameters used to guarantee comfort and the approaches to improve the quality of the indoor environment were often studied separately [1]. Furthermore, comfort in indoor spaces can be ensured through the control of all the environmental factors, which include indoor air quality, thermal comfort, lighting, and acoustical quality [2]. In this context, the indoor environmental quality (IEQ) that includes all these aspects, assumes a fundamental role in the determination of the conditions of comfort in buildings. Since the time that humans spend indoors has largely increased, several studies have been carried out in order to enhance comfort in indoor spaces, especially in offices, schools, hospitals, etc. The focus of researchers was often based on possible improvements of air quality [3] or on the reduction of the impact of pollutants in indoor air [4]. Visual quality and acoustical comfort have been also studied in workplaces and in educational rooms [5–8], since they may have a great impact on the focus and on the performance of the occupants. Finally, thermal comfort has been often studied in relation to the characteristic of the envelope and of the internal structures [9], as the thermal behavior of the building can largely influence the environmental conditions indoors, which also has an impact on the comfort and performance of people.

In sport facilities, these four aspects have been studied in order to improve the comfort and the performance of the athletes. In particular, light has been recognized as one of the most important factors to ensure the correct practice of the sport [10,11], while indoor air quality has been considered fundamental for ensuring health of the athletes [12]. Then, since swimming pools and sport halls are environments in which noise level and speech intelligibility can determine the comfort or the safety of the occupants, research on the assessment of acoustic conditions and on the use of acoustic treatments to improve the quality of the environment have been carried out [13].

The thermal environment is probably the most important parameter that should be considered when performing sports, as it can determine the safety and the performance of the athletes. In moderate environments, defined as spaces in which it is possible to reach the condition of thermal well-being, only few studies have been carried out. These studies were focused on the monitoring the thermal conditions according to thermal comfort indices [14,15], on the interventions to improve thermal comfort [16,17], on the comparison between objective and subjective measurements [18,19], on the assessment of thermal comfort to balance energy use [16,20] and on the association between thermal comfort and physiological responses during exercise [21,22]. However, there was no standardization in the measurement methodologies or in the models that were used to predict thermal sensation in sport facilities. Even the norms regarding the perception of the thermal environment do not often consider the parameters that should be maintained in sport halls and swimming pools and only in some cases Sports Federations provide these values, even if they are often incomplete.

The main purpose of this article is to consolidate the existing knowledge regarding the thermal environment and its application to sport facilities. In order to achieve this result, it was necessary to use a multidisciplinary approach that considers all the aspects of thermal comfort, from the engineering approach, which treats man as a heat engine, to the physiological and psychological ones, which also play a key role in the perception of the thermal environment, especially during sport practice. Then, in order to consider the practical aspects of the prediction of the thermal sensation of the athletes, the main standards and Federations' norms have been reviewed, as well as the models used by researchers to evaluate thermal comfort in sport facilities. Finally, the methodologies developed to assess thermal comfort in these environments have been studied. This paper lays the groundwork for future research on the determination of indoor thermal comfort in sport facilities located in moderate environments, as in these spaces thermal conditions have a fundamental role in the performance and in the health of the athletes.

## 2. Thermal Comfort Approaches

The perception of the thermal environment can be considered dependent on several factors, derived from different fields of research. In particular, three main approaches have been identified: engineering, physiological, and psychological approaches. The engineering approach is based on the representation of the humans as 'heat engines', who can exchange heat with the environment and it implies that the thermal sensation is dependent on the heat balance of the human body. The physiological approach considers instead the mechanisms with which the body responds to the thermal environment (e.g., thermoregulatory responses). Finally, the psychological approach concerns the psychological phenomena regarding the individuals' perception of a certain environment [23].

### 2.1. Engineering Approach

In the engineering approach, the human body is represented as a heat engine, which can give or receive heat from the environment through conduction, convection, radiation, and evaporation. The heat in the body is produced by the metabolic processes occurring during human life. The heat exchange between the body and the environment can be determined through the heat balance equation [23]

$$M - W = C_k + C + R + E + S \tag{1}$$

where M is the metabolic rate of the body (W), W is the mechanical work (W), $C_k$ is the heat transfer by conduction (W), C is the heat transfer by convection (W), R is the heat transfer by radiation (W), E is the heat transfer by evaporation (W), and S is the heat storage (W).

In conditions of thermal equilibrium, the heat storage is null (S = 0) and the heat balance equation can be written as

$$M - W - C_k - C - R - E = 0 \tag{2}$$

Note that this equation is generally applied to steady state conditions and it should be carefully adopted during sport practice, as exercise is usually performed under transient conditions. Figure 1 reports the mechanisms of heat transfer during exercise.

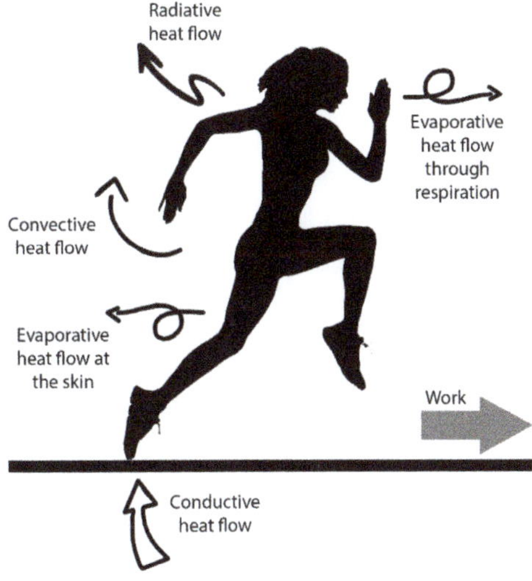


**Figure 1.** Heat transfer mechanisms during sport activity. The body can exchange heat with the environment through conduction, convection, radiation, and evaporation. The production of external work through muscular activity leads to an increase of the heat that has to be dispersed in the environment.

### 2.1.1. DuBois Area

The heat produced by the body flows through the body surface. A method for the calculation of the nude body surface area is given by DuBois formula [24]

$$A_{DB} = 0.2025 \, W^{0.425} \, H^{0.725}, \tag{3}$$

where W is the weight (kg) and H is the height (m) of the body. Generally, the value of $A_{DB} = 1.8 \ \text{m}^2$ is assumed.

### 2.1.2. Heat Exchange through Conduction

Generally, the heat exchange between the body and the environment through conduction is limited, as it involves small parts of the body. Therefore, the conductive effects are often neglected, or included in the convective effects [25]. However, conduction must be considered in the heat balance when the body is in contact with large surfaces. In this case, the heat loss or gain is dependent on factors such as the body and surface temperatures, the area of contact and the conductivity of the surface and of the body tissues [26]. During sport activity, heat exchange through conduction can occur for example in running, when the athlete is running on a hot road, or in cycling, when the athlete is in contact with the seat of the ridden bicycle.

### 2.1.3. Heat Exchange through Convection

Heat transfer through convection totals up to 15% of the whole heat loss in stationary conditions, but even more when the air is moving over the body surface [26]. During sport activity, convection can occur due to the body movement, which generates air (e.g., in running, riding, etc.) or water (e.g., swimming) currents or due to the air movement (e.g., wind). The air movement around the skin is responsible for convective cooling.

Heat transfer by convection is given by [25]

$$C = h_c \, (T_{sk} - T_a) \, A_c \, f_{cl},\qquad(4)$$

where $h_c$ is the convective heat transfer coefficient (W/m$^2$ K), $T_a$ is the air temperature (K), $T_{sk}$ is the mean skin temperature (K), $A_c$ is the body surface involved in the heat exchange through convection (m$^2$) ($A_c \approx A_{DB}$) and $f_{cl}$ is the clothing area factor.

The clothing area factor ($f_{cl}$) can be calculated as [27]

$$f_{cl} = 1.00 + 0.28 \, I_{cl},\qquad(5)$$

where $I_{cl}$ (clo) is the thermal insulation of clothing, whose values are provided for everyday garments in the tables reported in the ISO 9920. Movement tends to let the insulating characteristics of the clothing and of the boundary air layer decrease. In warm environments, where convective heat loss has a positive effect, fabrics are developed in order to let the air to flow between the body and the garments. Conversely, in cold environments, clothing is designed in order to minimize the air movement, preventing convective heat transfer and maintaining body warmth [28].

The convective heat transfer coefficient ($h_c$) is a function of several parameters such as the velocity of the currents, density, and viscosity of the fluid involved and the shape of the exposed surface. An approximate value of $h_c$ is given by [25]

$$\begin{aligned} h_c &= 3.5 + 5.2 \, V_{ar}, &\text{for } V_a \leq 1 \text{ m/s} \\ h_c &= 8.7 \, V_{ar}^{0.6}, &\text{for } V_a > 1 \text{ m/s} \end{aligned}\qquad(6)$$

where $V_a$ is the air velocity (m/s), $V_{ar}$ is the resultant air velocity (m/s) considering the environmental air velocity and the movement of the person and it can be calculated as [25]

$$V_{ar} = V_a + 0.0052 \, (M - 58),\qquad(7)$$

where M is the metabolic heat production (W/m$^2$), with the condition that it is considered M = 200 W/m$^2$ when M exceeds the value of 200 W/m$^2$.

The influence of the human body's movement on heat exchange can be considered through the calculation of the convective heat transfer coefficient. Several studies have been carried out on this topic, analyzing standing and seating postures and different air speeds occurring due to the movements of mannequin simulating walking and running [29,30] or to the wind [31]. Further research has been developed using computational fluid dynamics to assess the convective heat transfer of individual body segments for cyclist positions [32]. Moreover—since water convection is the only important heat transfer mechanism—in the past, several studies have been performed in order to determine $h_c$ analytically [33], or on a heated copper manikin located in the water [34], or detected on experimental data on humans [35].

### 2.1.4. Heat Exchange through Radiation

Thermal radiation is considered to be one of the factors that can influence the most the heat exchange during sport activity [26]. Only in water sports, the component of radiative heat loss is usually negligible [36]. Since body temperature during exercise is generally higher than the air temperature, there is a loss of radiative heat energy from the body. Only in warm environments, where the air temperature may be higher than the skin temperature, the body can gain heat through radiation.

The heat loss through radiation is given by [25]

$$R = h_r \, (T_{sk} - T_r) \, A_r \, f_{cl},\tag{8}$$

where $h_r$ is the radiative heat transfer coefficient (W/m$^2$K), $T_r$ is the mean radiant temperature (K), $T_{sk}$ is the mean skin temperature (K), $A_r$ is the effective radiation area of the body (m$^2$) and $f_{cl}$ is the clothing area factor.

$h_r$ can be calculated as

$$h_r = 4\sigma\varepsilon_{sk}\left(\frac{T_r + T_{sk}}{2}\right)^3,\tag{9}$$

where $\sigma = 5.67 \times 10^{-8}$ W/m$^2$K$^4$ is the Stefan-Boltzmann coefficient, $\varepsilon$ is the emissivity of the body (for the skin $\varepsilon = 0.97$–$0.98$).

$A_r$, the effective radiation area of the body is given by [25]

$$A_r = (A_r/A_{DB}) \, A_{DB},\tag{10}$$

where $A_r/A_{DB} = 0.67$ (for squatting position)—0.70 (for sitting position)—0.77 (for standing position).

### 2.1.5. Heat Exchange through Evaporation

Heat loss through evaporation can occur through skin (by passive diffusion or sweating) and respiratory system (by breathing). Under steady state conditions, it accounts 10% to 25% of the total heat loss and it depends on factors such as relative humidity of the environment, air and skin temperature, air velocity, and clothing [37]. During sport activity, thermoregulation depends mainly on the heat loss through evaporation of sweat and it can arrive to account up to 90% of the total heat loss [38]. In water sports, evaporation cannot be considered as a mechanism of heat exchange [36].

Heat exchange through evaporation can be calculated as [25]

$$E = h_e \, (P_{skH2O} - P_{aH2O}) \, A_e \, F_{pcl},\tag{11}$$

where $h_e$ is the evaporative heat transfer coefficient (W/m$^2$ Pa), $P_{aH2O}$ is the water vapor pressure in the environment (Pa), $P_{skH2O}$ is the water vapor pressure in saturated air at $T_{sk}$ (Pa), $A_e$ is the evaporative surface (m$^2$), and $F_{pcl}$ is the clothing permeability factor.

$h_e$ can be calculated as [25]

$$h_e = k \, h_c,\tag{12}$$

with $k = 16.7$ K/Pa

$A_e$ can be calculated as

$$A_e = (A_e/A_{DB}) \, A_{DB} = w \, A_{DB},\tag{13}$$

where w is the skin wittedness, which is a physiological index defined as the ratio between the actual sweating rate and the maximum sweating rate that occurs when the skin is completely wet. w can range from 0.06, when the evaporative heat loss is caused only by passive diffusion, to 1, when the skin surface is completely wet.

### 2.1.6. Strategies Adopted from Athletes using Heat Transfer Mechanisms to Support Thermoregulation

The heat transfer mechanisms may support thermoregulation and improve sport performance [39]. Conduction is often used when an athlete is warm to decrease his body temperature. In particular, possible solutions are to put him in contact with cold surfaces or to let him wear special clothing such as ice vests. When the athlete is cold, wearing sport garments that present good thermal insulation may prevent the heat flow from the skin to the environment. In fact, conduction is particularly important when designing sport equipment, especially when it is composed by conductive materials, as for example the baseball bats or the motor racing seats. In this case, it is important to maintain the equipment at a temperature that is safe for the athletes.

Convection is an effective method to decrease body temperature and it can be supported by the use of fans that increase the heat flow and cool him down. On the contrary, the use of wind-breaker jackets may prevent excessive body cooling when the athlete is cold.

Heat loss through radiation can be increased by increasing the skin exposure to the environment or decreased by exposing the athlete to the sun or to other radiation sources.

Finally, evaporation is a fundamental heat transfer mechanism during exercise, as the body can produce a great amount of sweat. If an athlete is warm, pouring the water over the body can be an efficient way to decrease his temperature, as it leaves more water on the skin to evaporate. Conversely, if an athlete is cold, solutions to prevent the sweat evaporation include removing the water from the skin, removing wet clothes, or wearing additional clothes.

### 2.1.7. Use of Sport Garments to Control Heat Transfer Mechanisms

The presence of clothing on the human body has several implications on the heat balance. In particular, when considering sport garments, the selection of certain materials and design play a key role in the performance of the athletes. In particular, sport clothing must provide thermo-physiological comfort, supporting the wearer's thermoregulation, keeping the wearer at a comfortable temperature and maintaining the micro-climate between skin and textile as dry as possible [40]. For this reason, understanding the requirements of each sport is an essential step in the design of sport apparel and different studies have been performed for specific sports as for example baseball [41], snowboarding [42], rowing [43], athletics [44], or fitness [45]. The importance of sport clothing is evident when the protection from the environment is required for survival (e.g., mountain sports), but even in common applications it can have a fundamental role in athletes' performance. The aim of sports garments is in fact to provide a comfortable microclimate for the athletes, since comfort may affect their sport performance as it can avoid them from using more reserves in order to maintain the heat balance with the environment [46]. The aim of the study on sport garments is particularly relevant as, in some sports, a uniform is required (e.g., fencing), which does not allow the athletes to modify their conditions, and to adapt to the thermal environment.

The aspects that must be considered with regard to the thermal performance of clothing are the thermal and the moisture management. The early versions of performance garments consisted of a three-layer system, constructed with a base layer with the function of managing the moisture, a middle layer necessary for the insulation and a protective outer layer [47]. Even if this system was primarily used for outdoor apparel; nowadays, it is often adopted in the production of garments for indoor sports, with specific adaptations in order to improve performance. In fact, the necessities of the athletes may be different according to the environment in which they are exercising, as in moderate climates the heat production is high and heat generally flows from the body to the ambient, while in severe environments (both cold and hot) the mechanisms of heat exchange may be different. Moreover, professional athletes may have additional requirements for their apparel, as they have to exchange a great amount of heat with the environment, due to their high metabolic rates. For this reason, the composition of the garments is particularly relevant, as it can determine the heat exchange preventing the heat transfer by conduction, maintaining still air in the clothing, and managing moisture.

The mechanisms of heat transfer through clothing are shown in Figure 2. Regarding the heat exchange through conduction, the presence of garments on the body reduces the amount of heat loss and the characteristics of the clothing may affect the way in which the heat is exchanged [48]. Substantially, the heat transfer through conduction occurs between the inside and the outside of the clothing. In hot environments, apparel should be composed by high conductive materials, in order to let the heat flow from the body to the surroundings, while in cold conditions they should present low conductivity and they are required to have layer that can trap air, in order to improve their thermal resistance [28]. During sport activity, convection may occur between the apparel (or the skin) and the environment. In warm conditions, the main necessity is to cool down the body, therefore apparel is designed to allow the air flow between the body and the garments, while in cold environments the movement of air should be minimal [28]. Considering the sport of swimming, the optimization of the convective flux has not only the function of improving the heat exchange, but it can enhance the performance thanks to the better hydrodynamics [49]. With regard to the heat exchange through radiation, thermal insulation of the clothing can lead to a decrease of this heat flux, through the reduction of the temperature difference between the skin and the surrounding area. In particular, color and texture of sport clothing are specifically relevant to the heat gain and loss through radiation [28]. For this reason, several studies have been performed in order to produce clothing with metallic coatings or finishing technology that can shield infrared radiation [49]. Finally, since evaporative heat loss is specifically relevant during sport activity, it is important to take it into consideration when designing sport garments. In particular, in warm climates, clothing should be conceived to transport the sweat on their outer layer in order to let it evaporate, thus reducing the body temperature. In cold environments the athletes must also sweat, even if the evaporation of moisture is not intended, thus fibers that draw sweat to the outside of the apparel or to the internal microclimate of clothing are generally used [28].

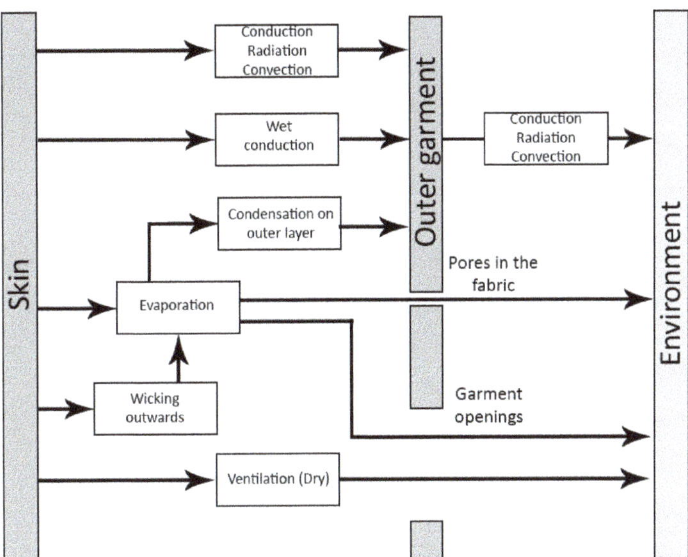


**Figure 2.** Heat transfer through clothing (modified from [50]). Heat flows from the internal to the external part of the clothing through conduction and then it is exchanged with the environment through conduction, convection, radiation, and evaporation.

*2.2. Physiological Approach*

The physiological approach concerns the mechanisms with which the body reacts to the thermal environment (e.g., thermoregulatory responses).

2.2.1. Heat Production during Exercise

The estimation of the heat produced by the metabolism is fundamental to assess the human thermal environment. The free energy necessary for living processes comes from the food and is then converted in the body cells thanks to the ATP-ADP cycle (adenosine triphosphate–adenosine diphosphate cycle) for ensuring life processes and for producing internal and external work. Internal mechanical work consists of the processes that take part in the body, such as the blood circulation, the movement of the air through the lungs or the work of the heart, while external mechanical work is a consequence of muscular contraction [51]. The metabolic heat is a waste product of the metabolism and it must be dispersed in the environment, in order to maintain the body temperature constant (Figure 3).

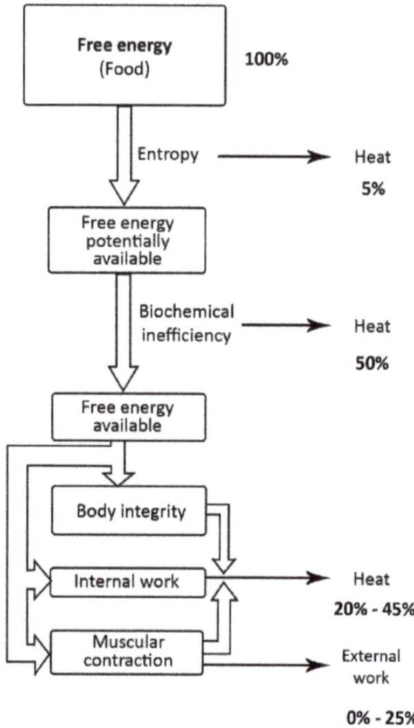


**Figure 3.** Transformation of free energy into work and heat (modified from [51]). The free energy coming from food is transformed into internal and external work and it is fundamental to guarantee body integrity. However, due to the inefficiency of the systems, part of the energy is converted into heat, which must be released in the environment.

The energy expenditure is distinguished in basal metabolism (at rest) and energy metabolism (when muscular work occurs). The basal metabolism is used for vital processes such as cerebral, circulatory, or respiratory activities and it is dependent on several factors—including sex, age, or hormonal activity—while the energy metabolism occurs during muscular work and it depends on the intensity of the work, the speed and the duration of the muscular contraction. The measurement of the metabolic activity is performed through the detection of the oxygen consumption. In particular, 1 l of O2 corresponds to 21.1 kJ if carbohydrates are oxidized or to 19.6 kJ if lipids are oxidized [51].

During sport activity, muscular contraction takes place thanks to the energy released by ATP. However, the ATP reserves are sufficient only for work lasting about one second. For this reason, the body presents some energetic mechanisms able to re-synthetize ATP. There are three main energy systems for ensuring these mechanisms [52]:

- anaerobic alactacid metabolism
- anaerobic lactacid metabolism
- aerobic metabolism

Of these, it is important to consider the power (maximum amount of energy produced), capacity (total amount of energy produced), latency (time necessary to obtain the maximum power), and resting time (time necessary for the reconstitution of the system). Table 1 shows the characteristics of each of these energy systems.

**Table 1.** Characteristics of the three energy systems to re-synthetize ATP.

| | Anaerobic Alactacid Metabolism | Anaerobic Lactacid Metabolism | Aerobic Metabolism |
|---|---|---|---|
| **Power** | High (60–100 Kcal/min) | Medium (50 Kcal/min) | Low (20 Kcal/min) |
| **Capacity** | Very low (5–10 Kcal) | Medium (40 Kcal) | High (2000 Kcal) |
| **Latency** | Minimum | Medium (15–30 s) | High (2–3 min) |
| **Resting time** | Rapid | Subordinate to the elimination of lactic acid in the muscles | Long (36–48 h) |

In general, these energy systems during sport activity do not occur separately, but they intervene together. As the number of sport activities is vast, it can be fundamental to describe the kind of mechanism occurring through the time necessary to perform certain movements (Table 2). For example, a basketball match has a duration of 40–48 min; therefore, considering what previously stated, it would mean that the metabolic system involved is the aerobic. However, in basketball rapid and intense movements occur, which involve also the anaerobic systems. Therefore, basketball and other sports such as fencing, baseball, football, golf, hockey, tennis, or volleyball involve both aerobic and anaerobic phases. In other sports such as swimming, running, skiing, or rowing, the energetic system used depends on the duration of the competition. For example, as swimming 200 m and running 800 m require the same time, the energetic system used by the body will be the same.

**Table 2.** Duration of the performance in relation to the energy system (modified from [52]).

| Time | Energy System | Sport |
|---|---|---|
| Less than 30 s | Anaerobic alactacid | Running 100 m |
| 30–90 s | Anaerobic alactacid + lactacid | Running 200–400 m, swimming 100 m, skating |
| 90 s–3 min | Anaerobic lactacid + Aerobic | Running 800 m, combat sports (2–3 min matches) |
| Over 3 min | Aerobic | Marathon, jogging, cross-country skiing |

The measurement unit of the metabolic rate is the Met; 1 Met corresponds to the metabolic rate at rest.

$$1 \text{ Met} = 58 \text{ W/m}^2$$

Standards report the metabolic rate for different activities and the way in which they can be calculated [53] while the 2011 Compendium of Physical Activities [54] shows the metabolic rate for different sports. Table 3 displays the metabolic rate corresponding to different activities.

**Table 3.** Typical metabolic rates for different activities (modified from [54]).

| Activity | Metabolic Rate (Met) |
|:---:|:---:|
| **Resting** | |
| Sleeping | 0.8 |
| Seating, quiet | 1.0 |
| Standing, relaxed | 1.2 |
| **Sport and Activities** | |
| Archery | 4.3 |
| Badminton | 5.5 |
| Basketball | 8.0 |
| Bicycling | 7.5 |
| Boxing | 12.8 |
| Calisthenics | 3.5 |
| Dancing | 7.8 |
| Fencing | 6.0 |
| Fishing | 3.5 |
| Football | 8.0 |
| Gymnastics | 3.8 |
| Hockey | 8.0 |
| Running | 7.0 |
| Skiing | 7.0 |
| Swimming | 4.8–13.8 |
| Tennis | 7.3 |
| Volleyball | 4.0 |

### 2.2.2. Thermoregulation during Exercise

In humans, temperature is regulated through control systems that ensure homeostasis through behavioral and physiological mechanisms of thermoregulation. The first includes all the tools that humans can use to support their thermal comfort, such as the choice of an appropriate clothing or the adjustment of the indoor environmental conditions (opening/closing a window, use HVAC systems, etc.). The second consists of several physiological mechanisms which can intervene to maintain homeostasis, which are the vasomotor response (vasoconstriction or vasodilation), sweating, and shivering. The physiological thermoregulation is a feedback system: temperature receptors are located in the skin and they are connected to the hypothalamus, which has the function of providing homothermia and can activate the mechanisms of thermoregulation through nervous pathways (Figure 4). The physiological field of thermoregulation is generally wider than the zone of thermal neutrality which also represents the zone of thermal comfort. Therefore, behavioral thermoregulation occurs earlier than physiological thermoregulation.

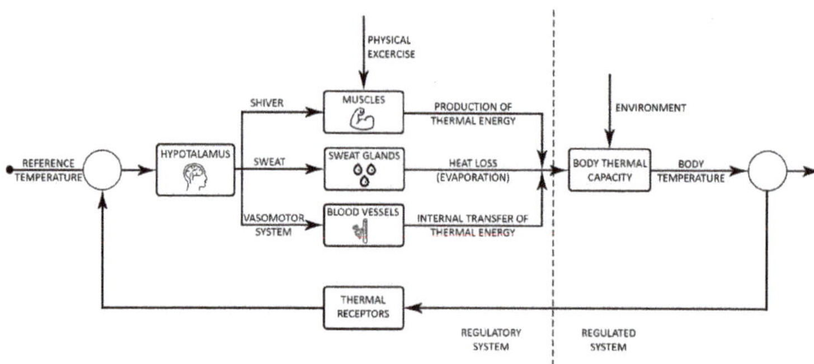


**Figure 4.** Human thermoregulatory system (modified from [55]). Temperature receptors located in the skin are connected with the hypothalamus, which compares the temperature of the body with a reference temperature and it can activate thermoregulatory systems in order to maintain body temperature constant.

During sport practice, the heat production of the body may exceed 1000 W, thus the body temperature tends to increase. In fact, only a modest part of the heat produced by muscles is initially transferred to the environment and most of it increases body's internal temperature. For example, it has been demonstrated that during intense cycle exercise, the temperature can rise up to 1 °C/min [56]. This heat storage cannot be maintained for long periods, or the athletic performance would be compromised due to the overheating and heat exhaustion. When the temperature reaches a certain limit, the thermoregulatory systems occur and the heat is dissipated though vasodilation and sweat. These mechanisms do not occur only in warm environments, but in every condition when the physical exercise is intense enough. In fact, muscular usage generally increases body temperature, usually resulting in temperatures higher than the environmental temperature.

Training can improve thermoregulation during sport practice, leading to an increase of sweat rate and skin blood flow. In fact, elite performers usually present an augmented sweat secretion that occurs in the early phase of exercise, leading indeed to a fast dehydration. Furthermore, professional athletes show an increase of blood total volume and maximal cardiac output, resulting in a better heat dissipation through vasodilation that leads the temperature decrease thanks to the convective cooling [57]. Acclimatization plays also a crucial role in sport performance, especially when competitions take place in different environments from where athletes are used to train.

Gender and age differences can also play a key role in thermoregulation, since physiological properties (e.g., sex hormones, exercise capacity, etc.), anthropometric characteristics (e.g., body mass and size), body composition (e.g., muscles and body fat), and physical activity level may be different. In general, women and elderly people have a lower sweat capacity and a higher core temperature than men [58] and therefore they present less endurance because of heat exposure. Actually, it has been observed that the human response to exercise depends mostly on the aerobic capacity ($VO_{2max}$), which is also correlated to factors such as gender and age [59].

Finally, another aspect that may affect thermoregulation is clothing, as it represents an additional layer that may delay the heat transmission through conduction or prevent the sweat evaporation. However, research has shown that in warm and in moderate environments, garments do not have any effect on the thermoregulatory response. In fact, the addition of layers or the fabric characteristics seem not to affect the physiological responses of the body [60]. On the contrary, in cold conditions clothing may influence thermoregulation and in this case the ideal clothing can block air movement but allows the passage of water vapor when the production of sweat occurs [61].

### 2.2.3. Body Temperature during Exercise

The human body can be considered divided in two parts, a core (inner part) and a shell (outer part). The temperature of the central core in stationary conditions is maintained stable by the thermoregulatory systems activated by the hypothalamus that maintains homeostasis, as the internal temperature cannot exceed certain limits, or the vital organs would result compromised. The shell temperature is usually defined by the mean temperature on the skin and it can vary according to the environmental conditions. The variation of the body temperature depends on several factors such as the environmental conditions, the thermoregulatory system and the metabolic rate, which determine the heat production of the body. During sport activity, the core temperature can rise up to 40 °C, due to muscular strain which determines a large amount of metabolic heat production. Temperatures higher than 40 °C may cause performance break-down or, in extreme cases, health problems [62]. Skin temperature has instead a different trend that it is usually inversely proportional to the exercise intensity, at least at the onset of exercise [63]. Recent studies on runners showed that in the first phase of exercise the body presents a decrease of skin temperature due to the vasoconstriction while, as soon as the core temperature reaches the threshold values, the warmer blood is directed to the shell, leading to an increase of skin temperature and to a decrease of core temperature [64]. However, since different sports involve the use of specific muscles, in the non-active regions the skin vasoconstriction is particularly evident, while in the active ones the skin temperature increases earlier due to the thermal conduction from the active muscles to the skin surface above them [65].

The evaluation of body temperature during exercise is fundamental to ensure a good performance and healthy conditions to the athletes. In extreme cases, when the body temperature gets too high (hyperthermia) or too low (hypothermia), accidents may occur, as it can happen in warm and humid environments or in cold spaces. Warm and humid environments are particularly critical for athletes [52], since their thermoregulatory system cannot properly operate (high temperature prevents heat transfer by convection and radiation and high humidity levels do not allow the sweat evaporation). For this reason, the assessment and the control of core and skin temperatures have a primary importance in the performance and in the safety of the athletes.

### 2.3. Psychological Approach

The thermal sensation perceived by humans derives from the sensory experience, therefore it cannot be based only on physical or physiological approaches. In fact, often the environmental factors are not always the cause of thermal dissatisfaction in buildings [66]. Even people's expectations may influence their satisfaction, as occupants can be satisfied with a certain environment because they do not expect any better condition, or be dissatisfied because they would expect a different environment. For example, elite athletes may have different expectation than other athletes, as they are more used to high quality environmental conditions. Thus, a certain environment could be considered satisfactory or not according to the personal experience of the single athlete and not for the physiological or environmental conditions.

### 2.3.1. Thermal Sensation

Thermal sensation is related to physical and physiological aspects, but also to psychological features, as it is related to how humans feel in a certain environment. It is important to distinguish how a person feels and how he or she would like to be (warmer/colder) or how a certain environment can be described. In particular, physical exercise can affect the thermal response of the athletes, as they may feel warm in environments in which they would perceive cold feelings in conditions of rest. The research of McIntyre [67] shows that usually cold sensations are determined by mean skin temperature, while warm feelings occur initially due to skin temperature and then due to core temperature and they are closely related to the skin wettedness. Furthermore, it has been demonstrated [68] that skin temperature, which may affect the thermal sensation of the athletes, is closely related to environmental conditions for a large range of exercise (from 30% to 70% $VO_{2max}$). Acclimatization can also play a fundamental role in the perception of the thermal environment, reducing the warmth sensation up to 70–80% during sport practice where air temperature was maintained 50 °C [68].

### 2.3.2. Thermal Discomfort

Thermal discomfort during exercise has an important role in sport performance, since it may affect the sense of effort of the athletes. Warm discomfort appears when physiological mechanisms such as vasodilation and sweat secretion occur, but it is also dependent on factors like body temperature and skin wittedness [69]. In particular, it was shown that during exercise in thermal equilibrium the level of skin wittedness providing a sensation of comfort rises from 0–10% to 20–25%, showing the influence of physical activity on human feelings. Moreover, when considering steady-state exercise, warm sensations are generally reduced when the athletes are well trained; otherwise, thermal discomfort was usually associated with the sweat rate and to the vasomotor response [68]. On the contrary, cold discomfort occurs due to vasoconstriction and to the consequent reduction of skin temperature [69]. However, this situation is generally related to sport practice in cold environments, since the majority of heat produced by the body during exercise leads usually to warm sensations.

## 3. Standards

Thermal environments are divided into moderate, hot, and cold. For moderate environments, the reference standard is the UNI EN ISO 7730 [70], which reports the indices that can be used to evaluate the perception of the thermal environment in indoor spaces. Furthermore, general standards explain the main concepts regarding the thermal environment, including the calculation of clothing insulation, the assessment of metabolic rate for different activities, the methods to conduct objective, and subjective measurements and the explanation of the physiological responses of humans. Figure 5 shows an overview on the current normative regarding thermal comfort, even if there is to consider that in this field, the normative is continuously evolving, as technical and scientific knowledge is rapidly developing.

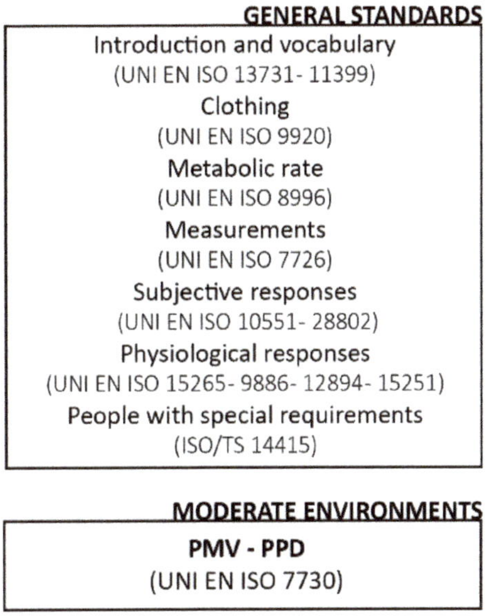

**GENERAL STANDARDS**

Introduction and vocabulary
(UNI EN ISO 13731- 11399)

Clothing
(UNI EN ISO 9920)

Metabolic rate
(UNI EN ISO 8996)

Measurements
(UNI EN ISO 7726)

Subjective responses
(UNI EN ISO 10551- 28802)

Physiological responses
(UNI EN ISO 15265- 9886- 12894- 15251)

People with special requirements
(ISO/TS 14415)

**MODERATE ENVIRONMENTS**

**PMV - PPD**
(UNI EN ISO 7730)

**Figure 5.** Standards for thermal environments: an overview (modified from: [55]). General standards can be applied to all environmental conditions, while the others can be used for the calculation of specific indices developed for different thermal environments.

With regard to sport facilities, generally they can be very heterogeneous, therefore specific literature and standards for the regulation of the parameters that should be maintained in these environments are needed. In fact, often different activities are carried out in these spaces as they are usually multifunctional buildings in which diverse sports are performed, hence the difficulty of finding a unique norm that handles all these aspects. Since the legislation regarding construction and maintenance of sport facilities is often related to hygienic conditions, the normative varies among countries. For example, in the USA the first standard concerning the thermal aspects in indoor environments is the ASHRAE 55/2004, regarding "Thermal Environmental Conditions for Human Occupancy" [71]. However, this standard does not provide any specific value to be maintained in sport facilities and usually different States present diverse regulations on these aspects. In Russia, the SNIP 31-112-2004 "Physical Training and Sports Halls" [72] reports the values of air temperature, relative humidity, and air velocity that should be maintained in an "ordinary sports hall", where different activities can be carried out. Therefore, this standard does not provide any specific information regarding the values to be maintained in different sport facilities, in relation to the sport performed. In Europe, a unique standard regarding thermal comfort in sport facilities does not exist and the regulations may vary among the countries, as it happens in the USA. In particular, in Italy, the most important standard managing thermal comfort in sports halls is the guideline of CONI [73], which defines indications about air quality, thermal, lighting, and acoustic environment in sport halls and swimming pools. Table 4 shows the main values to be maintained in indoor sports halls and swimming pool, according to CONI's guidelines.

**Table 4.** Environmental parameters for sports halls and natatorium facilities (modified from [73])

|  | Air Temperature (°C) | Relative Humidity (%) | Ventilation Rate (Air Exchange/h) [2] | Maximum Air Velocity (m/s) | Environment |
|---|---|---|---|---|---|
| **Indoor Sports Halls** | 16–20 | 50 | (3) | 0.15 | Playing field |
|  | 20–22 | 50 | (3) | 0.15 | Pre-athletic spaces |
|  | 18–22 [6] | 50 | 5 | 0.15 | Changing rooms |
|  | 22 [7] | 70 | 8 | 0.15 | Showers |
|  | 22 | 60 | 5–8 | 0.15 | Sanitary facilities |
|  | 20 | 50 | 2.5 | 0.15 | First aid |
|  | 20 | 50 | 1.5 | 0.15 | Offices |
|  | 20 | 50 | 1 | 0.20 | Halls |
|  | 16 | 50 | 0.5–1 | 0.25 | Storage rooms |
|  | 20 | 50 | 0.5 | 0.20 | Other spaces |
| **Swimming Pools** | (8) (5) | ≤70 [8] | (8) (4) | ≤0.10 [8] | Poolside |
|  | 28 | 70 | 3 | 0.15 | Pre-athletic spaces |
|  | ≥20 [8]–24 [6] | 60 | ≥4 [8]–5 | 0.15 | Changing rooms |
|  | 24 [7] | 70 | 8 | 0.15 | Showers |
|  | ≥20 [9] | 60 | ≥4 [8]–5–8 | 0.15 | Sanitary facilities |
|  | ≥20 [8]–22 | 50 | ≥4 [8] | 0.15 | First aid |
|  | 20 | 50 | 1.5 | 0.15 | Offices |
|  | 20 | 50 | 1.5 | 0.20 | Halls |
|  | 20 | 50 | 0.5–1 | 0.25 | Storage rooms |
|  | 20 | 50 | 0.5 | 0.20 | Other spaces |

Notes: (1) In the table are reported only the values concerning the thermal environment. The complete table can be found in [73]. (2) The values refer to the case of artificial ventilation. (3) At least 20 $m^3$/hour/person at maximum crowding for the spectator's area; 30 $m^3$/hour/person for the space occupied by the athletes. (4) Values to be established in relation to the thermo-hygrometric characteristics to be achieved. (5) For the water temperature in the pools, specific values are given by CONI's guidelines. (6) The temperature of the air in the changing rooms (excluding those of the swimming facilities) is appropriate to be 2–4 °C higher than that of the sport room. (7) The temperature of the water in the showers, must not be lower than 37 °C and not higher than 40–48 °C. (8) The thermo-hygrometric, ventilation and lighting engineering requirements must conform to what is indicated in the Agreement of 16 January 2003—between the Minister of Health, the Regions and the autonomous provinces of Trento and Bolzano on the sanitary aspects for the construction, maintenance, and supervision of the swimming pools.

However, since different sports present diverse requirements, International Federations often provide not only the rules regarding the game and the materials, but also standards concerning environmental parameters such as air temperature, relative humidity, and air velocity that should be maintained indoors for each sport. In Table 5 the environmental parameters provided for indoor sport facilities by federations recognized by the International Olympic Committee (IOC) are reported. It can be noticed that most federations show at least the values of air temperature that should be maintained in the playing area. However, several sport do not present any environmental value (e.g., boxing, fencing, etc.), therefore specific studies should be performed in order to define these parameters and to establish more precise values for the existing ones, since several sport federations report only the value of air temperature and they do not consider, for example, relative humidity, and air velocity.

**Table 5.** Environmental parameters to be maintained in the playing area given by sport federations.

| Federation | Air temperature (°C) | Relative Humidity (%) | Maximum Air Velocity (m/s) |
| --- | --- | --- | --- |
| Aquatics (FINA) [74] | 2 °C higher than water temperature (water temperature 25–28 °C) | - | - |
| Badminton (BWF) [75] | 18–30 | - | <0.2 |
| Basketball (FIBA) [76] | 16–20 | <50 | - |
| Boxing (AIBA) | - | - | - |
| Curling (WCF) [77] | 6–7 | controlled | No constant air movement |
| Fencing (FIE) | - | - | - |
| Gymnastics (IFG) [78] | Humidex = 22–38 | | - |
| Handball (IHF) [79] | 15–22 (heated halls) 18–24 (cooled halls) | - | <1 |
| Ice Hockey (IIHF) [80] | 6 | <70 | - |
| Ice skating (ISU) [80] | 6–12 | <70 | - |
| Judo (IJF) [81] | 17–26 | 30–40 | - |
| Table tennis (ITTF) [82] | 12–25 | - | <0.1 |
| Taekwondo (WT) | - | - | - |
| Tennis (ITF) [83] | 13–17 (winter) 6–8 below the external temperature (summer) >10 | 55–60 | - |
| Volleyball (FIVB) [84] | 16–15 (for official competitions) | - | - |
| Weightlifting (IWF) | - | - | - |
| Wrestling (UWW) [85] | 18–22 | - | - |

## 4. Thermal Comfort Models Applied to Sport Facilities

Sport facilities present high complexity, due to the different activities that are carried out in these environments. For this reason, experimental methods have been developed by researchers in order to predict or to assess the thermal conditions in swimming pools and sport halls. However, only a few studies has been developed, as shown by literature review displayed in Table 6.

**Table 6.** Literature review of the scientific papers regarding thermal comfort in sport facilities.

| Reference | Investigation | Monitoring Duration | Environment |
| --- | --- | --- | --- |
| Stamou et al., 2008 [86] | Evaluation of the thermal comfort in the Galatsi Arena in Athens through a CFD analysis. | - | Indoor stadium |
| Rajagopalan and Luther, 2013 [16] | Thermal comfort prediction with the use of Fanger's indices, adaptive comfort models and questionnaires. | 1 week | Sport hall within an aquatic center |
| Revel and Arnesano, 2014 [18] | Thermal comfort prediction with the use of Fanger's indices corrected for non-conditioned buildings in warm environments and questionnaires. | 4 days | Swimming pool + Gym |
| Revel and Arnesano, 2014 [20] | Development of a monitoring technology that includes thermal comfort calculated with the use of Fanger's indices corrected for non-conditioned buildings in warm environments to obtain information on how energy is used in sport facilities. | 4 days | Swimming pool + Gym |

**Table 6.** *Cont.*

| Reference | Investigation | Monitoring Duration | Environment |
| --- | --- | --- | --- |
| Kisilewics and Dudzinska, 2015 [14] | Measurements of the six basic parameters for the assessment of Fanger's indices. Calculation of the operative temperature. | 2 days | University sports hall |
| Zhai et al., 2015 [21] | Thermal comfort assessment with the use of questionnaires and its relation to air movement. Estimation through questionnaires of thermal responses, perceived air quality and Rate of Perceived Exertion (RPE). | - | Climate chamber |
| Cheng et al., 2016 [87] | Parametric modeling to predict and control thermal comfort and ventilation, based on the adaptive approach. | - | University multisport facility |
| Cianfanelli et al., 2016 [19] | Thermal comfort prediction with the use of Fanger's indices and questionnaires. | 2 months | Swimming pool + Polyvalent sport center |
| Khalil and Al Hababi, 2016 [88] | Investigation of thermal comfort in a gymnastic hall with the use of CFD for the calculation of PMV under different conditions. | - | Gymnastic sports hall |
| Lebon et al., 2017 [15] | Numerical analysis for the assessment of thermal comfort in an indoor swimming pool, through the calculation of Fanger's indices and Humidex index. | - | Swimming pool |
| Zora et al., 2017 [22] | Investigation on the relation between Fanger's index Predicted Mean Vote (PMV) and Rate of Perceived Exertion (RPE). Detection of skin and core temperatures. | - | Climate chamber |
| Bugaj and Kosinski, 2018 [17] | Thermal comfort prediction with the use of Fanger's indices and evaluation of possible improvements to improve the conditions of comfort. | 3 days | Indoor tennis court |
| Berquist et al., 2019 [89] | Assessment of thermal comfort using questionnaires | 5 months (12 h per day) | Gymnastic center |

## 4.1. Human Thermal Physiological Models

In order to determine the thermal comfort, several physiological models have been developed, from the one-node model, representing the complete human body as one node [90] to the more complex and realistic ones, representing the body with a multi-elements model using finite elements [91]. In sport applications, the most used models are [92]:

- Gagge's model, used for quasi-stationary conditions;
- Stolwijk's model, used for non-stationary situations;

The model of Gagge consists of a two-nodes model of the human body, representing the core and the shell, while the model of Stolwijk is a four-node model representing trunk, arms, hands, legs, and feet (Figure 6).

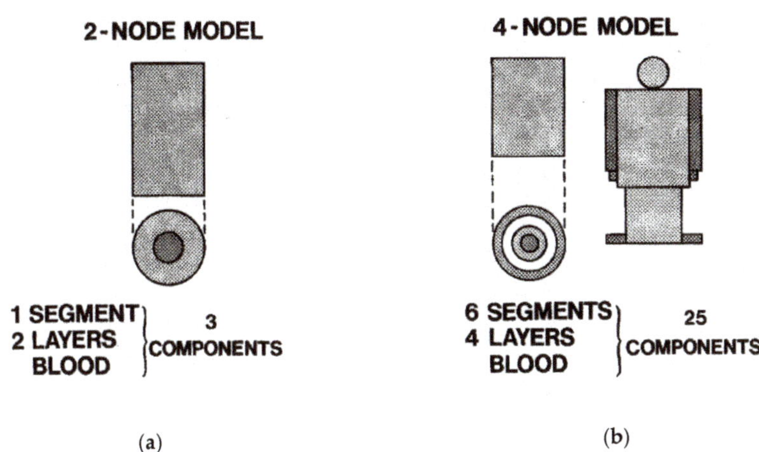



**Figure 6.** Human thermal physiological models: (**a**) Gagge's model; (**b**) Stolwijk's model [92].

The model of Gagge is used for most of the activities, while the model of Stolwijk is generally used only for the description of a wet swimmer coming out from the water. These models define the relation between individual parameters such as the metabolic rate and the clothing insulation with the environmental parameters, usually described by air temperature, relative humidity, mean radiant temperature, and air velocity.

For swimmers, thermal comfort depends on the time, due to the heat loss occurring because of the evaporation of the water on the skin, as swimmers are subjected to a drying process, which can be divided in two intervals [92].

- Interval 1: starting as soon as the swimmer leaves the pool and ending 10 min later;
- Interval 2: when the body is dry (it is a steady-state condition);

In particular, the adaptation of the model of Stolwijk to the wet swimmer (during interval I) includes the addition of an evaporative term [92]

$$E(I) = (p_{skin}(I) - p_a)\, 2.2\, h_c(I)\, (10\, v_a)^{1/2}\, f_{pcl}(I)\, S(I) \tag{14}$$

where $E(I)$ is the evaporative heat loss of segment I of a wet body (W), $p_{skin}$ is the saturated water vapor pressure at the skin of segment I (Pa), $p_a$ is the water vapor partial pressure of the air (Pa), $h_c$ is the convective heat transfer coefficient (W/m$^2$K), $v_a$ is the air velocity (m/s), $f_{pcl}(I)$ is the Nishi's permeation efficiency factor of segment I and $S(I)$ (m$^2$) is the surface area of segment I.

For the calculation, usually the segment I is assumed equal to the nude part of the swimmer's body. The drying process is a transient condition, therefore the evaporative heat loss starts from an initial amount and ends when $E(I) = 0$ and the skin surface is dry. If the condition is when the drying process is just started (swimmer gets out of the pool), the $p_{skin}$ can be calculated for a temperature equal to the pool temperature [18]. For the dry swimmer (Interval II) the evaporative term should not be considered [92].

### 4.2. Predictive Indices used to Assess Thermal Comfort

In moderate climates, the thermal environment is defined through six basic parameters: four environmental (air temperature $t_a$ (°C), relative humidity RH (%), mean radiant temperature $t_r$ (°C), and air velocity $v_a$ (m/s)) and two individual (clothing insulation $I_{cl}$ (clo) and metabolic rate M (Met)). It is the interaction of these factors which determines the thermal sensation of humans [23]. However, considering sport facilities, the conditions cannot be considered 'standard', as the metabolic activity can be very high and clothing insulation may vary according to the sport apparel worn and to the increased air velocity due to the body movement. For this reason, in these environments the correct estimation of the individual parameters has a fundamental role in the prediction of the thermal sensation of the athletes.

### 4.2.1. Fanger's Indices PMV and PPD

Most of the studies were based on the calculation of Fanger's indices, predicted mean vote (PMV) and predicted percentage of dissatisfied (PPD), derived from field measurements [14,16,17,19,22] or from simulations obtained through computational fluid dynamics (CFD) models [86,88].

PMV index is defined as the vote of an average individual regarding the thermal environment and it is a function of the six basic parameters, shown in Table 7. The calculation of this index requires an iterative method; therefore, it is generally performed through software, or directly by data loggers. The purpose of Fanger's indices is to correlate environmental and individual parameters to the subjective feeling of humans. Thus, Fanger proposed an experiment on 1296 individuals who, after remaining in a thermal chamber, had to give an answer regarding the thermal environment on a seven-point sensation scale, defined by ASHRAE (from -3 cold to +3 hot). Based on this survey, Fanger proposed the equation [93]

$$PMV = (0.303 \ e^{-0.036M} + 0.028) \ S \tag{15}$$

where M is the metabolic rate and S is the heat storage.

**Table 7.** Environmental and individual parameters used for the definition of PMV and PPD.

| Parameter | | Symbol | Unit |
|---|---|---|---|
| Environmental Parameters | Air Temperature | $t_a$ | °C or K |
| | Mean radiant temperature | $t_r$ | °C or K |
| | Partial pressure of water vapor | $p_a$ | Pa |
| | Air velocity | $v_a$ | m/s |
| Individual Parameters | Metabolic rate | M | $W/m^2$ or Met (1 Met = 58.2 $W/m^2$) |
| | Thermal insulation of clothing | $I_{cl}$ | $m^2$ K/W or Clo (1 Clo = 0.55 $m^2$ K/W) |

PPD index is defined as the predicted percentage of dissatisfied with regard to a certain environment, considering dissatisfied a person who, subjected to a certain thermal environment, express a rating of +3, +2, −2, or −3 on the thermal sensation scale. The relation between the PMV and PPD is [93]

$$PPD = 100 - 95 \ e^{0.03353 \ PMV^4 + 0.2179 \ PMV^2} \tag{16}$$

Note that the condition in which every subject is satisfied does not exist, and the minimum value of PPD is 5%.

However, the PMV method presents some limits, as indicated in the UNI EN ISO 7730 [70]. In fact, this method is applicable only when the environment can be defined moderate (PMV is less than 2 in absolute value) and when the six basic parameters stay within the limits shown in Table 8. This leads to the consideration that this method can present several problematics in the application to sport facilities, as the metabolic rate often exceeds the value of 4 Met.

**Table 8.** Range of applicability of environmental and individual parameters for the calculation of PMV [70].

| Parameter | | Range | Unit |
|---|---|---|---|
| Environmental Parameters | $t_a$ | +10–+30 | °C or K |
| | $t_r$ | +10–+40 | °C or K |
| | $p_a$ | 0–2700 | Pa |
| | $v_a$ | 0–1 | m/s |
| Individual Parameters | M | 0.8–4 | $W/m^2$ or Met (1 Met = 58.2 $W/m^2$) |
| | $I_{cl}$ | 0–2 | $m^2$ K/W or Clo (1 Clo = 0.55 $m^2$ K/W) |

### 4.2.2. Fanger's Indices PMV Corrected for Warm and Humid Environments (ePMV)

In some other cases [18,20] the PMV index was calculated considering the correction for non-air-conditioned buildings in warm climates, provided by Fanger and Toftum [94].

This model introduces the expectancy factor (e) shown in Table 9 that should be multiplied for the PMV in order to obtain the real thermal sensation of the athletes. This index is used to describe the perception of non-conditioned-buildings' occupants, as they may feel sensations of warmth less severe than the one predicted by PMV, due to their low expectations and factors related to metabolic activity [94].

**Table 9.** Expectation factor (e) for non-air-conditioned buildings in warm climates [18].

| Expectation | Classification of the Building | Expectation Factor e |
|---|---|---|
| High | Non-ventilated buildings located in regions where air-conditioned buildings are common. Warm period occurring briefly during the summer season. | 0.9–1.0 |
| Moderate | Non-ventilated buildings located in regions with some air-conditioned buildings. Warm summer season. | 0.7–0.9 |
| Low | Non-ventilated buildings located in regions with few air-conditioned buildings. | 0.5–0.7 |

In particular, the ePMV can be calculated as

$$ePMV = e\,PMV \tag{17}$$

In the research of Revel and Arnesano [18,20], the expectancy factor was considered equal to 0.7. Furthermore, for the evaluation of the PMV in the swimming pool, the condition of the swimmer just coming out from water was considered and the evaporative term described by [14] has been added for the calculation of the PMV.

### 4.2.3. Adaptive Comfort Model

The adaptive thermal comfort model has been applied for the parametric modeling used to predict and control of thermal comfort in a university multisport facility [87] and for spectators of a sport hall within an aquatic center [16], on the basis of the model of de Dear and Brager. In this model, it is taken into consideration that in general people in warm climate zones prefer warmer indoor temperatures than others living in cold climates [95]. This study consists of a statistical analysis which showed that occupants in naturally ventilated buildings present a wider tolerance with regard to the range of temperatures that can be recorded indoors.

### 4.2.4. Operative Temperature

The operative temperature has been calculated for the assessment of the thermal environment in an indoor sport hall during summer period [14]. The operative temperature is often used for the assessment of the thermal environment, even if it is not considered in the six basic parameters and it depends on radiative and convective exchanges. It can be calculated as [55]

$$t_o = \frac{h_r \cdot t_r + h_c \cdot t_a}{h_r + h_c} \qquad (18)$$

where:

$h_c$ = unitary convective conductance (W/m$^2$ K)
$h_r$ = unitary radiative conductance (W/m$^2$ K)
$t_a$ = air temperature (°C)
$t_r$ = mean radiant temperature (°C)

Since there are some difficulties in evaluating the operative temperature with this equation, two simplified expressions for its calculation exist. The first provides a value of the operative temperature dependent on the relative air velocity by means of a coefficient A,

$$t_o = A\, t_a + (1 - A)\, t_r \qquad (19)$$

where A = 0.5 when the relative air velocity is lower than 0.2 m/s, A = 0.6 when the relative air velocity is between 0.2 and 0.6 m/s, and A = 0.7 when the relative air velocity is between 0.6 and 1.0 m/s.

The second expression is an arithmetic mean of values of the two temperatures from which the operative temperature depends

$$t_o = (t_a + t_r)/2 \qquad (20)$$

### 4.2.5. Humidex

The humidity index (humidex) has been used to assess thermal comfort in a swimming pool [15]. This index is a dimensionless number which allows the evaluation of air temperature and humidity on an average person. It has been proposed for the evaluation of environments which presented high humidity as an alternative to the Fanger's indices. The value of humidex is given by [93]

$$\text{Humidex} = t_a + [5.555\,(Pa - 1.013)] \qquad (21)$$

where:

$t_a$ = air temperature (°C)
pa = partial vapor pressure (kPa)

Humidex identifies 4 different thermal levels, from the comfort level (level 0) to the definition of possible health risks (level 4).

### 4.3. Subjective Judgements Used to Assess Thermal Comfort

Even if this aspect is often underestimated, the perception of a thermal environment is strongly related to the psychological conditions. For this reason, most of the researchers in their studies assessed the thermal sensation of the athletes with the use of questionnaires [18,19,21,89]. In particular, with regard to the subjective perception of the thermal environment, the research of Revel and Arnesano [18] compared the predictive models to the subjective responses in order to determine the impact of high metabolic rates and sport garments on systematic errors in the prediction of Fanger's indices.

For the design of the questionnaires, UNI EN ISO 10551 Standard [96] regulates the subjective evaluation of the thermal environment, which consists of judgement scales regarding perception, comfort, and thermal preference and, in some cases, personal acceptability and tolerance. In particular, the scale of perception in moderate environments is represented on a seven-point scale showing the thermal sensation vote (TSV), which can be compared to the PMV. Furthermore, the evaluative scale shows the feeling of comfort of the subjects on a unipolar scale from 0 to 4, in which points 3 and 4 are characterized by an increasing level of discomfort and can be considered dissatisfied. For this reason, this scale can be compared to Fanger's PPD.

### 4.4. Thermal Environment and Performance: Correlation between PMV and RPE

The perception of the thermal environment has a great influence on the performance, which is related directly to the kind of activity that has to be carried out in terms of concentration, physiological effort, etc. Several studies have been performed on the problems related to the heat stress of the athletes in hot and humid environments, but little knowledge is available in moderate environments. In fact, even if in these environments the safety of the athletes is not usually compromised, their performance may be affected by certain ambient parameters. For this reason, in sport facilities, the impact of the thermal environment on the athletes has been considered, as better conditions could improve the efficiency of a training session, or even the performance in a competition. In particular, studies were carried out in order to correlate the Fanger's index PMV to the rate of perceived exertion (RPE), in order to find an association between the thermal and the physiological responses of athletes during exercise [22]. Furthermore, Zhai et al. [21] studied the effect of air movement for comfort during exercise at different levels of metabolic rate and the relation between the thermal sensation and the perceived exertion. In these researches, physiological characteristics such as the oxygen consumption, skin, and core temperatures were also detected. It resulted that PMV is related to RPE, thus increasing the condition of thermal comfort in sport facilities may have positive results also on the performance of the athletes.

## 5. Thermal Comfort Assessment in Practice

The assessment of thermal comfort in sport facilities presents some difficulties due to the complexity of these environments. In fact, often sport halls are multifunctional buildings, used for the practice of several sports. Furthermore, in sport facilities the activities carried out are not stationary, thus the six basic parameters cannot be assessed as in other environments such as offices, school buildings, etc. In this section, a review of the methodologies used to assess thermal comfort in sport facilities is reported.

### 5.1. Monitoring Duration

The monitoring duration was 2 days at minimum and 5 months at maximum. Data were recorded with a minimum of 15 seconds to a maximum of 15 min. The time of acquisitions varies largely in relation to the kind of environment investigated. When the conditions change rapidly, the acquisition time must be shorter (every 15 sec–1 min), otherwise data can be recorded every 15 min. Any standardized procedure was revealed by the review of the existing research on thermal comfort in sport facilities and the measurements were carried out in relation to the problematics of the case of study.

### 5.2. Assessment of the Individual Parameters

#### 5.2.1. Metabolic Rate

The metabolic rate can be determined by UNI EN ISO 8996 standard [53], which contains the data to calculate it with tables provided for different activity levels. However, more accurate estimation can be provided by studies concerning the level of metabolic activity of the sport considered [54]. For varying metabolic rates, standard UNI EN ISO 7730, suggests a time-weighted average that should be estimated during the previous 1 h of activity [70].

In the research of Revel and Arnesano [18], three phases were examined: transitory phase, steady state, and recovery state. These phases should be considered in the calculation of PMV, as they imply exercise levels. In fact, also the intensity of the exercise is fundamental for the metabolic production. In particular, sedentary activities are characterized by values around 1.0–1.5 Met, light intensity 1.6–2.9 Met, moderate intensity 3–5.9 Met, and vigorous intensity > 6 Met [54]. The correct evaluation of the metabolic rate is fundamental for the calculation of PMV and PPD, as mistakes in the assessment may lead to uncertainties. For this reason, researchers should focus on the determination of this parameter, in order to perform a correct evaluation of the thermal sensations.

### 5.2.2. Clothing Insulation

Standard UNI EN ISO 9920 [27] provides a procedure for the evaluation of the clothing insulation with tables reporting $I_{cl}$ values ($m^2$ K/W) for several clothing types. However, often sport garments are not included in this standard, thus specific research on the values of their insulation have been used by researchers, in order to provide a more precise estimation.

In sport facilities the clothing worn can be the same for all the athletes (in case in which a particular uniform is worn) or, more commonly, the garments may change according to the personal preference. For this reason, researchers calculated values of PMV and PPD for different clothing ensembles [17] or they were asking in questionnaires provided to users which were the garments worn during sport activity at the time of the test [18]. In the cases in which the movement of the human body was relevant and it modified the thermal insulation of the clothing due to the pumping effect, the correction proposed by UNI EN ISO 9920 was considered [27]. This correction is a function of air velocity and metabolic rate and it involves a decrease of the thermal insulation of the garments, which allows a greater heat transfer.

### 5.3. Measurement of the Environmental Parameters

The measurement of the four environmental parameters is also fundamental for the determination of the Fanger's indices. In buildings such as offices, where stationary activities are carried out, the probes are usually located close to the workstation or, in any case, in the proximity of the locations occupied by building users [55]. In sport facilities, athletes do not occupy a fixed position, thus the procedure for performing the measurements is more complex. However, in the review of the existing research, it resulted that the location of the probes in the sport facilities was standardized, as they were situated at a height varying from 0.6 m for the sitting position to 1.1–1.7 m for the standing position, usually in the center of the hall or, in the case of the swimming pool, close to the water [20].

## 6. Conclusions

In sport facilities, ensuring thermal comfort is particularly relevant, as it may affect the performance and health of the athletes. Thermal environments can be seen as a combination of physical, physiological, and psychological factors, and the interaction between these three aspects determines the thermal sensation of humans. In particular, in spaces in which physical activity is carried out, the physiological component is particularly relevant, as the metabolic rate is high and therefore the body produces a consistent amount of heat, which must be dispersed in the environment.

In order to predict the thermal sensation of the athletes performing in indoor sport facilities, most researchers focused on the calculation of Fanger's indices PMV and PPD. However, even if a correlation between the PMV corrected for warm and humid environments and the real sensation determined through questionnaires was found, the high metabolic rate occurring during sport practice may lead to an overestimation of the thermal sensation of the athletes. Furthermore, since sport facilities are multifunctional buildings in which several activities are carried out, the difficulties that have to be faced also concern the determination of a thermal environment which is comfortable for all the occupants, from the athletes to the spectators.

From the literature review, it results that only little knowledge is available on the determination of thermal comfort in indoor sport facilities and on the standardization of a measurement protocol to be applied in these spaces. Moreover, there is a lack in standards concerning the environmental parameters that should be maintained in sport halls. For this reason, further research should be developed on this topic, as performing in a comfortable environment may improve the performance of athletes and ensure healthy and pleasant conditions.

**Author Contributions:** All the authors contributed in equal parts to the research activity and to the paper writing.

**Funding:** This research received no external funding.

**Conflicts of Interest:** No conflict of interest or personal relationships that could have appeared to influence the work reported in this paper.

## References

1. Bluyssen, P.M. Management of the Indoor Environment: From a Component Related to an Interactive Top-down Approach. *Indoor Built Environ.* **2008**, *17*, 483–495. [CrossRef]
2. Bluyssen, P.M. *The Indoor Environment Handbook. How to Make Buildings Healthy and Comfortable*, 1st ed.; Routlege: London, UK, 2009; pp. 45–91.
3. Bluyssen, P.M. Towards an integrative approach of improving indoor air quality. *Build. Environ.* **2009**, *44*, 1980–1989. [CrossRef]
4. Bluyssen, P.M.; De Richemont, S.; Crump, D.; Maupetit, F.; Witterseh, T.; Gajdos, P. Actions to reduce the impact of construction products on indoor air: Outcomes of the European project healthy air. *Indoor Built Environ.* **2010**, *19*, 327–339. [CrossRef]
5. Leccese, F.; Salvadori, G.; Rocca, M. Visual discomfort among university students who use CAD workstations. *Work* **2016**, *55*, 171–180. [CrossRef] [PubMed]
6. Leccese, F.; Rocca, M.; Salvadori, G. Fast estimation of Speech Transmission Index using the Reverberation Time: Comparison between predictive equations for educational rooms of different sizes. *Appl. Acoust.* **2018**, *140*, 143–149. [CrossRef]
7. Leccese, F.; Salvadori, G.; Öner, M.; Kazanasmaz, T. Exploring the impact of external shading system on cognitive task performance, alertness and visual comfort in a daylit workplace environment. *Indoor Built Environ.* **2019**, in press. [CrossRef]
8. Zhang, D.; Tenpierik, M.; Bluyssen, P.M. Interaction effect of background sound type and sound pressure level on children of primary schools in the Netherlands. *Appl Acoust.* **2019**, *154*, 161–169. [CrossRef]
9. Leccese, F.; Salvadori, G.; Asdrubali, F.; Gori, P. Passive thermal behaviour of buildings: Performance of external multi-layered walls and influence of internal walls. *Appl. Energy* **2018**, *225*, 1078–1089. [CrossRef]
10. Fantozzi, F.; Leccese, F.; Salvadori, G.; Rocca, M.; Garofalo, M. LED Lighting for indoor sports facilities: Can its use be considered as sustainable solution from a techno-economic standpoint. *Sustainability* **2016**, *8*, 618. [CrossRef]
11. Di Pede, M.; Leccese, F.; Salvadori, G.; Di Ciolo, E.; Piccini, S. On the vertical illuminance in indoor sport facilities: Innovative measurement procedure to verify international standard requirements in fencing halls. In Proceedings of the Conference Proceedings—17th IEEE International Conference on Environment and Electrical Engineering and 2017 1st IEEE Industrial and Commercial Power Systems Europe, Milan, Italy, 6–9 June 2017. [CrossRef]
12. Andrade, A.; Dominski, F.H. Indoor air quality of environments used for physical exercise and sport practice: Systematic review. *J. Environ. Manag.* **2018**, *206*, 577–586. [CrossRef]
13. Nowicka, E. *The Index Method of Acoustic Design of Sports Enclosures*; EuroNoise: Maastricht, The Netherlands, 2015.
14. Kisilewicz, T.; Dudzińska, A. Summer overheating of a passive sports hall building. *Arch. Civ. Mech. Eng.* **2015**, *15*, 1193–1201. [CrossRef]
15. Lebon, M.; Fellouah, H.; Galanis, N.; Limane, A.; Guerfala, N. Numerical analysis and field measurements of the airflow patterns and thermal comfort in an indoor swimming pool: A case study. *Energy Effic.* **2016**, *10*, 527–548. [CrossRef]

16. Rajagopalan, P.; Luther, M.B. Thermal and ventilation performance of a naturally ventilated sports hall within an aquatic centre. *Energy Build.* **2013**, *58*, 111–122. [CrossRef]

17. Bugaj, S.; Kosinski, P. Thermal comfort of the sport facilities on the example of indoor tennis court. In Proceedings of the IOP Conference Series: Materials Science and Engineering 415, Krakow, Poland, 11–13 September 2018. [CrossRef]

18. Revel, G.M.; Arnesano, M. Perception of the thermal environment in sports facilities through subjective approach. *Build. Environ.* **2014**, *77*, 12–19. [CrossRef]

19. Cianfanelli, C.; Valeriani, F.; Santucci, S.; Giampaoli, S.; Gianfranceschi, G.; Nicastro, A.; Borioni, F.; Robaud, G.; Mucci, N.; Romano Spica, V. Environmental Quality in Sports Facilities: Perception and Indoor Air Quality. *J. Phys. Educ. Sport Manag.* **2016**, *3*, 57–77. [CrossRef]

20. Revel, G.M.; Arnesano, M. Measuring overall thermal comfort to balance energy use in sport facilities. *Measurement* **2014**, *55*, 382–393. [CrossRef]

21. Zhai, Y.; Elsworth, C.; Arens, E.; Zhang, H.; Zhang, Y.; Zhao, L. Using air movement for comfort during moderate exercise. *Build. Environ.* **2015**, *94*, 344–352. [CrossRef]

22. Zora, S.; Balci, G.A.; Colakoglu, M.; Basaran, T. Associations between Thermal and Physiological Responses of Human Body during Exercise. *Sports* **2017**, *5*, 97. [CrossRef]

23. Parsons, K.C. *Human Thermal Environments: The Effects of Hot, Moderate, and Cold Environments on Human Health, Comfort and Performance*, 2nd ed.; Taylor & Francis Group: London, UK, 2003.

24. DuBois, D.; DuBois, E.F. A formula to estimate surface area if height and weight are known. *Arch. Intern. Med.* **1916**, *17*, 863. [CrossRef]

25. American Society of Heating, Refrigerating and Air-Conditioning Engineers (ASHRAE). Thermal Comfort. In *ASHRAE Handbook Fundamentals*; Owen, M.S., Ed.; ASHRAE: Atlanta, GA, USA, 2009.

26. Hardy, J.D.; Du Bois, E.F.; Soderstrom, G.F. The Technic of Measuring Radiation and Convection. *J. Nutr.* **1938**, *15*, 461–475. [CrossRef]

27. International Organization for Standardization. *ISO 9920, Ergonomics of the Thermal Environment. Estimation of Thermal Insulation and Water Vapour Resistance of a Clothing Ensemble*; International Organization for Standardization: Geneva, Switzerland, 2009.

28. Blair, K.B. Materials and design for sports apparel. In *Materials in Sports Equipment*, 1st ed.; Subic, A., Ed.; Woodhead publishing: Philadelphia, PA, USA, 2007; Volume 2, pp. 60–86. [CrossRef]

29. Oliveira, M.V.A.; Gaspar, A.R.; Francisco, S.C.; Quintela, D.A. Convective heat transfer from a nude body under calm conditions: Assessment of the effects of walking with a thermal manikin. *Int. J. Biometeorol.* **2011**, *56*, 319–332. [CrossRef]

30. Luo, N.; Weng, W.G.; Fu, M.; Yang, J.; Han, Z.Y. Experimental study of the effect of human movement on the convective heat transfer. *Exp. Ther. Fluid Sci.* **2014**, *57*, 40–56. [CrossRef]

31. De Dear, R.J.; Arens, E.A.; Zhang, H.; Oguro, M. Convective and radiative heat transfer coefficients for individual human body segments. *Int. J. Biometeorol.* **1997**, *40*, 141–156. Available online: https://escholarship.org/uc/item/9hn3s947 (accessed on 10 October 2019). [CrossRef]

32. Defraeye, T.; Blocken, B.; Koninckx, E.; Hespel, P.; Carmeliet, J. Computational fluid dynamics analysis of drag and convective heat transfer of individual body segments for different cyclist positions. *J. Biomech.* **2011**, *44*, 1695–1701. [CrossRef]

33. Rapp, G.M. Convection coefficients of man in a forensic area of thermal physiology: Heat transfer in underwater exercise. *J. Physiol.* **1971**, *63*, 392–396.

34. Witherspoon, M.; Goldmarn, F.; Breckenndgje, R. Heat transfer coefficients of humans in cold water. *J. Physiol.* **1971**, *63*, 459–462.

35. Boutelier, C.; Bougues, L.; Timbal, J. Experimental study of convective heat transfer coefficient for the human body in water. *J. Appl. Physiol.* **1977**, *42*, 93–100. [CrossRef]

36. Holmér, I.; Bergh, U. Thermal Physiology of Man in the Aquatic Environment. In *Bioengineering, Thermal Physiology and Comfort*, 1st ed.; Cena, K., Clark, J.A., Eds.; Elsevier: Amsterdam, The Netherlands; Oxford, UK; New York, NY, USA, 1981; Volume 10, pp. 145–156. [CrossRef]

37. Luginbuehl, I.; Bissonnette, B.; Davis, P.J. Thermoregulation: Physiology and perioperative disturbances. In *Smith's Anesthesia for Infants and Children*, 7th ed.; Motoyama, E.K., Davis, P.J., Eds.; Elsevier: Philadelphia, PA, USA, 2006; pp. 154–176. [CrossRef]

38. Sportsengine. Available online: https://lancerselmbrookschools.sportngin.com/page/show/629570-heat-information (accessed on 12 October 2019).

39. Pdhpe. Available online: https://www.pdhpe.net/sports-medicine/what-role-do-preventative-actions-play-in-enhancing-the-wellbeing-of-the-athlete/environmental-considerations/temperature-regulation/ (accessed on 20 October 2019).

40. Bartels, V.T. Improving comfort in sports and leisure wear. In *Improving Comfort in Clothing*, 1st ed.; Guowen, S., Ed.; Woodhead Publishing: Philadelphia, PA, USA, 2011; pp. 385–411. [CrossRef]

41. Sherwood, J.; Drane, P. Design and materials in baseball. In *Materials in Sports Equipment*, 1st ed.; Subic, A., Ed.; Woodhead Publishing: Philadelphia, PA, USA, 2007; Volume 2, pp. 159–184. [CrossRef]

42. Subic, A.; Kovacs, J. Design and materials in snowboarding. In *Materials in Sports Equipment*, 1st ed.; Subic, A., Ed.; Woodhead Publishing: Philadelphia, PA, USA, 2007; Volume 2, pp. 185–202. [CrossRef]

43. Filter, B.K. Design and materials in rowing. In *Materials in Sports Equipment*, 1st ed.; Subic, A., Ed.; Woodhead Publishing: Philadelphia, PA, USA, 2007; Volume 2, pp. 271–295. [CrossRef]

44. Linthorne, N. Design and materials in athletics. In *Materials in Sports Equipment*, 1st ed.; Subic, A., Ed.; Woodhead Publishing: Philadelphia, PA, USA, 2007; Volume 2, pp. 296–320. [CrossRef]

45. Caine, M.; Yang, C. Design and materials in fitness equipment. In *Materials in Sports Equipment*, 1st ed.; Subic, A., Ed.; Woodhead Publishing: Philadelphia, PA, USA, 2007; Volume 2, pp. 321–338. [CrossRef]

46. Abreu, M.J.; Catarino, A.P.; Cardoso, C.; Martin, E. Effects of sportswear design on thermal comfort. In Proceedings of the AUTEX 2011 Conference, Mulhouse, France, 8–10 June 2011; Available online: https://core.ac.uk/download/pdf/55616300.pdf (accessed on 9 October 2019).

47. McCann, J. Material requirements for the design of performance sportswear. In *Textiles in Sport*; Shishoo, R., Ed.; Woodhead Publishing Series in Textiles: Philadelphia, PA, USA, 2005; pp. 44–69. [CrossRef]

48. Ogulata, R.T. The Effect of Thermal Insulation of Clothing on Human Thermal Comfort. *Fibres Text. East. Eur.* **2007**, *15*, 61–72. Available online: http://www.fibtex.lodz.pl/pliki/Fibtex_(ert6akuc1tje52dn).pdf (accessed on 5 October 2019).

49. Rossi, R.M. High-performance sportswear. In *High-Performance Apparel*; McLoughlin, J., Sabir, T., Eds.; Woodhead Publishing: Philadelphia, PA, USA, 2018; pp. 341–356. [CrossRef]

50. Havenith, G.; Richards, M.G.; Wang, X.; Bröde, P.; Candas, V.; Den Hartog, E.; Holmér, I.; Kuklane, K.; Meinander, H.; Nocker, W. Apparent latent heat of evaporation from clothing: Attenuation and "heat pipe" effects. *J. Appl. Physiol.* **2008**, *104*, 142–149. [CrossRef]

51. De Capua, A. La Termoregolazione. Available online: https://www.unirc.it/documentazione/materiale_didattico/597_2007_48_747.doc (accessed on 15 October 2019).

52. Fox, E.L. *Fisiologia Dello Sport*; Editoriale Grasso: Bologna, Italy, 1986.

53. International Organization for Standardization. *ISO 8996, Ergonomics of the Thermal Environment—Determination of Metabolic Rate*; International Organization for Standardization: Geneva, Switzerland, 2005.

54. Ainsworth, B.; Herrmann, S.; Meckes, N.; Bassett, D.; Tudor-Locke, C.; Greer, J.; Vezina, J.; Whitt-Glover, M.; Leon, A. 2011 Compendium of Physical Activities: A Second Update of Codes and MET Values. *Med. Sci. Sports Exerc.* **2011**, *43*, 1575–1581. [CrossRef]

55. D'Ambrosio Alfano, F.R.; Piterà, L.A. *Qualità Globale Dell'ambiente Interno*; Editoriale Delfino: Milano, Italy, 2014; pp. 71–142.

56. Gleson, M. Temperature Regulation During Exercise. *Int. J. Sports Med.* **1998**, *19*, 96–99. [CrossRef] [PubMed]

57. Reilly, T.; Drust, B.; Gregson, W. Thermoregulation in elite athletes. *Curr. Opin. Clin. Nutr. Metab. Care* **2006**, *9*, 666–671. [CrossRef] [PubMed]

58. Kaciuba-Uscilko, H.; Grucza, R. Gender differences in thermoregulation. *Curr. Opin. Clin. Nutr. Metab. Care* **2001**, *4*, 533–536. Available online: https://www.ncbi.nlm.nih.gov/pubmed/11706289 (accessed on 9 October 2019). [CrossRef] [PubMed]

59. Davies, C.T.M. Thermoregulation during exercise in relation to sex and age. *Eur. J. Appl. Physiol. Occup. Physiol.* **1979**, *42*, 71–79. [CrossRef] [PubMed]

60. Gavin, T.P.; Babington, J.P.; Harms, C.A.; Ardelt, M.E.; Tanner, D.A.; Stager, J.M. Clothing fabric does not affect thermoregulation during exercise in moderate heat. *Med. Sci. Sports Exerc.* **2001**, *33*, 2124–2130. Available online: https://insights.ovid.com/pubmed?pmid=11740309 (accessed on 5 October 2019). [CrossRef] [PubMed]

61. Gavin, T.P. Clothing and thermoregulation during exercise. *Sports Med.* **2003**, *33*, 941–947. [CrossRef]

62. Cosinuss. Available online: https://www.cosinuss.com/2015/11/10/body-temperature-in-sports/ (accessed on 10 October 2019).

63. Nielsen, M. Die Regulation der Korpoertemperatur bei Muskelarbeit. *Skand. Arch. Physiol.* **1938**, *79*, 193–230. [CrossRef]

64. Torii, M.; Yamasaki, M.; Sasaki, T.; Nakayama, H. Fall in skin temperature of exercising man. *J. Sport Med.* **1992**, *26*, 29–32. [CrossRef]

65. Tanda, G. Total body skin temperature of runners during treadmill exercise. *J. Anal. Calorim.* **2018**, *131*, 1967–1977. [CrossRef]

66. Robertson, A.S.; Burge, P.S.; Hedge, A.; Sims, J.; Gill, F.S.; Finnegan, M.; Pickering, C.A.C.; Dalton, G. Comparison of health problems related to work and environmental measurements in two office buildings with different ventilation systems. *Br. Med. J.* **1985**, *291*, 373–376. [CrossRef]

67. McIntyre, D.A. *Indoor Climate*; Applied Science: London, UK, 1980.

68. Gagge, P.; Stolwijjk, J.; Saltin, B. Comfort, thermal sensation and associated physiological responses during exercise at various ambient temperature. *Environ. Res.* **1969**, *2*, 209–229. [CrossRef]

69. Gonzalez, R.R. Exercise Physiology and Sensory Responses. In *Bioengineering, Thermal Physiology and Comfort*, 1st ed.; Cena, K., Clark, J.A., Eds.; Elsevier: Amsterdam, The Netherlands; Oxford, UK; New York, NY, USA, 1981; Volume 10, pp. 123–144. [CrossRef]

70. International Organization for Standardization. *ISO 7730. Ergonomics of the Thermal Environment—Analytical Determination and Interpretation of Thermal Comfort Using Calculation of the PMV and PPD Indices and Local Thermal Comfort Criteria*; International Organization for Standardization: Geneva, Switzerland, 2006.

71. American Society of Heating, Refrigerating and Air-Conditioning Engineers. *ASHRAE 55/2004. Thermal Environmental Conditions for Human Occupancy*; ASHRAE: Atlanta, GA, USA, 2004.

72. SNIP 31-112-2004, The National Building Regulation of Russia. Considerations for the design and construction of buildings—Part 1 Sports halls. 2004. Available online: http://docs.cntd.ru/document/1200040660 (accessed on 2 October 2019).

73. CONI. *CONI1379/2008, Norme Coni per L'impiantistica Sportiva*; Comitato Olimpico Nazionale Italiano: Rome, Italy, 2008.

74. FINA's Regulations, FINA Facilities Rules (Part IX). 2017. Available online: https://www.fina.org/sites/default/files/rules-print-pdf/8458.pdf (accessed on 1 October 2019).

75. BWF's Regulations, Specifications for International Standard Facilities. 2018. Available online: http://www.badmintonpanam.org/wp-content/uploads/2018/12/3.3.4-Specs-for-Intl-Standard-Facilities-Nov2018.pdf (accessed on 1 October 2019).

76. FIP's Regulations, Regolamento Relative All'impiantistica Sportiva in cui si Pratica il Gioco Della Pallacanestro. 2015. Available online: http://www.fip.it/public/statuto/regolamento%20impianti%20sportivi%20ultimo%202015.pdf (accessed on 1 October 2019).

77. WCF's Regulations. Technical Requirements for Good Playing and Environmental Conditions in a New Curling-Rink. 2013. Available online: https://www.teamusa.org/-/media/USA_Curling/Documents/GD/WCF-Technical-Requirements.pdf?la=en&hash=8B30631A753CAEDF882D0B449EB49FB88575401 (accessed on 1 October 2019).

78. IFG's Regulations, Technical Regulations 2019. 2019. Available online: https://www.gymnastics.sport/publicdir/rules/files/en_Technical%20Regulations%202019.pdf (accessed on 1 October 2019).

79. IHF's Regulations, Recommendations and Guidelines for the Construction of Handball Playing Halls. 2008. Available online: http://www.handball.ee/g84s4107 (accessed on 1 October 2019).

80. IIHF's Regulations, Ice Rink Guide. 2016. Available online: https://www.iihf.com/IIHFMvc/media/Downloads/Projects/Ice%20Rink%20Guide/IIHF_Ice_Rink_Guide_web_pdf.pdf (accessed on 1 October 2019).

81. IJF's Regulations, JJIF Handbook 2009—Medical Section. 2009. Available online: http://www.jjif.info/fileadmin/JJIF/Documents/MEDICAL_Handbook_09.pdf (accessed on 1 October 2019).

82. FITET's Regulations, Regolamento Per L'omologazione Degli Impianti Sportivi per il Tennistavolo. 2018. Available online: https://www.fitet.org/la-federazione/regolamenti/regolamento-impianti.html?download=5600:regolamento-impianti-sportivi-per-il-tennistavolo (accessed on 1 October 2019).

83. ITF's Regulations, Facilities Guide. 2019. Available online: https://www.itftennis.com/technical/facilities/facilities-guide/indoor-structures.aspx (accessed on 1 October 2019).

84. FIVB's Regulations, Official Volleyball Rules. 2016. Available online: https://www.fivb.org/EN/Refereeing-Rules/documents/FIVB-Volleyball_Rules_2017-2020-EN-v06.pdf (accessed on 1 October 2019).

85. UWW's Regulations, Requirements for the Organisation of World Cups. 2018. Available online: https://unitedworldwrestling.org/sites/default/files/2018-09/senior_world_cups_en.pdf (accessed on 1 October 2019).

86. Stamou, A.I.; Katsiris, I.; Schaelin, A. Evaluation of thermal comfort in Galatsi Arena of the Olympics "Athens 2004" using a CFD model. *Appl. Eng.* **2008**, *28*, 1206–1215. [CrossRef]

87. Cheng, Z.; Li, L.; Bahnfleth, W.P. Natural ventilation potential for gymnasia—Case study of ventilation and comfort in a multisport facility in northeastern United States. *Build. Environ.* **2016**, *108*, 85–98. [CrossRef]

88. Khalil, E.E.; AlHababi, T. Numerical Investigations of Flow Patterns and Thermal Comfort in Air-Conditioned Gymnastic Sport Facility. In Proceedings of the 54th AIAA Aerospace Sciences Meeting, San Diego, CA, USA, 4–8 January 2016. [CrossRef]

89. Berquist, J.; Ouf, M.; O'Brien, W. A method to conduct longitudinal studies on indoor environmental quality and perceived occupant comfort. *Build. Environ.* **2019**, *150*, 88–98. [CrossRef]

90. Givoni, B.; Goldman, R. Predicting metabolic energy cost. *J. Appl. Physiol.* **1971**, *30*, 429–433. [CrossRef]

91. Li, Y.; Li, F.; Liu, Y.; Luo, Z. An integrated model for simulating interactive thermal processes in the humane clothing system. *J. Therm. Biol.* **2004**, *29*, 567–575. [CrossRef]

92. Lammers, J.T.H. Human Factors, Energy Conservation, and Design Practice. Ph.D. Thesis, Technische Hogeschool Eindhoven, Eindhoven, The Netherlands, 1978. [CrossRef]

93. Inail, La Valutazione del Microclima. 2018. Available online: https://www.inail.it/cs/internet/docs/alg-pubbl-valutazione-del-microclima.pdf (accessed on 1 October 2019).

94. Fanger, P.O.; Toftum, J.R. Extension of the PMV model to non-air-conditioned buildings in warm climates. *Energy Build* **2002**, *34*, 533–536. [CrossRef]

95. De Dear, R.J.; Brager, G.S. Developing an adaptive model of thermal comfort and preference. *ASHRAE Trans.* **1998**, *104*, 145–167.

96. International Organization for Standardization. *ISO 10551. Ergonomics of the Thermal Environment—Assessment of the Influence of the Thermal Environment Using Subjective Judgement Scales*; International Organization for Standardization: Geneva, Switzerland, 2002.

*Article*

# Analysis of the Impact of the Construction of a Trombe Wall on the Thermal Comfort in a Building Located in Wrocław, Poland

**Jagoda Błotny and Magdalena Nemś ***

Faculty of Mechanical and Power Engineering, Wroclaw University of Science and Technology, Wybrzeże Wyspiańskiego 27, 50-370 Wrocław, Poland; jagoda.blotny@gmail.com
* Correspondence: magdalena.nems@pwr.edu.pl; Tel.: +48-71-320-4826

Received: 10 November 2019; Accepted: 27 November 2019; Published: 29 November 2019

**Abstract:** Changes in climate, which in recent years have become more and more visible all over the world, have forced scientists to think about technologies that use renewable energy sources. This paper proposes a passive solar heating and cooling system, which is a Trombe wall located on the southern facade of a room measuring 4.2 m × 5.2 m × 2.6 m in Wrocław, Poland. The studies were carried out by conducting a series of numerical simulations in the Ansys Fluent 16.0 environment in order to examine the temperature distribution and air circulation in the room for two representative days during the heating and cooling period, i.e., 16 January and 15 August (for a Typical Meteorological Year). A temperature increase of 1.11 °C and a temperature decrease in the morning and afternoon hours of 2.27 °C was obtained. Two options for optimizing the passive heating system were also considered. The first involved the use of triple glazing filled with argon in order to reduce heat losses to the environment, and for this solution, a temperature level that was higher by 8.50 °C next to the storage layer and an increase in the average room temperature by 1.52 °C were achieved. In turn, the second solution involved changing the wall material from concrete to brick, which resulted in a temperature increase of 0.40 °C next to the storage layer.

**Keywords:** Trombe wall; thermal comfort; passive heating systems; heat accumulation

---

## 1. Introduction

Progressive changes in the climate have forced scientists to take an interest in passive heating systems that will reduce the amount of pollutants released into the atmosphere by reducing heat demand. Particular attention should be paid to heating systems, which are currently the main source of toxic compounds in the atmosphere in Poland. The proposed Trombe wall system uses solar radiation energy by accumulating it in a material with a high heat capacity, such as concrete. It enables the heat to be slowly released into a room, both during the day and also at night. The mechanism of heat exchange, in this case, involves radiation and conduction. The ventilated version of this storage wall includes air circulation, with heat also being provided by convection. This solution includes air ducts in the upper and lower part of the wall. Cold high-density air flows from the room into the Trombe wall air gap through the lower vent, heats up as a result of solar radiation, reduces its density, and then rises up. The resulting density difference causes natural circulation in the system under the influence of buoyant forces. The operating principle is illustrated in Figure 1.

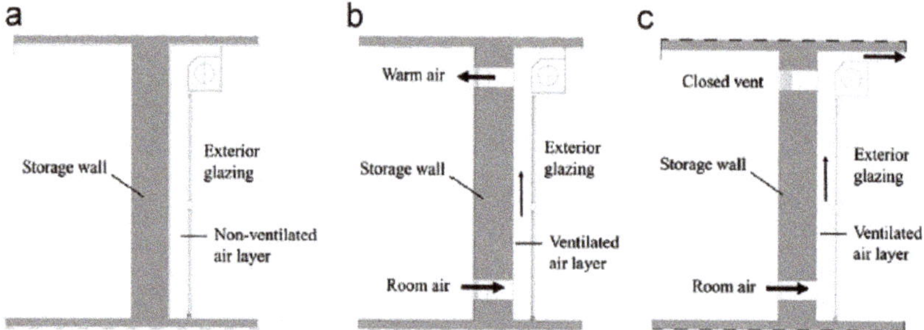


**Figure 1.** Principle of operation of a classic Trombe wall [1]; (**a**) non-ventilated version; (**b**) ventilated version working in the winter mode; (**c**) ventilated version working in the cooling (summer) mode.

The Trombe wall in its classic solution is already a well-known technology in countries with high solar gains during the day and low temperatures at night, and therefore especially in dry and hot climates. This is due to the fact that this element can function in heating and cooling modes. Table 1 contains a summary of the results of research conducted by scientists on real objects, and also numerical simulations that were carried out to optimize the given solution.

**Table 1.** Summary of the parameters subjected to optimization.

| Element | Parameter | Optimal Solution/Value |
|---|---|---|
| Glazing | Number of layers | 1 or 2, climate dependent selection [2] |
| | Material | Filling with noble gas (argon) Low emission coating [3] |
| Surface | $\alpha$ | 37% [4] |
| Storage wall | Material, colour | Concrete/brick/aerated concrete [5]/PCM in the darkest possible colour |
| | Thickness | 30 cm–40 cm/37 cm [6,7], for PCM the key is to position the capsules correctly [8] |
| | Isolation level | Depending on the solution [2,9,10] |
| Air duct | The ratio of depth to duct height | 1/10 [11] |
| Shading devices | External elements | Hoods [12,13], blinds [14], shutters [12,15], curtains [16,17] |
| | Internal elements (inside a duct) | Double-sided blinds—on one side covered with a material with a high radiation absorption coefficient, and on the other with a material with high reflectivity [18], placed inside the duct at a distance of 0.09 m from the glazing—in the case of a 0.14 m wide duct—regulation of the amount of radiation [14] |

The classic Trombe wall solution is still being modified and developed in order to increase its efficiency in the heating and cooling operation mode. This is conducted by the addition and modification of structural elements. A list of solutions on which research works are conducted is presented in Table 2.

**Table 2.** List of Trombe wall configurations with additional technology elements; designations: S—south, E—east, W—west.

| Technology | Heat Accumulating Material | Orientation and Location | Equipment | Possible Hazards | Additional Advantages |
|---|---|---|---|---|---|
| Classic | Concrete/brick | S Northern Hemisphere | Hood, blinds | Reverse heat flow at night and on cloudy days | Simple construction |
| Water [19] | Water | S Poland [19] | Water purification [20] | Water freezing | High heat capacity, aesthetic value, lower heat losses [21] |
| Zig-zag [22] | Concrete/brick | S S-E S-W USA | Additional sections forming a V-shaped wall | It needs to be considered at the stage of constructing a room—special arrangement of the walls | Light supply in the morning hours, the heat is given back to the room for longer |
| Hybrid [23] | Porous ceramic material, concrete | S Spain | Wetting nozzles, ceramic material Hood and blinds | Difficult construction that requires a special frame | Simultaneous cooling and humidifying of the air in summer |
| Composite | Concrete/brick | S France | Air gap acting as external insulation, internal insulation, thermal diode [24] | Risk of reverse thermo-circulation (foil in the ventilation opening is necessary [24]), operation only in the heating mode | Prevents inverted heat flow |
| PCM-Trombe [25] | Concrete + PCM(capsules)Plaster + PCM | S France [25], Turkey [26] | Triple NTG glazing [26], fans, electric heaters | Too fast dissipation of heat to a room in the winter season | A thinner wall and a smaller loading of the structure, more energy gains |
| Fluidized [27] | Storage bed with low density and good absorptive properties | S Turkey | Filters, fan | Storage bed particles could get inside a room | Good heat exchange due to the direct contact of particles with the heated air |
| Ribbed [28] | Concrete/brick | S Tunisia | Metal, ribs | The tested solution did not include the ventilated Trombe wall, limited use (no possibility of cooling in the summer season) | The ability to reduce the surface of the Trombe wall |
| PV-Trombe [29] | Concrete/brick | S Malaysia [29], China [30] | PV, fan | Heating of panels reduces their efficiency, and therefore there is a lower heating efficiency [30] | Additional generation of electricity |
| TC-Trombe [31] | Concrete/brick | S China | Catalyser | As the temperature increases, the efficiency of the oxidation process decreases | Air purification from volatile organic compounds (the driving energy of the reaction is used to heat the air) |
| Classic + solar chimney [32] | Concrete/brick | S Duct: E,W Iran | Spray nozzles, solar chimney | Requires closing of the chimney during the winter and the ensuring of adequate tightness and thermal insulation | It provides air circulation in the solar chimney in the absence of sunlight and also ensures an adequate cooling effect in summer |

## 2. Materials and Methods

In the conducted simulations, a ventilated Trombe wall, constituting the southern facade of a room measuring 5.2 m × 4.2 m× 2.6 m, was considered. The simplest and cheapest construction solutions were chosen; the 348 mm thick storage wall made of concrete, covered with a 2 mm thick black paint as an absorber. An air gap of 150 mm separates the absorber from the 4 mm thick glazing. The wall has three upper and three lower ventilation ducts with dimensions of 300 mm × 200 mm. The properties of the used materials are summarized in Table 3, which include: $d$—thickness, $\rho$—density, $Cp$—specific heat, $\lambda$—heat transfer coefficient, $a$—solar absorption coefficient, and $z$—refractive index of a material.

**Table 3.** Properties of the materials that were used during modeling.

| Element | $d$, mm | Material | $\rho$, kg/m$^3$ | $Cp$, J/kg K | $\lambda$, W/m K | $a$, 1/m | $z$,– |
|---|---|---|---|---|---|---|---|
| Medium | - | air | ideal gas | 1006.43 | 0.0242 | 0 | 1 |
| Storage wall | 398 | concrete | 2000 | 960 | 1.5 | 1.7 | 0 |
| Absorber | 2 | black paint | 2100 | 1050 | 1.6 | 0 | 0 |
| Glazing | 4 | glass | 2500 | 840 | 0.81 | 200 | 1.5 |

The external walls of the room with a thickness of 0.442 mm were modeled by designating the appropriate simulation conditions. In order to consider their insulation from the environment, the weighted average densities, specific heat, and heat transfer coefficient of the individual layers were determined. The properties of the individual layers are summarized in Table 4. The selection of insulation was conditioned by obtaining a heat transfer coefficient lower than 0.14 W/m K, which corresponds to that of energy-saving buildings. The calculations took into account the heat transfer resistance on both the inside and outside.

**Table 4.** Properties of the materials that form the outer wall.

| Material | $d$, m | $\lambda$, W/m K | $Cp$, J/kg K | $\rho$, kg/m$^3$ |
|---|---|---|---|---|
| Mineral-lime plaster | 0.005 | 0.76 | 840 | 1850 |
| Mineral wool | 0.230 | 0.035 | 750 | 130 |
| Aerated concrete | 0.200 | 0.35 | 840 | 800 |
| Cement-lime plaster | 0.007 | 0.82 | 840 | 1850 |
| Weighted average | - | 0.198 | 793.2 | 479.9 |

### 2.1. Numeric Mesh of the Room

In order to build a numerical mesh of the room, a 3D model of the considered geometry was made using CATIA software. This model should be divided into three elements: air, which is the medium in question, the storage wall, and the absorber. Glazing was simulated by appropriate settings of the boundary conditions in order to minimize the number of mesh elements. The obtained geometries were implemented in the Ansys Fluent 16.0 environment. Afterwards, a mesh was created (Figure 2a) with a bigger number of elements within the connections of the individual geometry elements, in particular at the contact point of the storage wall and the air in the room, where the largest changes in temperature distribution can be observed as a result of heat dissipation from the storage wall to the room. Individual meshes were checked with regards to their quality, obtaining in most elements an index value within the range of 0.7–1.0 (Figure 2b). The absorber, due to its small thickness, was the biggest challenge when creating the mesh. A high mesh quality could only be obtained by increasing the number of mesh elements, which in turn resulted in a significant increase in the calculation time. However, since the essential element of the experiment is the medium, the quality of the absorber mesh introduces an acceptable range of errors.

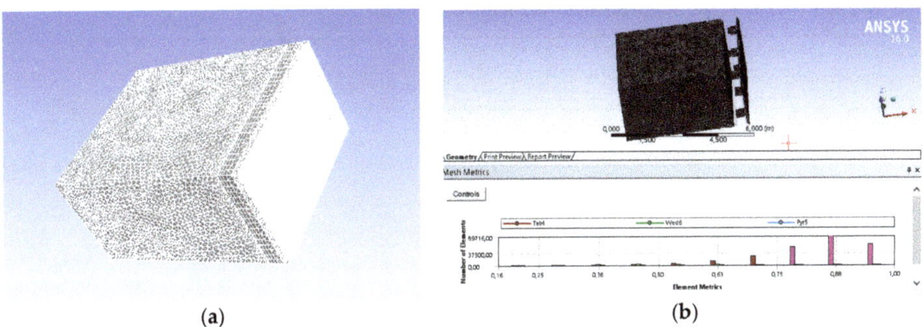

**Figure 2.** Numeric view of the mesh: (**a**) the room with Trombe wall elements; (**b**) air, together with an assessment of the quality of the elements.

## 2.2. Boundary Conditions

The boundary conditions for the implementation of the simulation for the heating period were determined on the outer walls of the room. For the ceiling and the north, east, and west walls, the determined value of the outside air temperature and the heat transfer coefficient were estimated and increased in order to also consider windy weather conditions. The temperature was adopted for a certain hour in accordance with meteorological data for Wroclaw. The same conditions were used for the floor, but the temperature corresponds to the ground temperature and the heat transfer coefficient changes. All the walls do not participate in solar radiation tracking, and calculations for heat transfer by radiation are also not conducted. For the southern wall, a condition was established on the external glass covering, and it includes air temperature and the heat transfer coefficient, and for the radiation calculations, it includes tracking of solar radiation with the semi-permeability condition. On the connections of the glass with the air gap, the gap with the absorber, the absorber with the storage wall, and also the walls with the air in the room, the condition of connecting the walls (interface, coupled wall) was determined. The replacement material and wall thickness were also specified in order to avoid the creation of additional meshes and connections between the walls. A list of conditions is given in Table 5. The heat transfer coefficient from the outside air to the wall was estimated by assuming the temperature of the layer adjacent to the wall and was then increased to take into account windier days.

**Table 5.** List of boundary conditions for the simulation.

| Element | $\alpha$, W/m$^2$K | $T$, °C | Type of Radiation |
|---|---|---|---|
| Floor | 3 | 1 | Not transparent |
| Walls and ceiling | 3 | $T_{amb}$ | Not transparent |
| Glazing | 3 | $T_{amb}$ | Translucent, with additional solar radiation tracking |

The air temperature $T_{amb}$ is the temperature for a given hour during the selected representative day on which the simulation is carried out. The lowest average temperature, according to meteorological data, was determined for the month of January with a value of −0.7 °C. Due to the fact that the proposed passive system is only intended to support the heating system, the selected day has daytime temperatures close to the average value. Choosing extremely low temperatures would not reflect the work of the system on most days when its role would be greater (heat losses to the environment increase at very low outside temperatures, and therefore, operation of the support system is necessary). Thus, 16 January (data for TMY [33]) was chosen as the representative day, for which the temperatures are summarized in Table 6.

**Table 6.** Ambient temperature for 16 January [33].

| Hour | 10:00 | 11:00 | 12:00 | 13:00 | 14:00 |
|---|---|---|---|---|---|
| $T_{amb}$, °C | −1.0 | −0.5 | 0.0 | 0.4 | 0.7 |

The ground temperature was assumed as a constant value for a given month at a depth of about 1 m [34]. For January, it is equal to around 1 °C.

### 2.3. Settings of the Module That Are Responsible for the Calculations

Carrying out the simulation requires taking into account relevant equations, which the program solves with a given number of iterations. The governing equations of Navier Stokes were linked to the RANS turbulence model. The used realizable two-equation turbulence model $k - \varepsilon$ is necessary due to the air circulation between the room and the air gap, which is caused by the resulting density difference. This condition also requires consideration of the effect of gravity by determining the Boussinesq model. Another equation included in the calculations, in order to observe the heat exchange between the external environment and the air inside the room, is the energy equation. This allows appropriate boundary conditions to be set that reflect the conducted simulation. The inclusion of radiation greatly complicates the calculations, often causes errors, and also significantly extends the calculations. Therefore, this model was simplified by excluding the consideration of radiation for individual elements, and only solar radiation tracking [35] was determined for the southern wall. This option requires entering the latitude and longitude of Wrocław and then orienting the solid in accordance with the selected directions. This model allows a specific day of the year and the hour with accuracy to the nearest minute to be set for the conducted simulation. A coupled calculation system based on pressure and speed was used. The simulations were made based on the research carried out in [36]. The mathematical notation [36] should take into account the mass and moment equation for three-dimensional turbulent flow.

### 3. Results for the Passive Heating System

The simulations were carried out for the classic Trombe ventilated wall solution operating in the heating mode. Afterward, the effect of exchanging the storage material from concrete to brick, and also exchanging the single glazing to triple glazing filled with argon were examined. The simulation results focus on the temperature distribution throughout the entire section of the room, and also air circulation, which are crucial for achieving thermal comfort. The results obtained for individual hours are summarized in tabular and graphic form.

### 3.1. The Basic Solution

Simulations were carried out for the heating period, and temperature values were read from 10:00 to 14:00. Between 7:00 and 10:00, solar radiation is not intense enough to achieve a heating effect, and after 14:00, the radiation is no longer recorded by the meteorological station. Due to the calculations in the set state, it is not possible to test the heat release to the room at later hours. The temperature distribution in the room was presented on several main planes that are shown in Figure 3a. In turn, the distribution of speed vectors is presented in Figure 3b.

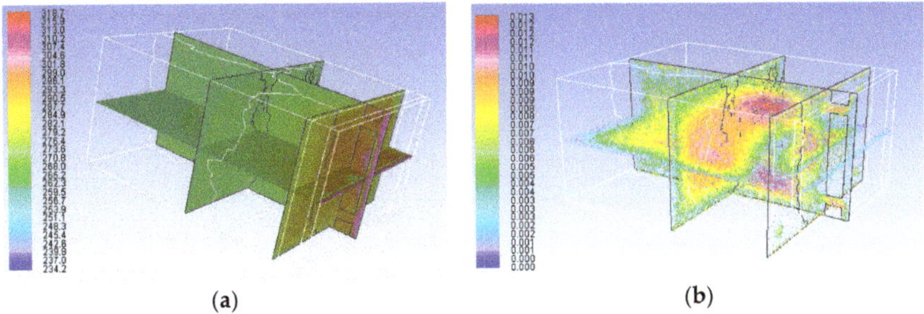

**Figure 3.** Simulation results for the basic solution: (**a**) temperature distribution; (**b**) distribution of speed vectors.

The result of each simulation consisted of a reading of the average temperature from the given plane. A summary of these values is given in Table 7, and it includes ambient temperature $T_{amb}$, the temperature in the middle of the room $T_{room}$, the temperature of air by the storage wall $T_{air\_wall}$, the temperature on the connection between the absorber, and the wall $T_{abs}$, the temperature of the storage wall from the room side $T_{wall\_ins}$, and the temperature of air in the gap $T_s$.

**Table 7.** Averaged temperatures on the individual planes for the basic solution.

| Hour | 10:00 | 11:00 | 12:00 | 13:00 | 14:00 |
|---|---|---|---|---|---|
| $T_{amb}$, °C | −1.00 | −0.50 | 0.00 | 0.40 | 0.70 |
| $T_{room}$, °C | 0.00 | 0.61 | 1.09 | 1.41 | 1.58 |
| $T_{air\_wall}$, °C | 7.04 | 12.51 | 15.59 | 16.80 | 16.26 |
| $T_{abs}$, °C | 12.72 | 19.60 | 23.44 | 24.80 | 24.04 |
| $T_{wall\_ins}$, °C | 8.40 | 14.98 | 18.63 | 20.03 | 19.33 |
| $T_s$, °C | 15.59 | 30.81 | 39.01 | 41.70 | 39.98 |

*3.2. The Brick Storage Wall*

An important element of the Trombe wall is the storage wall and its properties, which depend primarily on the used material. Other simulations were carried out for concrete, which is the most frequently chosen material in such constructions. An alternative suggestion may be the use of brick, which has a lower specific heat $Cp_{brick} = 880$ J/kg K, lower density $\rho_{brick} = 1900$ kg/m$^3$ and a thermal conductivity coefficient with the value of $\lambda_{brick} = 1.05$ W/m K. This material is able to accumulate less heat, but at the same time, it delays heat transfer to the room. The resulting temperature distribution for 12:00 is shown in Figure 4a, while the distribution of speed vectors is presented in Figure 4b.

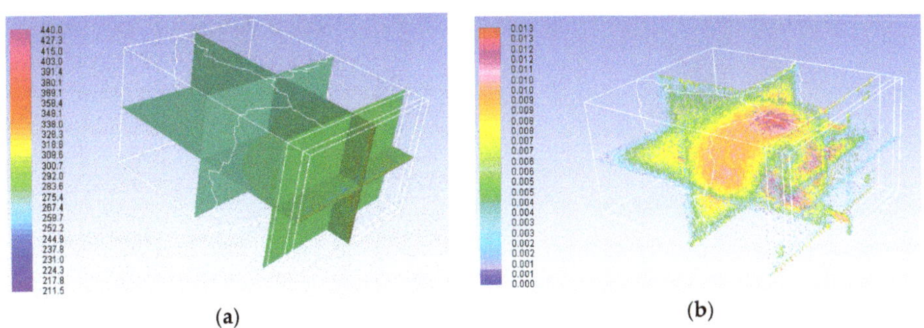

**Figure 4.** The results of simulations for the solution with the brick storage wall: (**a**) temperature distribution; (**b**) distribution of speed vectors.

The average temperature values in the individual planes are summarized in Table 8. Small differences of these results can be observed when compared to the simulation with the wall made of concrete.

**Table 8.** Average temperatures on the individual planes for the solution with the brick wall.

| Hour | 10:00 | 11:00 | 12:00 | 13:00 | 14:00 |
|---|---|---|---|---|---|
| $T_{amb}$, °C | −1.00 | −0.50 | 0.00 | 0.40 | 0.70 |
| $T_{room}$, °C | −0.01 | 0.60 | 1.10 | 1.43 | 1.59 |
| $T_{air\_wall}$, °C | 7.02 | 12.56 | 15.98 | 17.20 | 16.72 |
| $T_{abs}$, °C | 12.65 | 19.52 | 23.72 | 25.26 | 24.53 |
| $T_{wall\_ins}$, °C | 8.45 | 15.04 | 19.09 | 20.49 | 19.88 |
| $T_s$, °C | 15.66 | 30.90 | 39.17 | 41.91 | 40.16 |

### 3.3. Argon Filled Glazing

The basic solution of the Trombe wall is characterized by heat losses to the environment at the border of the glazing and the environment. This is associated with a large temperature difference between the heated air in the gap and the ambient temperature. The reduction of these losses, without significantly affecting the glass transmission, can be achieved by using glazing filled with a noble gas such as argon. Such a gas is characterized by a low thermal conductivity coefficient. Triple glazing with 4 mm thick glass and 14 mm thick argon filled spaces was considered in the simulation. Table 9 summarizes the values of the parameters of this glazing, which were calculated as a weighted average based on the thickness of the layer.

**Table 9.** Parameters of the triple glazing filled with argon.

| Material | $d$, mm | $\lambda$, W/m K | $Cp$,J/kg K | $\rho$, kg/m³ |
|---|---|---|---|---|
| Glass | 4 | 0.81 | 840 | 2500 |
| Argon | 14 | 0.017 | 519 | 1.7 |
| Total thickness and weighted average of the parameters for the triple glazing | 40 | 0.182 | 615.3 | 751.2 |

The temperature distribution for 12:00 in the individual planes is shown in Figure 5a, while the distribution of speed vectors is presented in Figure 5b.

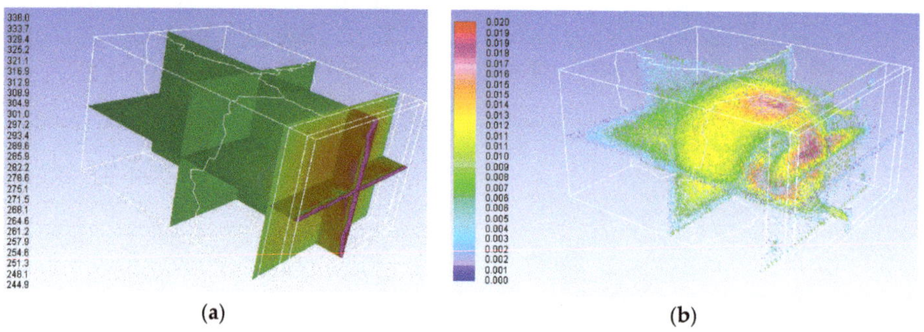

(a)         (b)

**Figure 5.** Simulation results for the solution with the triple glazing filled with argon: (**a**) temperature distribution; (**b**) distribution of speed vectors.

Analysis of the results contained in Table 10 confirms the achievement of a higher temperature level on each surface. An increase in temperature within the storage wall and the absorber is especially noticeable, which confirms the fact that there are fewer heat losses to the environment.

**Table 10.** Averaged temperatures on the individual planes for the solution with the glazing filled with argon.

| Hour | 10:00 | 11:00 | 12:00 | 13:00 | 14:00 |
|---|---|---|---|---|---|
| $T_{amb}$, °C | −1.00 | −0.50 | 0.00 | 0.40 | 0.70 |
| $T_{room}$, °C | 0.21 | 0.97 | 1.52 | 1.87 | 2.00 |
| $T_{air\_wall}$, °C | 10.42 | 18.97 | 23.71 | 25.30 | 24.30 |
| $T_{abs}$, °C | 17.30 | 27.98 | 33.91 | 35.86 | 34.58 |
| $T_{wall\_ins}$, °C | 12.55 | 22.84 | 28.51 | 30.37 | 29.14 |
| $T_{s}$, °C | 26.22 | 50.31 | 63.20 | 67.28 | 64.30 |

## 3.4. Discussion of Results

The heating level for the use of the selected technology is best reflected by the average temperature inside the room and the air temperature by the storage wall. The results for these quantities for the individual solutions are compiled on a common graph and shown in Figure 6. This figure also contains the temperature at the connection of the absorber and the storage wall and the ambient temperature in order to show the appropriate reference point. A slight increase of the values can be observed for the storage wall made of brick, for which the temperature distribution in the room almost coincides with the trend line for the wall made of concrete. A significant increase in the values is noticeable when using glazing filled with argon.

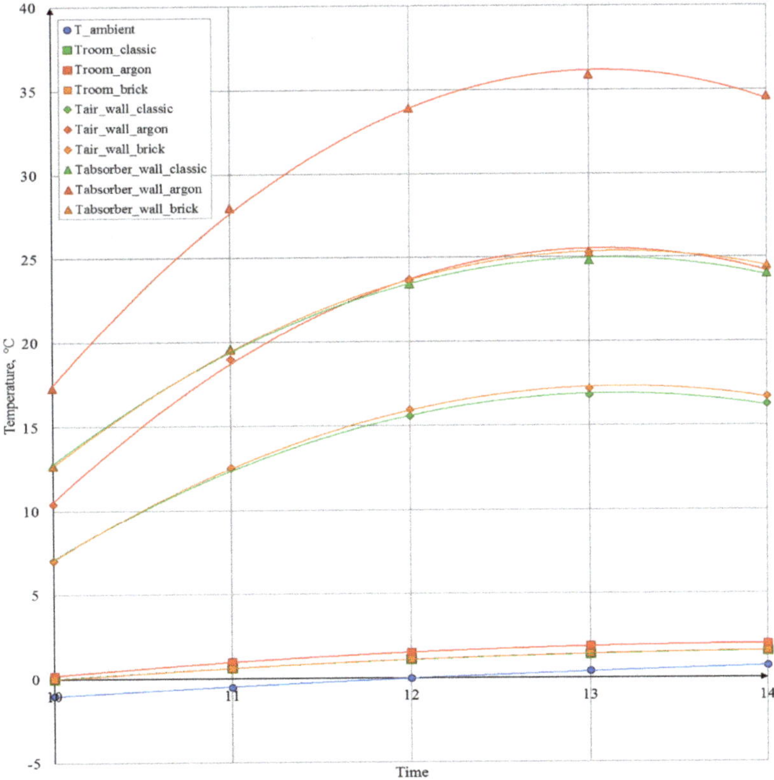



**Figure 6.** Summary of results in graphic form for the individual solutions.

When analyzing the obtained temperature distribution for the heating mode, an increase in the temperature in the middle of the room ranging from 1–2 °C can be observed. Taking into account the large dimensions of the room and the fact that the analysis was conducted under steady-state conditions, this result is considered as rational. The air zone just next to the storage wall is characterized by temperatures within the range of 20–30 °C, while the highest temperatures are achieved by the air in the gap, which receives heat by convection from the absorber. Higher values of temperature ranges were observed for the simulations with additional glazing, due to which heat losses to the environment are reduced. The impact of the change in the storage material is small, and the obtained results are only slightly higher than for the wall made of concrete.

The distribution of air speed vectors in the room when considering the heating mode enables a visible area of air circulation, caused by the difference in the density of cool and warm air, to be noticed. The maximum speed values are in the range of 0.013–0.020 m/s, which allows heat comfort to be maintained in the room. Higher velocity values were registered for the solution with argon filled glazing, which is associated with a greater temperature difference, and thus a greater difference in air density, which in turn is the driving force of the proposed solution.

## 4. The Trombe Wall Operating as a Cooling System

The operation of the Trombe wall in the cooling system, which has potential in the Polish climate, was also examined. For this purpose, the room geometry and simulation conditions were changed.

### 4.1. Changes in Geometry

New geometry and mesh elements were created for the cooling system (Figure 7a). Ducts, which are the inlets of cooler air, were added to the top of the north wall. A 10 cm thick insulation was placed on the inner side of the storage wall, the properties of which are listed in Table 11. The air was simulated according to the Boussinesq model, with the compressibility factor $\beta$ equal to 0.003 (Table 11).

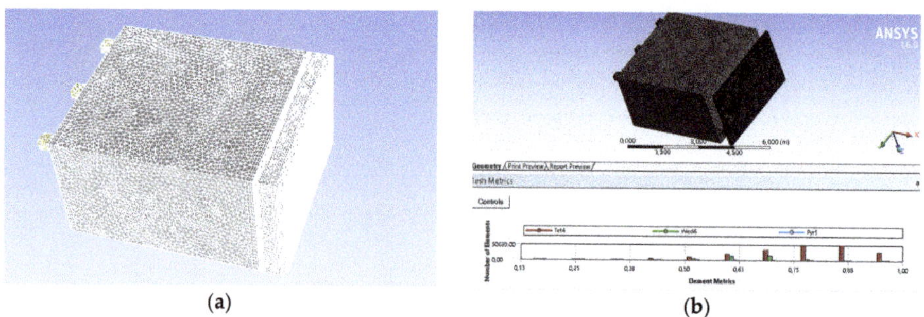

(a)      (b)

**Figure 7.** Numeric mesh view: (**a**) the room with the elements of the Trombe wall; (**b**) air, together with an assessment of the quality of the elements.

**Table 11.** Parameters of the changed properties of the medium and the added insulation in the cooling mode.

| Element | $d$, mm | Material | $\rho$, kg/m$^3$ | $C_p$, J/kg K | $\lambda$, W/mK | $a$, 1/m | $z$, – | $\beta$, – |
|---|---|---|---|---|---|---|---|---|
| Medium | - | air | 1.225 | 1006.43 | 0.0242 | 0 | 1 | 0.003 |
| Insulation | 100 | mineral wool | 130 | 750 | 0.030 | - | - | - |

The top ducts located in the upper part of the storage wall were closed, and warm air outlets located in the glazing were added. The air domain mesh with the quality scale of the elements is shown in Figure 7b. In order to simplify the simulation conditions, the glazing and the north wall were implemented by setting appropriate boundary conditions. Due to the greater complexity of the geometry, the authors resigned from the absorber, which was characterized by weaker elements of its mesh. Instead, the solar radiation absorption coefficient for the storage wall was determined with a value that corresponds to the absorber, i.e., 1.7 m$^{-1}$.

### 4.2. Boundary Conditions

The boundary conditions for the walls of the room when considering the cooling mode of operation were adopted in the same way as for the heating system. However, the glazing conditions were changed, which was included in the geometry. The heat transfer coefficient $\alpha$ was reduced to 1 W/m$^2$ K in order to obtain atmospheric conditions that hinder the system's operation and cause higher heating of the storage wall. The value of 3 W/m$^2$ K, which was confirmed by the calculations, was left for the remaining walls. These conditions are summarized in Table 12.

**Table 12.** Summary of the boundary conditions concerning the heat transfer in the cooling mode simulation.

| Element | $\alpha$, W/m$^2$ K | $T$, °C | Type of Radiation |
|---|---|---|---|
| Floor | 3 | 17.5 | Not transparent |
| Walls and ceiling | 3 | $T_{amb}$ | Not transparent |
| North wall | 3 | $T_{amb}$ | Not transparent |
| Glazing | 1 | $T_{amb}$ | Semi-translucent, with additional solar radiation tracking |

The outside air temperature divided into hours was adopted in accordance with TMY [33]. The highest average temperature, equal to 17.8 °C, was recorded for August. Due to the fact that in the summer the cooling system is necessary at an outside temperature higher than the comfort temperature, and that there is also a risk of overheating the room during this period, the day with the highest outside temperatures was chosen, i.e., 15 August. The temperatures in the hours for which the simulations were performed are listed in Table 13.

**Table 13.** Ambient temperature for 15 August [33].

| Hour | 10:00 | 11:00 | 12:00 | 13:00 | 14:00 | 15:00 | 16:00 | 17:00 |
|---|---|---|---|---|---|---|---|---|
| $T_{amb}$, °C | 28.6 | 29.8 | 30.9 | 30.4 | 29.9 | 29.4 | 28.5 | 27.5 |

The ground temperature was assumed as a constant value for a given month at a depth of about 1 m [34], and for August, it is around 17.5 °C. Due to the changed geometry of the room, two conditions regarding the air flow in the room were also added, which are presented in Table 14.

**Table 14.** Summary of boundary conditions concerning the air flow in the cooling mode simulation.

| Element | Condition | Radiation Tracking |
|---|---|---|
| Inlets | Speed: 0.5 m/s | Does not participate |
| Outlets | Relative pressure: 0 Pa | Participates |

On the ducts located on the north facade of the building, the air inlet condition was established with an inlet speed of 0.5 m/s, while on the ducts located in the glazing, the condition of the air outlet to the atmosphere was set at a relative pressure of 0 Pa. These conditions are shown in Table 13.

### 4.3. Simulation Results

The operation of this solution is based on the introduction of cooler air into the room from the north side. For the purpose of this simulation, a temperature of 5 °C lower than the ambient temperature, as well as the obtaining of air flow between the north and south side due to the difference in temperature, were assumed. The temperature distribution on the individual planes of the room (Figure 8a), read at 12:00, clearly shows the inflow of cooler air, which, due to its higher density, falls to the bottom of the room. A warmer air zone is visible in the upper part of the room.

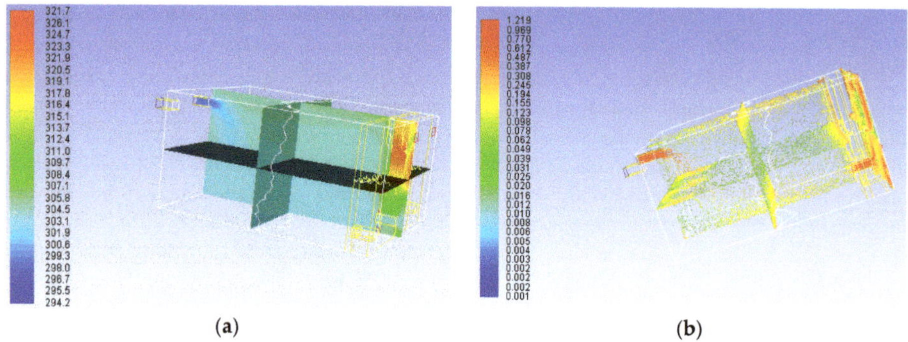

(a)       (b)

**Figure 8.** Simulation results for the solution in the cooling system: (**a**) temperature distribution; (**b**) distribution of speed vectors.

A flow of air into the room is visible, the stream of which is dispersed in the volume of air in the room, reaching higher velocity values at the edges of the geometry. The stream accelerates again in the ducts that connect the room to the gap and also in the air outlets to the environment.

The average temperature values were read for the individual planes located in the middle of the room $T_{room}$, at the connection of the storage wall and the absorber $T_{abs}$, and in the air gap $T_s$. The maximum average room temperature decrease of 2.27 °C can be seen, while between 12:00 and 14:00, temperature equalization and a slight increase in room temperature can be noticed. This is due to the increase in the temperature of the absorber and the air in the gap as a result of higher intensity radiation at that time. The results are summarized in Table 15.

**Table 15.** Average temperatures on the individual planes for the cooling system.

| Hour | 10:00 | 11:00 | 12:00 | 13:00 | 14:00 | 15:00 | 16:00 | 17:00 |
|---|---|---|---|---|---|---|---|---|
| $T_{amb}$, °C | 28.60 | 29.80 | 30.90 | 30.40 | 29.90 | 29.40 | 28.50 | 27.50 |
| $T_{room}$, °C | 27.52 | 29.35 | 30.90 | 30.80 | 29.99 | 29.10 | 27.24 | 25.23 |
| $T_{abs}$, °C | 31.97 | 35.97 | 39.89 | 41.92 | 43.67 | 34.54 | 31.74 | 27.91 |
| $T_s$, °C | 34.57 | 39.00 | 42.88 | 45.87 | 48.61 | 37.36 | 34.45 | 29.87 |

## 5. Conclusions

The carried out simulations indicate the potential of the Trombe wall as a passive heating and cooling system for the climate of Wrocław. The maximum increase in the average air temperature in the analyzed room of 1.52 °C was observed for the solution with argon filled glazing. The air temperature at the storage wall increases in relation to the ambient temperature within the range from 8.04–16.40 °C for the classic solution, from 11.42–24.90 °C for the solution with argon filled glazing, and from 8.02–16.80 °C for the solution with the brick storage wall. The maximum temperature values are reached at the highest solar radiation, which occurs at 13:00.

Air circulation between the room and the air gap was observed, which confirms the heat transfer by convection, and thus proves the correctness of the simulation. The maximum air speed in the room reaches 0.013 m/s for the classic solution and for the solution with an exchanged storage material, while for the version with argon filled glazing, it is equal to 0.020 m/s. These values ensure comfort for residents.

When considering the cooling system, a maximum temperature drop of 2.27 °C was achieved. In contrast, during the hours with the most sunshine, i.e., between 12:00 and 14:00, the highest temperatures of the absorber and the air in the gap are observed, which causes equalization and a slight increase in the average room temperature. This problem can be solved by improving the insulation of the wall or by increasing the air draft by using a solar chimney. The maximum recorded air speed in the gap was equal to 1.219 m/s, while the air speed in the room remained within the range of 0.020–0.245 m/s with an increase to 0.5 m/s around the ducts, which is determined by the inlet speed. These values enable the heat comfort of residents to be achieved.

Based on previous research, it can be concluded that in Polish climatic conditions, it is economically justified to use the simplest design solutions, for which the results of achieved room temperatures are comparable with more expensive technologies.

In further studies, it will be necessary to conduct analyses of the Trombe wall operation throughout the whole year in order to fully analyze the cycle of its charging and discharging.

**Author Contributions:** Conceptualization, J.B. and M.N.; methodology, J.B.; formal analysis, M.N.; investigation, J.B.; writing—original draft preparation, J.B.; writing—review and editing, M.N.; funding acquisition, M.N.

**Funding:** This research was funded by the Ministry of Science and Higher Education in Poland within the grant for Wrocław University of Science and Technology. Project No. 049 M/0014/19.

**Conflicts of Interest:** The authors declare no conflict of interest.

## References

1. Saadatian, O.; Sopian, K.; Lim, C.H.; Asim, N.; Sulaiman, M.Y. Trombe walls: A review of opportunities and challenges in research and development. *Renew. Sustain. Energy Rev.* **2012**, *16*, 6340. [CrossRef]
2. Stazi, F.; Mastrucci, A.; di Perna, C. The behaviour of solar walls in residential buildings with different insulation levels: An experimental and numerical study. *Energy Build.* **2011**, *16*, 217–229. [CrossRef]
3. Richman, R.C.; Pressnail, K.D. A more sustainable curtain wall system: Analytical modeling of the solar dynamic buffer zone (SDBZ) curtain wall. *Build. Environ.* **2009**, *44*, 1–10. [CrossRef]
4. Jaber, S.; Ajib, S. Optimum design of Trombe wall system in mediterranean region. *Sol. Energy* **2011**, *85*, 1891–1898. [CrossRef]
5. Ozbalta, T.G.; Kartal, S. Heat gain through Trombe wall using solar energy in a cold region of Turkey. *Sci. Res. Essays* **2010**, *5*, 2768–2778.
6. Agrawal, B.; Tiwari, G.N. *Building Integrated Photovoltaic Thermal Systems: For Sustainable Developments*; Royal Society of Chemistry: Delhi, India, 2011.
7. Fang, X.; Li, Y. Numerical simulation and sensitivity analysis of lattice passive solar heating walls. *Sol. Energy* **2000**, *69*, 55–66. [CrossRef]
8. Khalifa, A.J.N.; Abbas, E.F. A comparative performance study of some thermal storage materials used for solar space heating. *Energy Build.* **2009**, *41*, 407–415. [CrossRef]
9. Guohui, G. A parametric study of Trombe walls for passive cooling of buildings. *Energy Build.* **1998**, *27*, 37–43.
10. Zrikem, Z.; Bilgen, E. Theoretical study of a composite Trombe–Michel wall solar collector system. *Sol. Energy* **1987**, *39*, 409–419. [CrossRef]
11. Liping, W.; Angui, L. A numerical study of Trombe wall for enhancing stack ventilation in buildings. In Proceedings of the 23rd International Conference on Passive and Low Energy Architecture, Geneva, Switzerland, 6–8 September 2006.
12. Stazi, F.; Mastrucci, A.; di Perna, C. Trombe wall management in summer conditions: An experimental study. *Sol. Energy* **2012**, *86*, 2839–2851. [CrossRef]
13. Soussi, M.; Balghouthi, M.; Guizani, A. Energy performance analysis of a solar-cooled building in Tunisia: Passive strategies impact and improvement techniques. *Energy Build.* **2013**, *67*, 374–386. [CrossRef]
14. Hong, X.; He, W.; Hu, Z.; Wang, C.; Ji, J. Three-dimensional simulation on the thermal performance of a novel Trombe wall with venetian blind structure. *Energy Build.* **2015**, *89*, 32–38. [CrossRef]
15. Chel, A.; Nayak, J.K.; Kaushik, G. Energy conservation in honey storage building using Trombe wall. *Energy Build.* **2008**, *40*, 1643–1650. [CrossRef]

16. Ji, J.; Yi, H.; He, W.; Pei, G. PV-trombe wall design for buildings in composite climates. *J. Sol. Energy Eng.* **2007**, *129*, 431–437. [CrossRef]

17. Chen, B.; Chen, X.; Ding, Y.H.; Jia, X. Shading effects on the winter thermal performance of the Trombe wall air gap: An experimental study in Dalian. *Renew. Energy* **2006**, *31*, 1961–1971. [CrossRef]

18. He, W.; Hu, Z.; Luo, B.; Hong, X.; Sun, W.; Ji, J. The thermal behavior of Trombe wall system with venetian blind: An experimental and numerical study. *Energy Build.* **2015**, *104*, 395–404. [CrossRef]

19. Mohamad, A.; Taler, J.; Ocłoń, P. Trombe Wall Utilization for Cold and Hot Climate Conditions. *Energies* **2019**, *12*, 285. [CrossRef]

20. Tyagi, V.V.; Buddhi, D. PCM thermal storage in buildings: A state of art. *Renew. Sustain. Energy Rev.* **2007**, *11*, 1146–1166. [CrossRef]

21. Hordeski, M.F. *New Technologies for Energy Efficiency*; The Fair-mont Press: New York, NY, USA, 2011.

22. NREL. Building a Better Trombe Wall. Colorado: Department of Energy's Premier Laboratory for Renewable Energy & Energy Efficiency Research, Development and Deployment. 2012. Available online: https://www.nrel.gov/docs/legosti/fy98/22834.pdf (accessed on 2 June 2019).

23. Melero, S.; Morgado, I.; Neila, F.J.; Acha, C. Passive evaporative cooling by porous ceramic elements integrated in a Trombe wall. *Archit. Sustain. Dev.* **2011**, *2*, 267.

24. Zalewski, L.; Chantant, M.; Lassue, S.; Duthoit, B. Experimental thermal study of a solar wall of composite type. *Energy Build.* **1997**, *25*, 7–18. [CrossRef]

25. Zalewski, L.; Joulin, A.; Lassue, S.; Dutil, Y.; Rousse, D. Experimental study of small-scale solar wall integrating phase change material. *Sol. Energy* **2012**, *86*, 208–219. [CrossRef]

26. Kara, Y.A. Diurnal performance analysis of phase change material walls. *Appl. Therm. Eng.* **2016**, *102*, 1–8. [CrossRef]

27. Tunc, M.; Uysal, M. Passive solar heating of buildings using a fluidized bed plus Trombe wall system. *Appl. Energy* **1991**, *38*, 199–213. [CrossRef]

28. Abbassi, F.; Dehmani, L. Experimental and numerical study on thermal performance of an unvented Trombe wall associated with internal thermal fins. *Energy Build.* **2015**, *105*, 119–128. [CrossRef]

29. Irshada, K.; Habib, K.; Thirumalaiswamy, N. Performance evaluation of PV-Trombe wall for sustainable building development. *Procedia CIRP* **2015**, *26*, 624–629. [CrossRef]

30. Sun, W.; Ji, J.; Luo, C.; He, W. Performance of PV-Trombe wall in winter correlated with south facade design. *Appl. Energy* **2011**, *88*, 224–231. [CrossRef]

31. Yu, B.; He, W.; Li, N.; Wang, L.; Cai, J.; Chen, H.; Ji, J.; Xu, G. Experimental and numerical performance analysis of a TC-Trombe wall. *Appl. Energy* **2017**, *206*, 70–82. [CrossRef]

32. Rabani, M.; Kalantar, V.; Dehghan, A.A.; Faghih, A.K. Empirical investigation of the cooling performance of a new designed Trombe wall in combination with solar chimney and water spraying system. *Energy Build.* **2015**, *102*, 45–47. [CrossRef]

33. Website of the Republic of Poland, Data for A the Typical Meteorogical Year for Wrocław. Available online: https://www.gov.pl (accessed on 2 June 2019).

34. Website of Heat Pump Producer SOLIS. Available online: http://solis.pl/index.php/projektowanie_instalacji/instalacje_zrodla/temperatury_gruntu_w_zaleznosci_od_pory_roku_i_glebokosc (accessed on 2 June 2019).

35. Fluent, A.N. *Ansys Fluent Theory Guide*; ANSYS: Canonsburg, PA, USA, 2012.

36. Bajc, T.; Svorcan, J.; Todorovićc, M.N. CFD analyses for passive house with Trombe wall and impact to energy demand. *Energy Build.* **2015**, *98*, 39–44. [CrossRef]

*Article*

# Simulation Analysis of a Ventilation System in a Smart Broiler Chamber Based on Computational Fluid Dynamics

Shikai Zhang [1], Anlan Ding [1], Xiuguo Zou [1,2,3,4,*], Bo Feng [2,3,4], Xinfa Qiu [2,3,4], Siyu Wang [5], Shixiu Zhang [1,6], Yan Qian [1,6], Heyang Yao [1] and Yuning Wei [1]

[1] College of Engineering, Nanjing Agricultural University, Nanjing 210031, China; shikainjau@126.com (S.Z.); jsdinganlan@163.com (A.D.); zhangshixiu@yeah.net (S.Z.); qianyan@njau.edu.cn (Y.Q.); Charcy_yang@163.com (H.Y.); 18305180897@163.com (Y.W.)

[2] Key Laboratory of Meteorological Disaster, Ministry of Education (KLME), Nanjing University of Information Science and Technology, Nanjing 210044, China; 1191229236@163.com (B.F.); xfqiu135@nuist.edu.cn (X.Q.)

[3] Joint International Research Laboratory of Climate and Environment Change (ILCEC), Nanjing University of Information Science and Technology, Nanjing 210044, China

[4] Collaborative Innovation Center on Forecast and Evaluation of Meteorological Disaster (CIC-FEMD), Nanjing University of Information Science and Technology, Nanjing 210044, China

[5] School of Environmental Science and Engineering, Nanjing University of Information Science and Technology, Nanjing 210044, China; wangsiyu1984@163.com

[6] Jiangsu Province Engineering Laboratory of Modern Facility Agriculture Technology and Equipment, Nanjing 210031, China

* Correspondence: zouxiuguo@njau.edu.cn; Tel.: +86-25-5860-6585

Received: 3 May 2019; Accepted: 5 June 2019; Published: 6 June 2019

**Abstract:** In this paper, a CFD (computational fluid dynamics) numerical calculation was employed to examine whether the ventilation system of the self-designed smart broiler house meets the requirements of cooling and ventilation for the welfare in poultry breeding. The broiler chamber is powered by two negative pressure fans. The fans are designed with different frequencies for the ventilation system according to the specific air temperature in the broiler chamber. The simulation of ventilation in the empty chamber involved five working conditions in this research. The simulation of ventilation in the broiler chamber and the simulation of the age of air were carried out under three working conditions. According to the measured dimensions of the broiler chamber, a three-dimensional model of the broiler chamber was constructed, and then the model was simplified and meshed in ICEM CFD (integrated computer engineering and manufacturing code for computational fluid dynamics). Two models, i.e., the empty chamber mesh model and the chamber mesh model with block model, were imported in the Fluent software for calculation. In the experiment, 15 measurement points were selected to obtain the simulated and measured values of wind velocity. For the acquired data on wind velocity, the root mean square error (RMSE) was 19.1% and the maximum absolute error was 0.27 m/s, which verified the accuracy of the CFD model in simulating the ventilation system of the broiler chamber. The boundary conditions were further applied to the broiler chamber model to simulate the wind velocity and the age of air. The simulation results show that, when the temperature was between 32 and 34 °C, the average wind velocity on the plane of the corresponding broiler chamber (Y = 0.2 m) was higher than 0.8 m/s, which meets the requirement of comfortable breeding. At the lowest frequency of the fan, the oldest age of air was less than 150 s, which meets the basic requirement for broiler chamber design. An optimization idea is proposed for the age of air analysis under three working conditions to improve the structure of this smart broiler chamber.

**Keywords:** smart broiler chamber; ventilation system; wind velocity; age of air; computational fluid dynamics; simulation analysis

---

## 1. Introduction

In summer, a closed broiler house will produce many harmful gases, including ammonia, hydrogen sulfide, carbon monoxide, carbon dioxide, dust, etc. The ventilation system design of a broiler house plays an important role in improving the internal environment of the broiler house (such as temperature and air quality). In a closed broiler house designed with mechanical ventilation, air flow is influenced by the power and quantity of fans; the size, shape, and location of air inlets and outlets; the distribution of temperature; the difference between the indoor and outdoor temperature; and the flow of outside air [1]. In view of the ventilation of a broiler house applying fine breeding, this paper proposes a practical ventilation scheme. The ventilation system of the broiler house should be investigated to examine its design rationality [2].

CFD (computational fluid dynamics) simulation technology is a very effective method to predict and evaluate the performance of ventilation system design [3]. At present, CFD is widely applied to the ventilation simulation of broiler houses. Researchers have extensively employed CFD technology in simulating the environment of livestock houses using two-dimensional models or three-dimensional models under the conditions of natural ventilation or mechanical ventilation [4–6]. Kic et al. [7] used Fluent software to simulate the ventilation in broiler houses in summer and winter. They concluded that the accuracy of three-dimensional simulation was higher than that of two-dimensional simulation.

Some researchers used CFD to compare the strengths and weaknesses of different ventilation schemes for livestock houses and tried to select the optimal solution and further optimize it. Bjerg et al. [8] researched the influence of different modes of air intake on the airflow in a pig house using CFD technology. Norton et al. [9] investigated the improvement of eaves openings of livestock houses by CFD technology. Mostafa et al. [10] simulated and compared four different ventilation schemes of the broiler houses designed with duct ventilation in winter, and concluded that the four schemes could improve the uniformity of indoor wind velocity to 60–70%. Yao et al. [11] used CFD technology to simulate and analyze an airflow problem where a large amount of airflow diffused above the goose house and the ventilation effect on the ground was blocked, and then they proposed an optimized scheme. Li et al. [12] used CFD to simulate and analyze the influence of different opening angles, installation heights, and wind velocity of rectangular air inlets on the airflow distribution in a closed flat broiler house. Coradil [13] used CFD technology to simulate the temperature field in a broiler house heated by a stove.

In order to improve the accuracy of CFD in evaluating the ventilation conditions of a broilers house, some researchers have proposed appropriate settings for different boundary conditions or turbulence models. Blanes-Vidal et al. [14] studied the influence of different boundary conditions on the simulation accuracy in a broiler house, and compared the simulation results with experimental data. Cheng et al. [15] obtained the resistance coefficients under different conditions by comparing the different shapes and sizes of broiler house models, and then simplified the occupied area of caged hens as a porous medium for wind tunnel experiments. They concluded that the result by RNG (the renormalization group) k-ε is more accurate for the full geometry model.Sun et al. [16] applied PIV (particle image velocimetry) technology to the research of a broiler house. Seo et al. [17] simulated the ventilation conditions of the broiler houses with natural ventilation in winter by CFD technology, and improved the original ventilation scheme based on the simulation results. The numerical simulation results shown by the previous researchers have indicated that the relative error of the RNG k-ε turbulence model is the smallest among the turbulence models selected. Therefore, it is effective and feasible to employ CFD simulation technology to optimize the ventilation conditions of existing livestock houses [18].

With the advantages of low cost, high efficiency, and good repeatability, CFD simulation has been widely used in evaluating the ventilation environment of broiler houses. However, researchers have mostly focused on large-scale farms, and there is limited research on the simulation of the ventilation environment of fine breeding broiler house. In this paper, based on the on-site measurement of the structure of a smart broiler house, SolidWorks was used to establish a three-dimensional model of the broiler house. Then, ICEM CFD (the integrated computer engineering and manufacturing code for computational fluid dynamics) was used to mesh the model of the empty chamber and the model of the house with broilers. After that, CFD was introduced into Fluent software to numerically simulate the distribution and velocity of the airflow. The velocity contours of the empty chamber were compared with those of the chamber with broilers. In addition, the ageof airwas also simulated to check the ventilation efficiency of the broiler house. In summary, the design rationality of the ventilation system of the smart broiler house was analyzed and verified in a comprehensive manner.

## 2. Materials and Methods

### 2.1. Research Objects and Measurement Methods

#### 2.1.1. Layout of Experimental Site

Compared with large broilers farms, the small-scale broiler house in this experiment used a variety of sensors to detect the environmental parameters, which is easier to manage and more intelligent than large farms.The broiler house for the experiment is located in the Jinniuhu Subdistrict, Luhe District, Nanjing City, Jiangsu Province. The experiment started on 20 July 2018 and lasted for 21 days. The broiler house consists of two symmetrical chambers. The four walls around the broiler chamber are made of 55 mm-thick colored steel–polystyrene sandwich plates. Each chamber is 1.9 m in width, 2.9 m in length, with a total area of 5.51m$^2$. The roof is slope-shaped with a height of 1.88 m on the west and 1.77 m on the east, which is convenient for draining rainwater. The ventilation system in the chamber consists of an air inlet, an air outlet, and an internal circulation. The internal circulation refers to the part where the openings at both ends are connected through pipes. Fan A and Fan B were arranged at the air outlet and the internal circulation air inlet, respectively. A Pulin Leshi 400 axial flow negative pressure fan was selected, with a rated air volume of 9000 m$^3$/h and theoretical applicable area of 20–35 m$^2$. Frequency conversion controllers were used to change the frequencies at different temperatures to create an optimal ventilation environment for broilers. Since the internal structures of the two broiler chambers were completely consistent, this paper only analyzed the ventilation structure of one chamber. Figure 1 shows the internal environment of the broiler chamber. Figure 2 shows the structure and dimensions of the chamber. Figure 3 presents the three-dimensional chamber model established by SolidWorks.

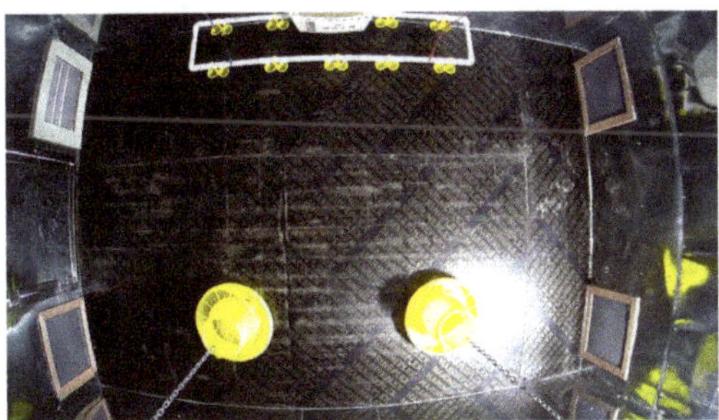

**Figure 1.** Internal picture of the experimental broiler chamber.

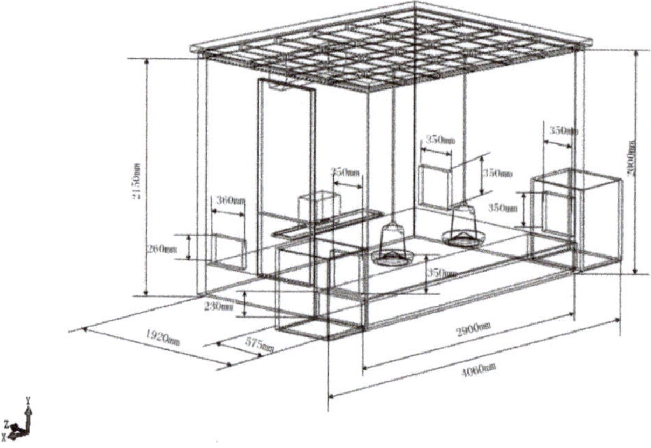

**Figure 2.** The structure and dimensions of the experimental broiler chamber.

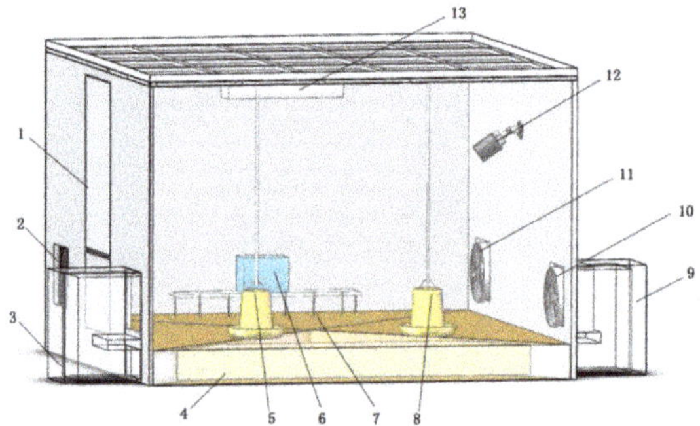

**Figure 3.** The three-dimensional chamber model was established by SolidWorks with (1) chamber door, (2) air inlet, (3) internal circulation front air bellow, (4) internal circulation pipeline, (5) Feeder A, (6) water tank, (7) drinking water pipe, (8) Feeder B, (9) internal circulation rear air bellow, (10) Fan B, (11) Fan A, (12) camera, and (13)air conditioner.

### 2.1.2. Selection of Measurement Points

According to the average height of broilers and a certain active area, the plane of 20cm above the ground was selected. The selected plane was divided into three lines, with five points in each line. Thus, a total of 15 measurement points was set, as shown in Figure 4. The wind velocity was measured using a Testo 405i handheld hot-wire anemometer with a measurement range of 0~30 m/s and a measurement error of ±(0.1m/s+5%). Since the broilers might interfere with the measurement work in the broiler chamber, the measured data were drawn from the empty broiler chamber after the breeding. The velocity of the fan inlet was acquired by taking an average of thenine points at the air inlet.The measurement points are shown in Figure 4b. Table 1 displays the measured wind velocity data of the nine points at the air inlet under three working conditions. The measured wind velocity data were transmitted to a mobile phone through Bluetooth, and they could be read and stored by the mobile phone.

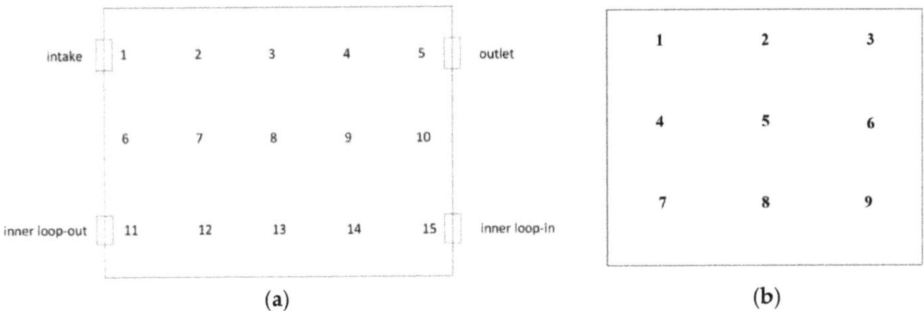


**Figure 4.** Point distribution: (**a**) The plane distribution position of the detection points; (**b**) The inlet distribution position of the detection points.

**Table 1.** The measured wind velocity values of the nine measuring points at the air inlet.

| Measuring Point Number | Measuring Velocity Values (m/s) | | |
|---|---|---|---|
| | Fan A = 50 HZ | Fan A = 30 HZ | Fan A = 20 HZ |
| 1 | 1.67 | 0.81 | 0.27 |
| 2 | 1.74 | 0.83 | 0.34 |
| 3 | 1.64 | 0.79 | 0.31 |
| 4 | 1.69 | 0.93 | 0.37 |
| 5 | 1.97 | 1.15 | 0.47 |
| 6 | 1.73 | 1.07 | 0.39 |
| 7 | 1.78 | 0.87 | 0.29 |
| 8 | 1.80 | 0.94 | 0.32 |
| 9 | 1.79 | 0.92 | 0.30 |
| Average value | 1.76 | 0.92 | 0.34 |
| Standard deviation | 0.091 | 0.11 | 0.05 |

### 2.1.3. Ventilation System

Fan A and Fan B are installed at two different air vents and work independently. They perform frequency modulation and temperature control according to the actual indoor temperature. The specific scheme is shown in Table 2. Since the experiment was conducted in summer in Jiangsu, the temperature at night is generally higher than 24 °C, so the 10 Hz working condition of Fan A is not met and, therefore, the simulation experiment did not study this. Because the broiler chamber is set outside the chamber, the sun shines directly on it, and sometimes the temperature is higher than 34 °C. To ensure the continuity of the experiment, the air conditioner was utilized to avoid high temperatures at which the broiler would have a severe thermal emergency response. At this time, the air conditioner is only used to reduce the indoor temperature, and this paper does not study this working condition. The simulation analysis in this paper is mainly based on five working conditions, as shown in Table 2.

**Table 2.** The experimental plan of fan frequency based on five working conditions. RPM = revolutions per minute.

| Number of Work Condition | Temperature | Fan A | RPM (Fan A) | Fan B | RPM (Fan B) | Air Conditioner |
|---|---|---|---|---|---|---|
| | >34 °C | 0 | 0 | 30 Hz | 840 | On (set 30 °C) |
| 1 | 32~34 °C | 50Hz | 1400 | 50 Hz | 1400 | Off |
| 2 | 30~32 °C | 50Hz | 1400 | 30 Hz | 840 | Off |
| 3 | 28~30 °C | 40Hz | 1120 | 20 Hz | 560 | Off |
| 4 | 26~28 °C | 30Hz | 840 | 10 Hz | 280 | Off |
| 5 | 24~26 °C | 20Hz | 560 | 0 | 0 | Off |
| | <24 °C | 10Hz | 280 | 0 | 0 | Off |

### 2.2. CFD Model

#### 2.2.1. CFD Control Equations

CFD is a kind of numerical simulation under the control of basic flow equations. In the numerical calculation, the air is considered as a continuous, steady, and incompressible Newtonian fluid. The continuity equation is also a mass conservation equation, and any flow movement must satisfy the Law of Conservation of Mass [19].

(1) Mass conservation equation

$$\frac{\partial(\rho u)}{\partial x} + \frac{\partial(\rho v)}{\partial y} + \frac{\partial(\rho w)}{\partial z} = 0 \tag{1}$$

where, $\rho$ is the fluid density, and $u$, $v$, and $w$ are the components of the velocity vector in $x$, $y$, and $z$ directions, respectively.

(2) Momentum conservation equation

$$\frac{\partial(\rho u_i)}{\partial t} + \frac{\partial\left(\rho u_i u_j\right)}{\partial x_j} = -\frac{\partial p}{\partial x_i} + \frac{\partial \tau_{ij}}{\partial x_i} + \rho g_i + F_i \tag{2}$$

where, $p$ is the static pressure on the fluid microelement body, $t$ is time, and $g_i$ and $F_i$ represent the gravity volume force and other external volume forces acting on the microelement body in the $i$ direction, respectively. $\tau_{ij}$ is the viscous stress tensor acting on the surface of the microelement body due to the molecular viscous effect.

#### 2.2.2. Mesh Division of the Broiler Chamber

In order to improve the calculation speed and meshing quality, a three-dimensional model was established in ICEM CFD15.0 according to the coordinates of each point. At first, the 1:1 broiler chamber model was simplified, and the steel frame structure at the top was removed. The internal structure was simplified without affecting the CFD simulation results. Non-structural meshes were selected to mesh the empty broiler chamber. The maximum side length of the mesh was set to 0.05 m. Since the main area of activity of broilers is located at the bottom of the broiler chamber, the ventilation condition on the ground surface is more worthy of our attention. The meshes were increased in density from a 0.6 m-high plane to the bottom of the empty chamber with a total cell number of 7,494,859, and the mesh inspection quality was greater than or equal to 0.34, which meets the calculation requirements. Mesh division of the broiler chamber in ICME CFD is shown in Figure 5.

**Figure 5.** Mesh division of the broiler chamber in ICME CFD (integrated computer engineering and manufacturing code for computational fluid dynamics).

### 2.2.3. Broiler Model

Simplification of Broiler

Cheng et al. [15] simplified a single broiler to three different degrees, including a body only model, a broiler body model, and an ellipsoid model. After simulation and comparison, they finally chose the body only model as the research model. Chen [20] used a block model for CFD research on the broiler chamber. For the model of this project, it was necessary to determine a suitable model. In this paper, 35 seven-week-old yellow feather broilers were studied in the broiler chamber, and the chamber space was small, so geometric modeling could be carried out on a single broiler to improve the simulation accuracy. At the same time, in order to improve the calculation efficiency and accuracy, the broiler was simplified to a certain extent according to the original volume, as shown in Figure 6, (a) shows the body only model, and (b) shows the block model.

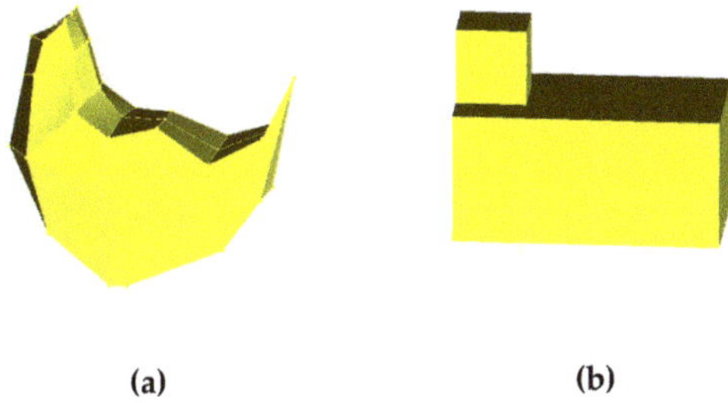

(a)                                (b)

**Figure 6.** Schematic diagram of broiler model: (**a**) body only model, (**b**)block model.

Efficiency Calculation and Selection

As shown in Figure 7, 35 measuring points were randomly selected from the position 15 cm away from the ground for comparison of model simulation results, and the corresponding velocity values of the 35 measuring points in the two models were drawn into the comparison chart of efficiency verification of wind velocity, as shown in Figure 8.

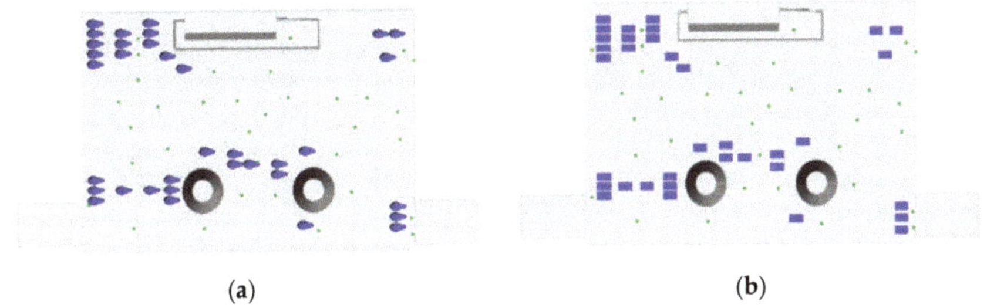

(a)                                (b)

**Figure 7.** Distribution of two models for broilers: (**a**) body only model, (**b**) block model.

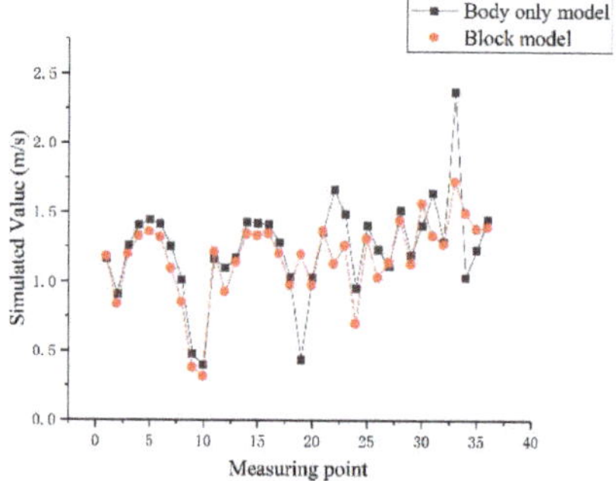


**Figure 8.** Comparison of efficiency verification wind velocity between the body only model and the block model.

From Figure 8, it can be seen that the simulation results of the body only model and the block model have good consistency, and the trend is consistent. The average absolute error between the body only model and the block model was 0.12, and the error between the two sets of simulation values was small, indicating that the impact on the wind field in the broiler chamber was approximate. Table 3 shows the efficiency comparison of different models.

**Table 3.** The efficiency comparison of different models.

| Selected Model | Time (Unit: s) | Number of Cells | Quality of Meshes | Time to Achieve Convergence (Unit: h) |
| --- | --- | --- | --- | --- |
| Empty chamber | 803 | 7,494,859 | >0.35 | 1.35 |
| Block model | 993 | 8,310,070 | >0.34 | 2 |
| Body only model | 4381 | 40,120,931 | >0.2 | 7.5 |

Under the distribution of two different broiler models, the mesh division time and Fluent numerical calculation time were counted, the mesh division number and convergence calculation time length were compared, and an appropriate broiler model was selected for the research in this paper. The computer processors were Intel ®Xeon E7-4830 @2.13 GHz and Intel ®Xeon E5-2660 V4 @2.0 GHz, respectively, with 64.0 GB of installed memory. The mesh division time of the block model and the body only model increased by 0.24 and 4.46 times, and the mesh number increased by 0.11 and 4.35 times, respectively. The mesh quality of the empty chamber was better than the mesh quality of the chamber with the block model, and both of them were better than the mesh quality of the chamber with the body only model. The calculation time of the block model and the body only model using Fluent increased by 1.48 and 5.5 times compared with the empty broiler chamber model. In terms of computational efficiency, the block model of broilers had greater advantages. Therefore, the block model was selected as the prototype to be studied in the next step.

### 2.3. Solver Parameter Settings

After the mesh model was completed, Fluent 15.0 software was used to perform the simulations of the empty broiler chamber, the broiler chamber with the block model, and the age of air in the broiler chamber with broilers under different working conditions. The simultaneous study of PIV and numerical simulation for a broiler house was conducted by Sun. Comparing the results of the two, it was found that the standard mean square error between the simulated values by the RNG k-ε turbulence model and the measured values was less than 0.25 and that the simulated values were more accurate [14]. Cheng et al. [17] explored the effects of flow resistance by considering the geometry of the broiler models (full geometry, ellipsoid, and maternal model in which the head, neck, and legs were ignored) and spatial distribution. In the wind tunnel experiment, five full geometric models were used to represent the broilers, and the model verification was performed using numerical simulation. Different turbulence models were evaluated, and the RNG k-ε model showed better performance than the other models. Specific parameters are given in Table 4.

**Table 4.** Specific parameters of solver that were used in Fluent.

| Parameters | Values |
| --- | --- |
| Simulated state | Steady state |
| Turbulence model | RNG (the renormalization group) k-ε |
| Air density/(kg/m$^3$) | 1.225 |
| Aerodynamic viscosity/(Pa·s) | $1.83 \times 10^{-5}$ |
| Dynamic mesh | Smoothing |

*2.4. Broiler Chambers*

In the simulation of the wind field in the broiler chamber, the location distribution of broilers had different effects on the wind field. From the video, we randomly selected an arrangement photo of the broiler positions, shown in Figure 9a. As the basis for the distribution of broilers, a simulated experimental arrangement of the broiler chamber with broilers was carried out, as shown in Figure 9b. This paper mainly observed the wind velocity and the wind velocity flow field arrangement in the random distribution area of broilers, compared the velocity of the wind field in the empty broiler chamber, and analyzed the influence of broilers on the wind field in the broiler chamber, so as to make better improvements and optimizations for future research.

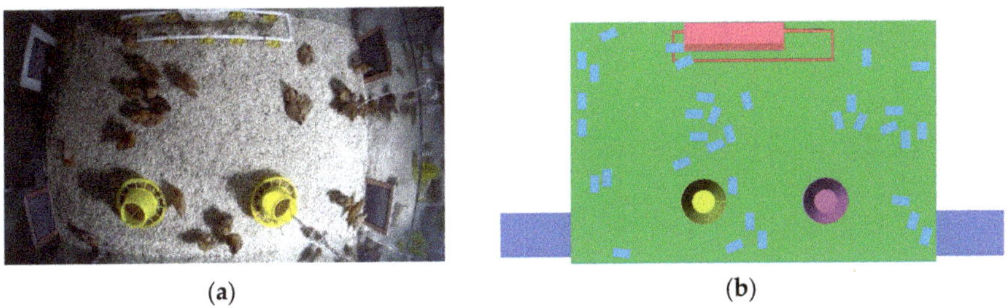

(a)      (b)

**Figure 9.** Experimental arrangement of the chamber with broilers was simulated by the block model. (a) one video frame of broiler distribution, (b) modeling of broiler distribution in ICEM CFD.

*2.5. RMS Error*

RMSE (Root mean square error) is very sensitive to extremely large or small errors in the two sets of data, so it can reflect the approximation of the two sets as well as the extent to which the simulated values deviate from the measured values. RMSE can be calculated by Equation (3).

$$\mathrm{RMSE} = \sqrt{\frac{\sum_{i=1}^{n}\left(X_{obs,i} - X_{model,i}\right)^2}{n}} \tag{3}$$

where, $n = 15$ is the number of measurement points, $i = 1,2, \ldots \ldots n$. $X_{obs,i}$ is the measured value of point $i$, and $X_{model,i}$ is the simulated value.

*2.6. Control Equation of Age of Air*

The age of air refers to the time taken for fresh air to travel from the entrance to each mesh cell, which can reflect the ventilation situation in the chamber and the residence time of airflow, thus reflecting the replacement velocity of indoor fresh air. Therefore, the age of air was used as an index to evaluate the air quality in the broiler chamber in this paper [21]. The control equation of steady-state age of air is given as Equation (4).

$$\nabla(\mu\tau) - \nabla(\Gamma\cdot\nabla\tau) = 1 \tag{4}$$

where, $\mu$ is the velocity of air (m/s), $\tau$ is the age of air (s), $\Gamma$ is the diffusion coefficient, and $\nabla = (\partial/\partial_x, \partial/\partial_y, \partial/\partial_z)$.

## 3. Results and Discussion

### 3.1. Wind Velocity Simulation of Empty Broiler Chamber under Different Working Conditions

#### 3.1.1. Numerical Comparison of Monitoring Points

Figure 10 shows the measured and simulated values of the 15 measurement points in the empty broiler chamber under different working conditions. The simulated values of the 15 measurement points were obtained by the momentum conservation equation in Fluent. Figure 10 shows the results in sequence, from working condition 1 to working condition 5. The maximum absolute error values and RMSE values are shown in Table 5 after the data were exported and compared.

**Table 5.** Error values under different working conditions.

| ErrorType | Fan A 20 Hz<br>Fan B 0 Hz | Fan A 30Hz<br>Fan B 10Hz | Fan A 40Hz<br>Fan B 20Hz | Fan A 50Hz<br>Fan B 30Hz | Fan A 50Hz<br>Fan B 50Hz |
|---|---|---|---|---|---|
| Maximum Absolute Error (Unit: m/s) | 0.1 | 0.15 | 0.24 | 0.24 | 0.27 |
| Root Mean Square Error (Unit: m/s) | 4.9 | 17.6 | 18.2 | 19.1 | 16.6 |

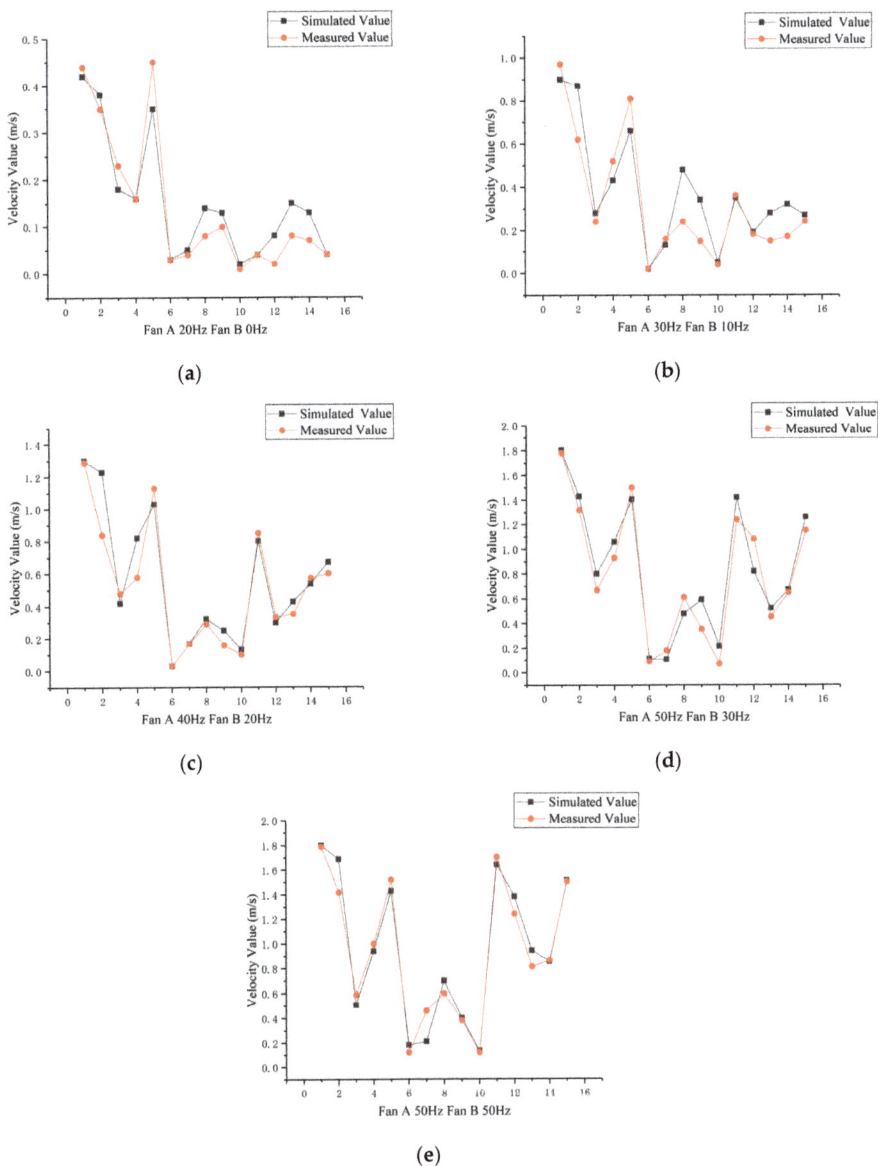

**Figure 10.** Numerical comparison of empty broiler chamber under (**a**)work condition 5; (**b**)work condition 4; (**c**)work condition 3; (**d**)work condition 2; and (**e**)work condition 1.

At the height of Y = 0.2 m, under different working conditions, the difference between the simulated values and the measured values is not large, and the variation rules are consistent. The RMSE in this research was 0.19 m/s, which is close to the result of 0.16 m/s by Yao [16], and the maximum absolute error was 0.27 m/s, which shows that the CFD model and the adopted boundary conditions are suitable for this smart broiler chamber model. Therefore, the boundary conditions corresponding to each working condition were obtained, and then they were applied to a broiler chamber for the simulation experiment.

3.1.2. Analysis of Velocity Contours

Air velocity contours of the empty chamber were arranged in sequence from working condition 1 to working condition 5. As shown in Figure 11, the research of air velocity on the Y-axis plane of the empty chamber was at the height of 0.2 m in order to compare the velocity relationship among different working conditions, the colormap of the contour was unified, and the color in the velocity contour corresponded to the velocity in the colormap. Therefore, the velocity range under the five working conditions was 0~3.5 m/s, and the change rule of air velocity was studied in sequence.

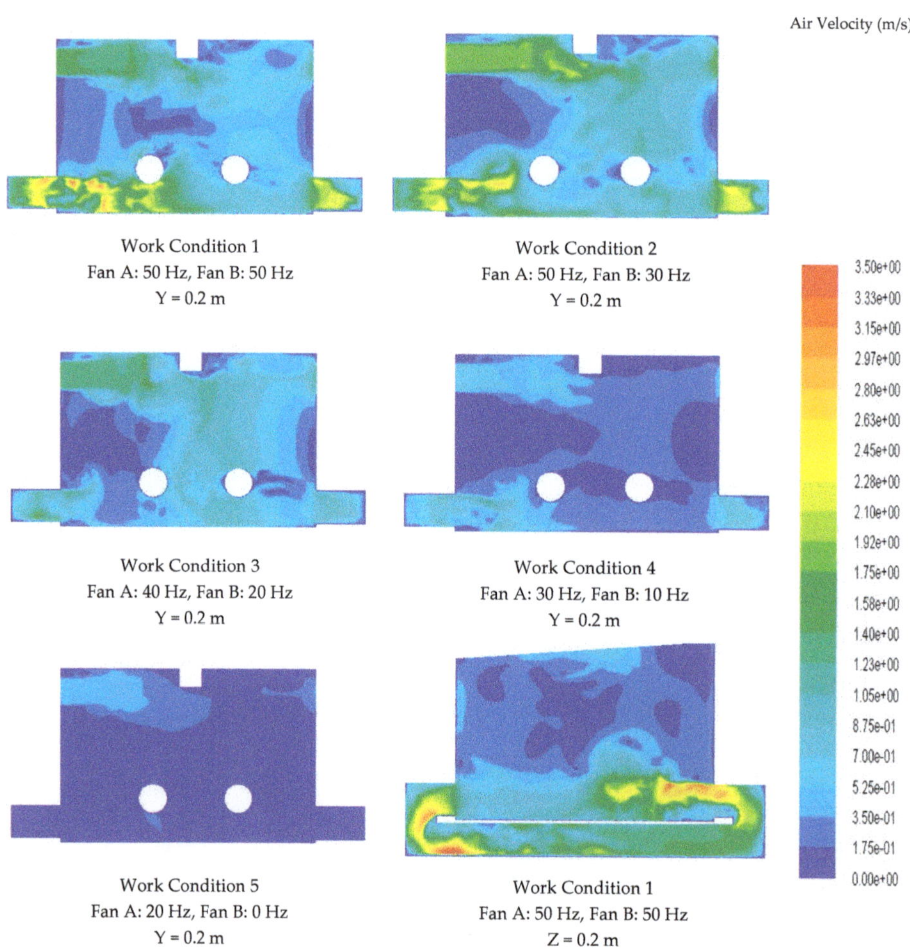


**Figure 11.** Simulated velocity contour of the empty chamber with the colormap on the left.

In Figure 11, according to the contour of the five working conditions, the law of the plane wind velocity field in the broiler chamber at the height of Y = 0.2 m was relatively consistent. The wind velocities at the inlet and outlet were large, and there was obvious convection at the two opposite outlets. The velocity at the inlet was large, and fresh air was fed into the chamber. The five contours all had obvious high-velocity wind areas at about x = 0.6 m from the entrance, which was due to the influence of the water pipe of the water tank, causing the airflow to rise or fall along the pipe wall. Working condition 5 did not open the internal circulation system, and Fan A frequency was low, which can allow a small amount of ventilation. There was no wind at the two corners of the internal circulation side. The wind field uniformity of the broiler chamber was not high, and the suitable living area for broilers was limited. Fan A and Fan B worked together when the internal circulation system was started, and fresh air flowing to the middle of the broiler chamber circulated to the two fan ports. The air flowing through Fan A was directly exhausted from the broiler chamber, while the air flowing through Fan B was blown out from the other side through the internal circulation pipeline. By comparing working conditions 1 and 2, when the frequency of Fan B increased, the weak wind area in the plane could be effectively reduced.

Under the working condition 1, the plane with Z = 0.2 m in the broiler chamber shows the contour of the wind velocity field on the internal circulation vertical plane. The wind was drawn into the ventilation duct from Fan B, and the airflow direction at the wall near the outlet side formed a certain angle with the wall, which made the wind vector line at the wall denser and improved the wind velocity at the top of the inlet wall. Therefore, a larger vortex was formed in the broiler chamber along the X direction, so that the upper airflow could be driven to the lower part to exit the broiler chamber through airflow circulation. Under these five working conditions, the higher the internal circulation frequency, the more uniform the wind velocity field as a whole, indicating that the design of internal circulation has a great effect on the wind-velocity uniform distribution of the broiler chamber.

*3.2. Simulation of the Broiler Chamber with Broilers under Three Common Working Conditions*

Since the simulated and measured values of different working conditions under the empty broiler chamber were close to each other, the determined boundary conditions under the empty broiler chamber were applied to the model of the broiler chamber with broilers. The temperature of the broiler chamber in which broilers are put was relatively higher than that of the empty chamber, so the working conditions 1 and 2 were commonly used during the day in summer and the working condition 4 was used at night. The following is a study on the wind velocity field of the broiler chamber with broilers under three common working conditions. The wind velocity contours are shown in Figure 12.

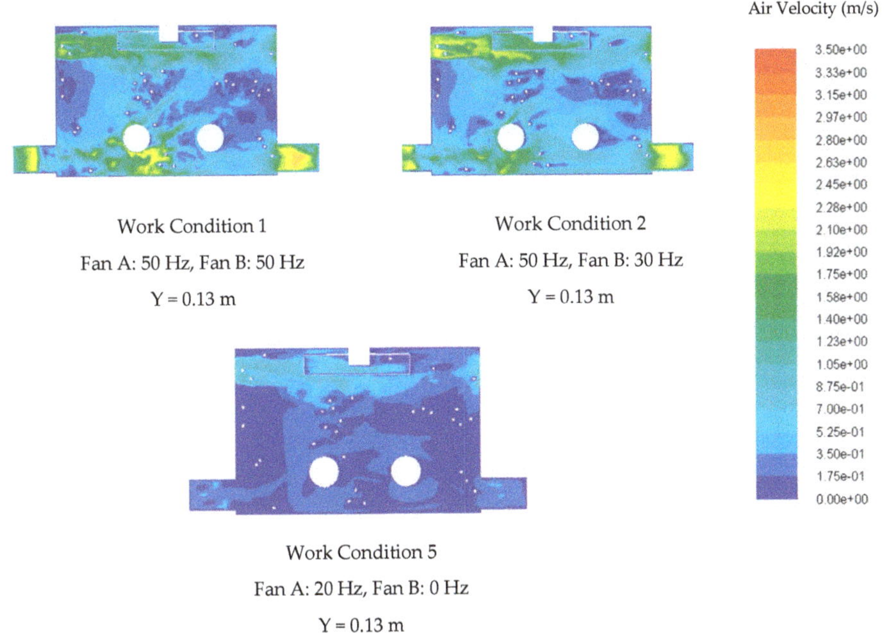


**Figure 12.** Simulated velocity contours of the broiler chamber with broilers.

The apparent temperature of broilers was generally analyzed by measuring the average skin temperature of side face, ears, comb, and lower leg, so the height of the velocity contour was selected as Y = 0.13 m (the average height of the middle part of the head in the broiler chamber). In the study from Yahav, the most suitable wind velocity was 2.0 m/s under a high-temperature environment (35 ± 1.0 °C). At that wind velocity, the body temperature was the lowest, and overhigh (3.0 m/s) or overlow (0.8m/s) wind velocities will raise the body temperature [21]. Zhang et al. [22] conducted experiments on 42-day-old broilers at 26 °C, 29 °C, and 32 °C with 0, 0.5 m/s, 1m/s, 1.5m/s, and 2.0m/s wind velocities. Under the condition of relatively high temperatures, considering the ventilation benefits, the optimum wind velocity was 1.2 m/s. According to Figure 12, the contours of working condition 2 and working condition 1 were more uniform in the wind velocity field due to the shielding and guiding effects of broilers on airflow compared with the contours of the same working condition of the empty chamber. The wind velocity at the vent can reach 2 m/s, which is favorable for the rapid emission of high-temperature gas. Therefore, when the weather temperature reaches above 30 °C, it is more suitable for broilers to gather near the vent. Most of the wind velocity values in the broiler chamber were above 0.875 m/s, which is close to the most suitable wind velocity, and there was almost no windless area, thus ensuring the normal living environment of broilers and avoiding a heat stress reaction. Working condition 5 corresponds to a temperature range of 24~26 °C. This condition mainly occurs at night when the air temperature is suitable for broilers and the fan is not necessary to increase the wind speed to cool the broilers, but needs to achieve continuous ventilation of the house.

### 3.3. Simulation of the Age of Air

The good ventilation environment in the broiler chamber not only cools the surface of the broiler and makes it feel comfortable, but also needs to provide sufficient fresh air to the broiler chamber. Therefore, the age of air under the broiler chamber was simulated. The age of air with Y = 0.2 m (broiler living plane), Z = 0.2 m (vertical section of internal circulation), and Z = 1.65 m (vertical section of air inlet and outlet) were simulated under three working conditions in turn. The contours of the age of air are shown in Figure 13.

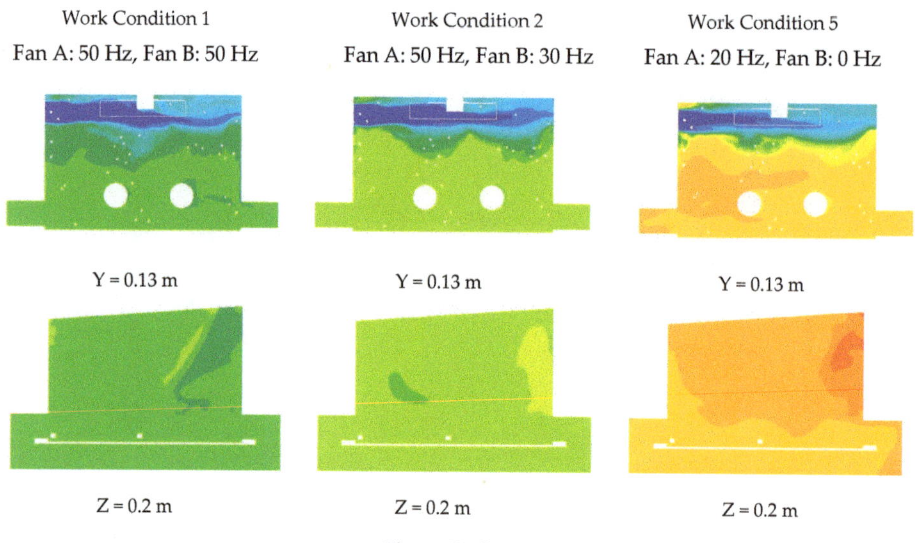


**Figure 13.** *Cont.*

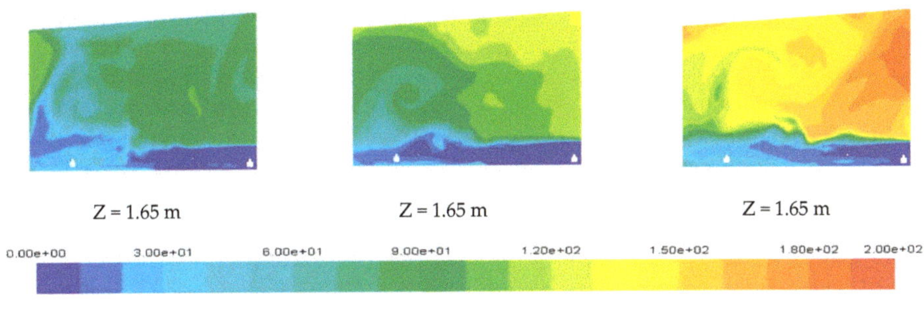



**Figure 13.** The contours of the age of air in the chamber with broilers.

In Figure 13, according to the contours, the ages of air were as follows: Working Condition 1 < Working Condition 2 < Working Condition 5, which proves that the increase of fan frequency greatly improves the ventilation efficiency of the broiler chamber. It can be seen from the contours of the plane at Y = 0.13 m that the convection area of the inlet was contrasted with the inner side. The ages of air at the inlet and outlet were younger, and fresh air changed faster. The age of air inside the broiler chamber was older than that at the air duct. Working conditions 1 and 2 need to use internal circulation. The age of air in the middle of the broiler chamber was younger than that in working condition 5. The internal circulation promotes the air circulation in the plane. The fresh air with the Z = 0.2 m section first reached the outlet and then reached the internal circulation outlet. The wall corner of the internal circulation inlet continued to move in the two directions of the internal circulation pipeline and wall climbing. The age of air was old, and there were vortexes and weak wind velocity, which is not conducive to the replacement of fresh air. The wall near the internal circulation inlet had an older age of air. At the bottom of section Z = 1.65 m, the air inlet and outlet realized convection ventilation, which was the most efficient area for ventilation in the broiler chamber. Through velocity analysis, the wind velocity was relatively high, and the vortex was generated upward through wall–floor collision, which made a great contribution to the renewal of the upper air.

Yu et al. [23] analyzed the age of air of the cabin kitchen with a length of 5.6 m and a width of 3 m. The air in the cabin was highly disturbed, and the air freshness was higher. Our results showed that the age was within 60 s, the oldest age of air was 255 s, which appeared in the corner, and the age of air in the air inlet and outlet of the broiler chamber was also within 60 s. The oldest age of air in the broiler living area was about 100 s under the working condition 1 and 150~180 s under the working condition 4, which was within a reasonable range. This showed that this ventilation system was conducive to timely updating of the fresh air inside the broiler chamber and to improving the living environment of the broilers.

## 4. Conclusions

The conclusions of this research are summarized as follows:

(1) Through actual measurement, the three-dimensional model of the broiler chamber was established and simulated in Fluent software. The comparison between the simulation results and the actual measurement results showed that the RMSE of the wind velocity at each measurement point was up to 19.1%, and the maximum absolute error was 0.27 m/s, which shows the effectiveness of the CFD model.

(2) The wind velocity and the age of air in the broiler chamber were simulated and analyzed. The design of setting different frequencies for the fan met the requirement of setting appropriate temperatures in different periods. In the high-temperature area (above 30 °C), the wind velocity at the air inlet and outlet was close to 2 m/s, which is beneficial to the cooling of broilers. The internal wind velocity could be regulated above 0.8 m/s, which meets the ventilation requirement in summer. The

age of air of the broiler chamber could be renewed within about 100s at the bottom of the chamber in a relatively high-temperature environment. When starting the working condition 5 at night, ventilation could be carried out in all parts of the chamber within 150~180 s. Compared with the residential ventilation using fans, the age of air of the broiler chamber in this research was much younger, so the design of the ventilation system in the broiler chamber is superior to the common residential standard.

(3) The height of the air inlet and the center of the fan in the broiler chamber were set to be the same as those in the empty chamber. When the broiler chamber is mechanically ventilated, a large amount of airflow circulates in the lower part, effectively improving the ventilation and cooling effect. As a result, more airflow passes through the surface of the broiler chamber. As the density of polluted gas is small, the polluted gas mostly gathers at the top, and needs to be circulated to the bottom through the vortex and then blown out from the air outlet, thus reducing the efficiency of pollutant emission. Based on this design, an exhaust fan can be applied to the upper part of the broiler chamber to improve the efficiency of pollutant emission.

**Author Contributions:** X.Z. and X.Q. conceived and designed the experiments; X.Z., S.Z. (Shikai Zhang), A.D., B.F., S.Z. (Shixiu Zhang), and Y.Q. performed the experiments and analyzed the data; S.W., H.Y., and Y.W. helped perform the data analysis; S.Z. (Shikai Zhang), A.D., and X.Z. wrote the paper.

**Funding:** This research was funded by the Fundamental Research Funds for the Central Universities of China (KYTZ201661), China Postdoctoral Science Foundation (2015M571782), Jiangsu Agricultural Machinery Foundation (GXZ14002), and University Student Entrepreneurship Practice Program of Jiangsu Province (No. 201810307010P).



## References

1. Park, G.; Lee, I.; Yeo, U.; Ha, T.; Kim, R.; Lee, S. Ventilation rate formula for mechanically ventilated broiler houses considering aerodynamics and ventilation operating conditions. *Biosyst. Eng.* **2018**, *175*, 82–95. [CrossRef]

2. Yao, H.; Sun, Q.; Zou, X.; Wang, S.; Zhang, S.; Zhang, S.; Zhang, S. Research of yellow-feather chicken breeding model based on small chicken chamber. *Agric. Eng.* **2018**, *56*, 91–100.

3. Du, L.; Yang, C.; Dominy, R.; Yang, L.; Hu, C.; Du, H.; Li, Q.; Yu, C.; Xie, L.; Jiang, X. Computational fluid dynamics aided investigation and optimization of a tunnel-ventilated poultry house in china. *Comput. Electron. Agric.* **2019**, *159*, 1–15. [CrossRef]

4. LEE, I.-B.; SASE, S.; SUNG, S.-H. Evaluation of CFD accuracy for the ventilation study of a naturally ventilated broiler house. *Jpn Agric. Res. Q.* **2007**, *41*, 53–64. [CrossRef]

5. Seo, I.H.; Lee, I.B.; Moon, O.K.; Kim, H.T.; Hwang, H.S.; Hong, S.W.; Bitog, J.P.; Yoo, J.I.; Kwon, K.S.; Kim, Y.H.; et al. Improvement of the ventilation system of a naturally ventilated broiler house in the cold season using computational simulations. *Biosyst. Eng.* **2009**, *104*, 106–117. [CrossRef]

6. Cheng, Q.; Li, H.; Rong, L.; Feng, X.; Zhang, G.; Li, B. Using CFD to assess the influence of ceiling deflector design on airflow distribution in hen house with tunnel ventilation. *Comput. Electron. Agric.* **2018**, *151*, 165–174. [CrossRef]

7. Kic, P.; Zajicek, M. A numerical cfd method for the broiler house ventilation analysis. *J. Inf. Technology in Agric.* **2011**, *1*, 1–7.

8. Bjerg, B.; Svidt, K.; Zhang, G.; Morsing, S.; Johnsen, J.O. Modeling of air inlets in cfd prediction of airflow in ventilated animal houses. *Comput. Electron. Agric.* **2002**, *34*, 223–235. [CrossRef]

9. Norton, T.; Grant, J.; Fallon, R.; Sun, D. Improving the representation of thermal boundary conditions of livestock during CFD modelling of the indoor environment. *Comput. Electron. Agric.* **2010**, *73*, 17–36. [CrossRef]

10. Mostafa, E.; Lee, I.; Song, S.; Kwon, K.; Seo, I.; Hong, S.; Hwang, H.-S.; Bitog, J.P.; Han, H.-T. Computational fluid dynamics simulation of air temperature distribution inside broiler building fitted with duct ventilation system. *Biosyst. Eng.* **2012**, *112*, 293–303. [CrossRef]

11. Yao, J.; Guo, B.; Ding, W.; Shao, X.; Shi, Z. Structure optimization and validation of goose house ventilation system based on airflow field simulation by CFD. *Trans. Chin. Soc. Agric. Eng.* **2017**, *33*, 214–220.

12. Li, W.; Shi, Z.; Wang, C. Numerical simulation of tunnel ventilation system of plane-raising enclosed henhouse. *J. China Agric. Univ.* **2007**, *12*, 80–84.

13. Coradi, P.C.; Costa, D.R.D.; Visser, E.M.; Martins, M.A.; Tinôco, I.D.F.F. Thermal comfort evaluation in a broiler house utilizing computational fluid dynamics (CFD) software. In Proceedings of the CIGR International Conference of Agricultural Engineering, Viçosa, Brazil, 3 August 2008.

14. Blanes-Vidal, V.; Guijarro, E.; Balasch, S.; Torres, A.G. Application of computational fluid dynamics to the prediction of airflow in a mechanically ventilated commercial poultry building. *Biosyst. Eng.* **2008**, *100*, 105–116. [CrossRef]

15. Cheng, Q.; Wu, W.; Li, H.; Zhang, G.; Li, B. CFD study of the influence of laying hen geometry, distribution and weight on airflow resistance. *Comput. Electron. Agric.* **2018**, *144*, 181–189. [CrossRef]

16. Sun, H.; Zhao, L.; Zhang, Y. Evaluating RNGk-εmodels using PIV data for airflow in animal buildings at different ventilation rates. *ASHRAE* **2007**, *113*, 358–365.

17. Seo, I.; Lee, I.; Moon, O.; Hong, S.; Hwang, H.; Bitog, J.P.; Kwon, K.; Ye, Z.; Lee, J. Modelling of internal environmental conditions in a full-scale commercial pig house containing animals. *Biosyst. Eng.* **2012**, *111*, 91–106. [CrossRef]

18. Zajicek, M.; Kic, P. Improvement of the broiler house ventilation using the CFD simulation. *Agron. Res.* **2012**, *1*, 235–242.

19. Khosravi Nikou, M.R.; Ehsani, M.R. Turbulence models application on CFD simulation of hydrodynamics, heat and mass transfer in a structured packing. *Int.Commun. Heat Mass Transf.* **2008**, *35*, 1211–1219. [CrossRef]

20. Jinru, C. Study of Indoor Environment Dynamic Simulation of Layer-House Based on the Model of Caged Pens of Porous Media. Master's Thesis, Hebei University of Technology, Wuhan, China, 2015.

21. Damin, Z.; Xiugan, Y. Numerical solution of age of air. *J. Beijing Univ. Aeronaut. Astronaut.* **1997**, *23*, 565–570.

22. Shaoshuai, Z.; Huajie, D.; Minhong, Z.; Jinghai, F.; Ying, Z.; Meng, L.; Xiumei, L. Effects of air velocity and moderate ambient temperatures on physiological, endocrine and immune indices of broilers. *Chin. J. Anim.Nutr.* **2017**, *29*, 69–79.

23. Yu, N. Numerical simulation of the air distribution of typical ship kitchen ventilation system. *Ship Ocean Eng.* **2018**, *47*, 89–92.

*Article*

# An IoT Integrated Tool to Enhance User Awareness on Energy Consumption in Residential Buildings

**Marco Dell'Isola [1], Giorgio Ficco [1], Laura Canale [1,*], Boris Igor Palella [2] and Giovanni Puglisi [3]**

[1]   Dipartimento di Ingegneria Civile e Meccanica, Università degli Studi di Cassino e del Lazio Meridionale, Via G. Marconi 10, 03043 Cassino, Italy; dellisola@unicas.it (M.D.); ficco@unicas.it (G.F.)

[2]   Dipartimento di Ingegneria Industriale Piazzale Vincenzo Tecchio 80, Università degli Studi di Napoli Federico II, 80125 Naples, Italy; palella@unina.it

[3]   ENEA Centro della Casaccia, Via Anguillarese 301, 00123 Rome, Italy; giovanni.puglisi@enea.it

*   Correspondence: l.canale@unicas.it

Received: 23 October 2019; Accepted: 23 November 2019; Published: 26 November 2019

**Abstract:** Unaware behaviors of occupants can affect energy consumption even more than incorrect installations and building envelope inefficiencies, with significant overconsumptions widely documented. Real time data and an effective and frequent billing of actual consumptions are required to reach an adequate awareness of energy consumption. From this point of view, the European Directive 2012/27/EU already imposed the use of metering and sub-metering systems, setting the minimum criteria for billing and related information based on real energy consumption data. To assess the ability of buildings to exploit new information and communication technologies (ICT) and sensitize both landlords and tenants to related savings, the new European Directive 2018/844/EU promotes the use of a smart readiness indicator. At the same time, basic information about indoor thermal comfort should be also gathered, aimed at avoiding that an excessive saving tendency can determine the onset of issues related to excessively low internal temperatures. In this paper, the authors address the problem of gathering, processing, and transmitting energy consumption and basic indoor air temperature data in the framework of an Internet of Things (IoT) integrated tool aimed at increasing residential user awareness through the use of consumption and benchmark indexes. Two case-studies in which thermal and electrical energy monitoring systems have been tested are presented and discussed. Finally, the suitability of the communication of energy consumption in terms of temporal, spatial, and typological aggregation has been evaluated.

**Keywords:** user awareness; energy consumption; individual metering; feedback strategies; N-ZEB; IoT

---

## 1. Introduction

Encouraging energy savings in residential buildings has been a topic of scientific interest since the 1970s, when the energy crisis made people aware of the possible exhaustion of fossil fuels. Almost 50 years later, it is clear that all intervention addressed to improve the energy efficiency of buildings should be combined with actions aimed at increasing the awareness and participation of end users, also through more frequent and detailed information on energy consumption [1]. In the absence of frequent information, two buildings with similar thermo-physical characteristics and energy performances, even designed consistently with N-ZEB (Net-Zero Energy Buildings) criteria, can consume one twice the other depending upon the occupants' behavior [2]. In recent years, smart devices and information and communication technologies (ICT) allowed the possibility to set up integrated systems to support decisions at the building, district, and city stages, but they are not common owing to the complexity of the problem, the reduced interoperability among the different systems, and high costs [3]. In addition, the effectiveness of user's feedback actions addressed at the energy savings is a still debated topic in the scientific literature.

On the basis of the results of 38 different studies addressing the effectiveness of interventions aimed at encouraging families to reduce their energy consumption [4], two macro-categories are identified, depending on the kind of information provided to families: (i) antecedent strategies; (ii) consequent strategies. Antecedent strategies include media campaigns, workshops, educational conferences, and energy audits for targeted and personalized information [5,6]. It is proven that antecedent strategies raise user awareness, but do not necessarily lead to behavioral changes or sure energy savings. This category includes any type of user feedback for example, real-time feedback, information presented via in-home displays, mobile apps, or online services [7–9]. Feedback actions can be direct, when learned directly from the instrument display (meter, sub-meter, and so on) or indirect, when information on data consumption is preliminary processed before reaching the user. The inconsistencies in behaviors related to the use of energy in families are the result of the following reasons [10]: (i) temporal coherence of decisions, (ii) difficulty in processing consumption data and in assuming simple decisions, and (iii) effects of presentation.

The available literature identifies three key problems related to the feedback: (i) the poor evidence of effectiveness, (ii) the need for involving users, and (iii) the potential occurrence of unwanted consequences. The main finding is that actual in-home displays could not be effective in orientating users' behaviors. Thus, it is necessary to develop and test novel feedback devices accounting for the degree of user involvement [11]. In a recent experimental campaign [12], many interviewed users reported difficulties in the interpretation of the units (kW, kWh) and poor feedback (e.g., lack in the corresponding economic value). This research also highlighted the usefulness of presenting disaggregated data for each device (sub-metering), at least for the most energy-consuming devices (stove, oven, dishwasher, washing machine, dryer, and so on) and of benchmarks with historical consumption.

In this scenario, low-income families, such as those living in public housing, are a particular category of users to be approached in a specific way. A recent experimental study in seven European Union (EU) countries highlighted several problems both in the implementation of smart-metering solutions and in the use of personalized feedback for low-income families in the Mediterranean region [13]. The experimental results proved that the use of smart-meters associated with in-home displays is not so effective. On the other hand, the monitoring of individual electrical devices, the allocation of consumption inside the dwelling, and suggestions for energy retrofits are appreciated. The joint implementation of these measures and the personalization of user feedback resulted in electricity consumption savings varying in the range from 22% to 27% [14]. Unfortunately, the adoption of energy saving strategies in social housing could lead to a potential worsening of comfort conditions [15]. For example, the reduction of the average winter indoor air temperature could result in condensation phenomena and mold. In the same research paper, the authors also point out that providing end-users with information about their energy consumption is more effective when people live in relatively energy-efficient dwellings, but is less useful for users living in public housing.

Rating the performances of energy devices and benchmarking energy consumption is a highly debated topic, as they could represent effective tools to help users in managing their energy costs. In an interesting guidance on energy consumption benchmarks on residential customers of the Australian Energy Regulator [16], indications about displaying the most appropriate electricity consumption benchmarks are given. In the same document, it is suggested that benchmarks tailored on household size, climatic zone, type of heating system, seasonal factors, and so on are more effective and better explain the variability of users' energy consumption. Benchmarking methodologies can be classified in four different categories [17,18]: (i) regression model-based, requiring extensive data sets of buildings having similar characteristics; (ii) points-based rating system, comparing the measured energy consumption to best practice standards; (iii) simulation model-based, in which energy consumption is modeled in the building energy simulation environment; and (iv) hierarchal and end-use metrics, in which set of performance metrics is developed for the underlying system performance data. Kavousian et al. [19] presented a method to rank residential buildings based on their appliance energy efficiency, as the definition of a scale indicating the position within a distribution is considered to be more definitive than a comparison to a simple value. A literature review on up to date energy benchmarking methods and their performance levels is also provided in the work of [20], where twelve methods for benchmarking building energy consumption, including six black box methods, two gray box methods, and four white box methods, are analyzed, highlighting that many methods, although simple, can still achieve satisfactory performance. In the work of [18], research projects, tools, and programs focused on energy benchmarking methods, energy rating procedures, and classification schemes for the building sector are also reviewed and discussed.

Moreover, these technologies may also result in a series of issues related to the access to confidential information on users' activities and habits, privacy, confidentiality, and availability of data (also considering the greater quantity and vulnerability of data). This is for authorized parties (e.g., utility companies, metering companies); unauthorized parties (e.g., competitors, thieves, real estate owners); and, finally, for final users who are often unable to access and use their own data. In addition, the remote billing brings up problems related to data security and integrity (e.g., the risk of deletion/modification of information). Different privacy preservation techniques may be based on information theory, multiple source energy engineering, and cryptographic network protocols.

It is thus clear that, to achieve energy saving through more frequent and detailed information to the user via individual metering devices and smart technologies, a strategic feedback design is needed.

In this framework, the paper is aimed at addressing the problem of measuring, processing, and transmitting energy consumption data to final users through a suitably designed feedback strategy, without the intention of investigating deeper the social/behavioral causes underlying the eventual energy consumption variation. The development of tailored benchmark indicators applicable to metering and sub-metering in residential buildings is designed and proposed with the aim of making the information about energy consumption to the user more understandable. Three case-studies for thermal and electrical energy are presented and discussed. Finally, the communication strategy about energy consumption in terms of temporal, spatial, and typological aggregation is evaluated.

## 2. Architecture and Data Transmission

The integrated Internet of Things (IoT) tool developed by the authors is based on three levels. The first level is represented by metering and sub-metering systems for gathering energy consumption data of electrical, thermal, and natural gas devices (nodes) of the relative plants. The second level is the data concentration by wireless personal area networks (ZigBee protocol) and remote transmission data with the home router connected to the Internet. Smart meters may also directly communicate with the cloud. The third level is the web-based data management providing parallel solutions for data entry, storage, analysis, and processing. In particular, in this latter level, data for user feedback are processed by creating reports (e.g., indirect feedback), as well as real-time displaying via dashboard (e.g., direct feedback). Therefore, the IoT tool combines and stores information and data, as follows:

- the measurement module, which collects data from different sources (electric, thermal, and gas energy consumption and production);
- the configuration module, which collects data from the different sources (i.e., energy prices, weather data, and end-users' behavior);

Figure 1 shows a simplified sketch of the developed IoT integrated tool.

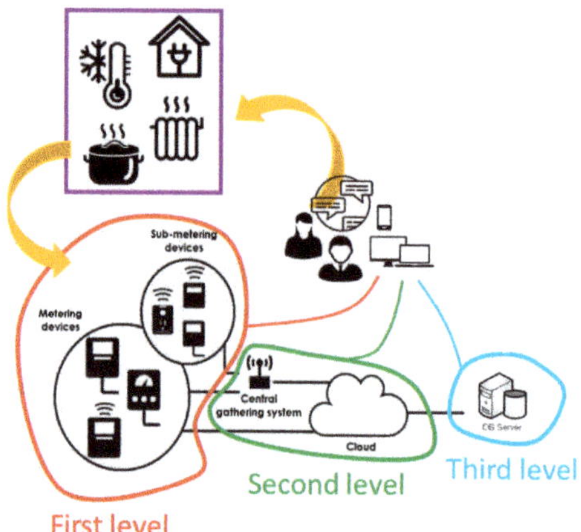


**Figure 1.** Internet of Things (IoT) integrated tool scheme.

In the following, for the sake of completeness, the metering and sub-metering systems generally employed in measuring energy consumptions within residential houses and buildings are further detailed.

*Level 1: Smart Metering and Sub-Metering Systems*

As well known, residential buildings are supplied by several energy carriers and their technical plants provide the energy necessary for the fundamental activities performed. For simplicity reasons, for classification of smart metering and sub-metering devices, the authors will refer to the "typical" (i.e., more widespread) configuration of technical plants supplying residential buildings in Italy [21,22]:

(i) heating system, consisting of a centralized or autonomous boiler supplied by natural gas for the production of hot water and of heating, through a hydronic system with radiators/fan coils emitters;

(ii) electrical system, for the electricity consumed inside the apartment by all the installed electrical appliances and devices and lighting systems;

(iii) natural gas system, supplying the main boiler and the cooking appliances.

Following the above-described classification, metering and sub-metering of the heating system allow the control and monitoring of energy consumption of heating and domestic hot water [23–28]. At the metering level, the smart direct thermal energy meter directly measures the consumptions of the boiler for heating and hot water production services at the building level. On the sub-metering level, the energy consumed by each apartment could be allocated through (smaller) thermal energy meters in the case of ring configuration of the plant. Otherwise, consumptions of single radiators, fan coils, or generic heating elements, when a vertical raising main configuration of the plant is present, are estimated through the so-called indirect accounting devices, such as two-sensors electronic heat cost allocators or insertion time counters. Additionally, the sub-metering of domestic hot water can be performed through a direct thermal energy meter or, alternatively, through a water meter suitable for hot water measurements.

Referring to the electrical system, different measuring devices are available at the metering level: smart electric energy meters, generally made up of a static type sensor with an associated processing system, are employed for fiscal purposes (i.e., metering and billing), but smart non-fiscal devices for monitoring current flows via the electric system phase are also available on the market. Electricity sub-metering multifunction devices also allow to monitor and control energy consumption of single appliances (e.g., refrigerator, dishwasher, oven, hairdryer) and they are often associated with the so-called smart plugs.

The natural gas system generally supplies natural gas for cooking purposes as well as to the main autonomous boiler for heating and, in some cases, the production of domestic hot water. Gas smart meters can be employed to measure the total gas volumes supplied to the house/building at the metering level. In this case, the meter is associated with temperature and pressure sensors, whose signals are processed by an electronic calculation module. On the market, "hybrid" smart gas meters equipped with electronic correction/transmission modules and static ultrasonic or thermal mass are also available. For sub-metering functions (e.g., cooking), small domestic gas meter can be used (e.g., class G2.5), but optimal operational conditions should be adequately considered, as the measured flow rates are often very low. Table 1 shows the technical specifications of the metering and sub-metering systems used by the authors in the case studies.

**Table 1.** Technical characteristics of metering and sub-metering (case studies): accuracy class as per measuring instrument directive (MID).

| Plant/Service | Function | Description | Accuracy | Reading | Range |
|---|---|---|---|---|---|
| Heating | Metering ** | Heat meter (Turbine) | Class 2 MID | 0.1 kWh | 240:1 |
| | Sub-metering | Two-sensor electronic heat cost allocator | n.a. | 0.001 UR | n.a. |
| | | Insertion time counter (comp. fluid temperature) | n.a. | 1 Wh | n.a. |
| Electric | Metering * | Static electricity meter | Class 1 MID | 1 kWh | n.a. |
| | Sub-metering ** | Current clump meter | n.a. | 1 Wh | n.a. |
| | Sub-metering ** | Smart Plug | n.a. | 1 Wh | n.a. |
| Natural Gas | Metering * | Hybrid gas meter | 1.5 MID | 1 dm$^3$ | 150:1 |
| | Sub-metering cooking ** | Hybrid gas meter | 1.5 MID | 0.1 dm$^3$ | 150:1 |
| | Sub-metering hot water ** | Hybrid gas meter | 1.5 MID | 0.1 dm$^3$ | 150:1 |
| Indoor air temperature monitoring | | Tlogger (thermal resistance) | 1 °C | 0.1 °C | −10/50 °C |

* fiscal; ** non-fiscal.

In this research project, basic information about indoor air temperature measurements in one or more zones of the apartments was also given. The aim was to allow the end-user to know in which way his energy saving behaviors were affecting, positively or negatively, the indoor air temperature, which was chosen in the present experimental campaign as a simple parameter that the end-user could directly relate to his perceived thermal comfort. In this way, the user could assess if eventual changes in his behavior toward energy savings (i.e., the closure of one or more radiators, no or excessive ventilation [29]) could be related to low indoor temperatures (e.g., below 18 °C), which represents an undesirable consequence to be avoided.

## 3. Methods

### 3.1. The Buildings Case-Studies

For the experimentation of feedback strategies on the consumption of thermal energy for heating, an experimental campaign is currently underway in two social housing buildings belonging to the Italian Territorial Agency for Social Housing (ATER), served by a centralized natural gas system and in a detached house all located in the district of Frosinone (Central Italy). The social housing buildings (building #1 and building #2), both built in the 1970s, have very low energy performance and would require relevant energy retrofit intervention, both to improve the insulation of the building envelope and to increase the efficiency of the heating plant. End-users are mostly low income and elderly, mainly living in single- or two-family units, with limited ability to interact with automation systems, and can be considered a representative sample of the typical Italian social housing user. Buildings #1 and #2 are almost identical in terms of constructive characteristics, inhabitants, floor areas, plan scheme, and heating system; the only exception is that one of the buildings consists of eight dwellings while the other has nine.

In each building, a thermal energy meter for the direct measurement of the thermal energy produced by the boiler (metering level) and two different indirect heat metering systems were installed (submetering level): (i) insertion time counters compensated with fluid temperature and thermostatic electronic valves controlled by programmable thermostat (building #1); (ii) two-sensor electronic heat cost allocators, mechanical thermostatic valves, and a programmable thermostat (building #2).

With regards to electrical energy consumption, an experimental campaign is currently underway in a detached house (building #3) located in the district of Frosinone (Central Italy) built in the first decade of the 2000s and inhabited by a family of four people. The house is a two-floor detached building, divided into two apartments, of which only one is actually inhabited by the family, but both are served by the main electrical energy meter with a maximum power installed of 4.5 kW. A current clamp meter was installed on the main power line of the sole inhabited apartment (metering level) in the fuse box of the house, whereas on the sub-metering level, two different devices were installed: (i) current clamp meter on the main light's powerline in the fuse box of the house; (ii) smart plugs on the more energy consuming electrical appliances.

In Figures 2 and 3, only two of the investigated buildings are depicted, as buildings #1 and #2 are almost identical.

**Figure 2.** Case-study building #1.

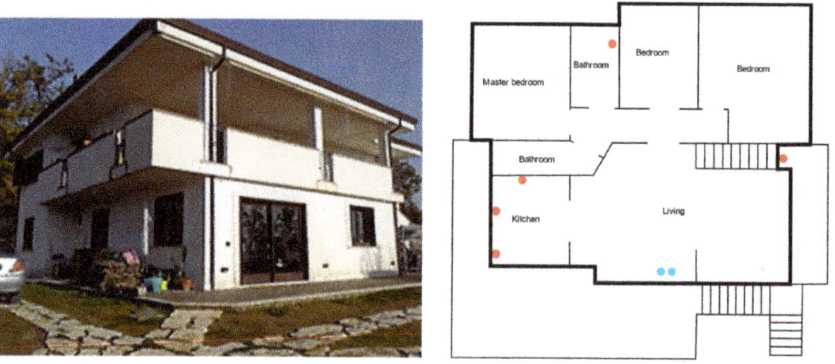


**Figure 3.** Case study building #3 with location of devices (red dots: smart-plugs, green dots: current clump meters).

Specific survey questionnaires were administrated to the inhabitants of the buildings, to assess user's attitude to adopt energy saving strategies and to interact with monitoring and control systems. In the survey, the user could assign a vote to seven statements, attributing a grade from 1 (not at all in agreement) to 10 (fully agreed), based on their level of concordance with the statement. For each question, there was also the option "I don't know". Each user was also given the opportunity to express general considerations in a special note field. The questions given to families are listed below:

A.  Overall, I feel satisfied with the installation of thermostatic valves and sub-metering devices in my apartment;
B.  I do often adjust the temperature using the chrono thermostat;
C.  During periods of absence from the apartment, I set the thermostat temperature to minimum to save energy;
D.  I think the installation of thermostatic valves and sub-metering devices in my apartment is helping me save on my gas bill;
E.  The temperature in my apartment is often too high and I am forced to open the windows;
F.  The temperature in my apartment is often too low;
G.  I use alternative systems to heat my apartment (for example, electric heaters and gas stoves).

A number of 18 family members (i.e., one for each of the families taking part to the experimental campaign) took part to the survey, namely, eight from building #1, nine from building #2, and one for building #3. In Figure 4, information about the composition of the investigated sample of families is given, for a calculated average number of family members of 2.6 people per dwelling (close to the national average of 2.4 [21,22]). Questionnaires were provided at the end of the first heating season following the installation of the metering and temperature control devices. Authors required answers from the householders (i.e., the person or the people managing the house).

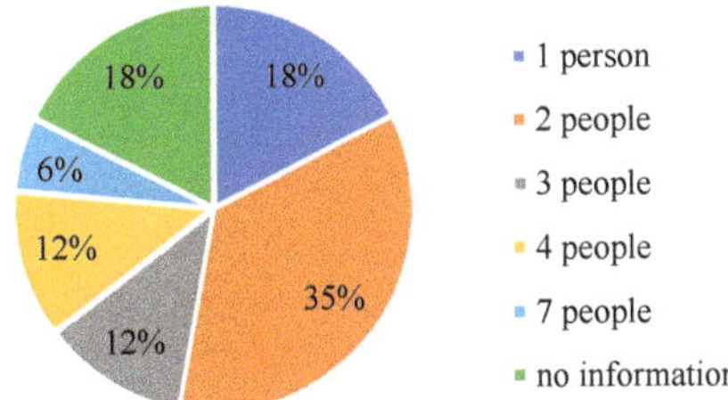


**Figure 4.** End-users living in investigated buildings, sample composition.

*3.2. Designing Information Strategies*

Effective strategies for improving end-users' awareness in "smart homes" are influenced by numerous aspects, such as the following [30]: (i) the quality of the perceived interaction (e.g., speed, brevity/easiness); (ii) information efficiency (e.g., accuracy and completeness); (iii) usability (e.g., ease of use, intuitiveness, user satisfaction); (iv) the aesthetics; (v) the usefulness (e.g., offered functions); and (vi) acceptability (e.g., low cost, number of potential users). The feedback of monitored data should reach end-users over time and the most adequate way to allow the full understanding of the phenomenon, before it is irreversible or no longer visible, linking it to specific retrofit actions [31]. To identify the most effective feedback, the authors analyzed the features shown in Table 2.

In the technical practice, the simplicity and/or cost of information system is sometimes favored, while in others, the completeness and/or the effectiveness of the information is favored. The different types of feedback can have very different costs and customer satisfaction levels, but a crucial issue should be the awareness and immediacy of information to lead users at performing higher energy savings. A unanimous judgment of end-users is the greater appreciation of a detailed, frequent, and actual feedback. Therefore, authors decided to differentiate between direct and indirect feedback: (i) by using frequent, synthetic, and immediate information in the case of direct feedback; (ii) by providing detailed and disaggregated information for each consumption area (i.e., bedrooms, living, bathroom, kitchen), for each energy carrier (i.e., thermal energy, electrical and natural gas), and for the device/system in the case of indirect feedback.

**Table 2.** Types of feedback.

| Characteristic | Description |
| --- | --- |
| Frequency | Continuous feedback (1/4 h, hourly, daily)<br>Deferred feedback (weekly, bi-monthly, half-yearly, yearly) |
| Content | consumption, kWh (absolute), % (relative)<br>costs, € (absolute), % (relative)<br>environmental impacts, $CO_2$ (absolute), % (relative) |
| Data aggregation | by location (e.g., room, living/sleeping area, apartment)<br>for use (e.g., heating, cooling, ventilation)<br>for plant/appliance (e.g., refrigerator, washing machine)<br>by energy carrier (e.g., electricity, heating, gas) |
| Presentation | Analog data (e.g., dashboard)<br>Numerical data (e.g., display)<br>Traffic lights, colors, and ideograms<br>Historical trend (e.g., trend, histograms)<br>Diagrams (e.g., pie, bar, ring) |
| Benchmark | Historical consumption<br>Consumption of other users (e.g., building average)<br>Expected theoretical consumption (e.g., based on climatic data, characteristics of energy systems, type of user) |
| Further information | Diagnosis (e.g., faults and malfunctioning)<br>Retrofit (e.g., indications and tips for rational use and efficiency) |

Other aspects positively evaluated in the technical literature are the diagnosis of faults and malfunctions, the comparison with historical consumption, simplicity, and effectiveness in understanding user information. Therefore, for indirect feedback and for each consumption area, the authors presented the following: (i) historical consumption benchmark, (ii) benchmark with average consumption of other users (building average), and (iii) theoretical expected consumption obtained on the basis of the specific characteristics of the user (e.g., characteristics of energy systems, type of user) and of climate data. To enhance the communication effectiveness, pie charts (for allocation) and bar charts (for comparisons with previous periods and with other users) were prepared.

*Benchmarking Indices and Energy Saving Tips*

Suitable tailored benchmarking indices were built by the authors as the ratio between the measured and the expected energy consumption of the room/appliance, as per Equation (1).

$$Benchmarking\ index = \frac{measured\ consumption - expected\ consumption}{measured\ consumption} \cdot 100 \tag{1}$$

In the following, the benchmarks designed for the case studies (i.e., heating and electricity services) are presented and discussed separately, owing to the corresponding peculiarities.

In the heating service case studies (buildings #1 and #2), the performance benchmarks were calculated as the ratio between the energy consumption measured at actual conditions of use (i.e., operational rating) and the estimated primary energy consumption adjusted to the actual conditions of use and climate (i.e., tailored rating), for each room, apartment, and the whole building [32].

A different approach was adopted for the electricity case study (building #3). In fact, it is well known that energy consumption of an appliance strongly depends on its use, which, in turn, relies on the number of family components, characteristics of the house (e.g., floor area and outdoor spaces), and on the end-user (e.g., income, work, age, presence of children and/or elderly people). These characteristics are then employed to build reference baseline values within certain descriptive distributions and adopted for benchmarking purposes. On the other hand, the benchmarking methodology proposed by the authors in this paper is based on a simplified hybrid approach. The expected electricity consumption is built with a "bottom-up" approach considering each electrical appliance installed in the house. Basic information is collected from the technical booklet of the appliances (i.e., energy consumption for cycle/energy labels/power) and then used to calculate the expected consumption through statistical data on the use of household appliances. In fact, for the purposes of the present research, it was necessary to use a simple representative indicator for the expected energy consumption of each electrical appliance. In this way, the methodology could be simply implemented in an IoT application for energy monitoring. To this aim, it was decided to consider the number of family components as the most representative parameter for the estimation of the expected energy consumption of electrical devices. This hypothesis, although introducing a certain level of simplification (e.g., for the calculation of consumption for lighting), is considered to be reliable for the calculation of the expected consumption of large electrical appliances, as the number of family members directly impacts their frequency of use. In this context, in order to determine the expected energy consumption of each electrical appliance, the authors made a preliminary analysis of statistical data about electrical energy use from the Italian National Institute of Statistics (ISTAT) and Italian National Agency for New Technologies, Energy, and Sustainable Economic Development (ENEA). In particular, regarding the Italian energy consumption, ISTAT provides statistical data about the following: (i) number of cycles per week for given electrical appliances for different numbers of family components; (ii) lights turn-on period; and (iii) expected expenditure for different numbers of family components.

In particular, Figure 5 shows the data about dishwasher and washing machine usage per week and the trend of the expenditure for electrical energy of different family sizes.

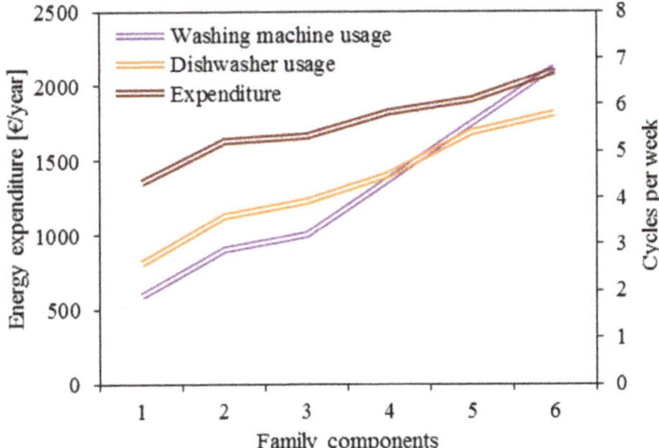


**Figure 5.** Data usage for main appliances [21,22].

Suitable usage coefficients were calculated by normalizing all the statistical data about energy use/expenditure with respect to reference number of family components (i.e., 2.4 for Italy [21,22]). This means that, for example, the expenditure coefficient for a family of four components was determined by dividing the expected statistical expenditure (estimated to be about 1908 €) for the expenditure expected for a family with 2.4 family components (i.e., about 1706 €). Usage coefficients were employed as a base to calculate, for each electrical appliance, the expected time of use (expressed in hours) in the reference period (year). The expected energy consumption is then built either per Equation (2) or per Equation (3), depending on the type of appliance and on the data available in the technical booklet.

$$expected\ consumption\ =\ num_{c,\,w} \cdot 52 \cdot cons_{cycle}, \tag{2}$$

$$expected\ consumption\ =\ UC \cdot num_{h,\,w} \cdot 52 \cdot power, \tag{3}$$

where $num_{c,\,w}$ is the number of weekly cycles of the given electrical appliance (calculated as a function of the usage coefficient [21,22]), $cons_{cycle}$ is the electrical energy consumption per cycle of the appliance declared by the manufacturer and retrievable in the technical booklet (expressed in kWh/cycle), *UC* is the usage coefficient of the appliance, $num_{h,\,w}$ is the expected weekly hours of use of the appliance, and *power* is the declared power of the appliance expressed in kW. Note that the use of Equation (2) was applied to all of the electrical appliances whose energy consumption per cycle was given in the technical booklet (e.g., dishwasher, washing machine, dryer), while Equation (3) was employed for all the other appliances. Number cycles/functioning hours per week were determined as the weighted average of the statistical data given in the work of [21,22] or, in the case in which data were not available, by making suitable assumptions.

In this context, the baseline values (i.e., expected consumption) employed to calculate the benchmarking indicators will be updated as soon as a change is made in electrical devices (i.e., replacement of an appliance with a new one, replacement of light bulbs), in the insulation of the building, or if the number of family components changes at some point.

For the present analysis, the tips presented in the indirect feedback were selected from a set of suggestions expressly provided by the ENEA as part of the dissemination actions about energy efficiency in residential buildings [33]. The set of selected tips is general advice targeting the correct management of each heating, cooling, lighting, and electrical systems, and was given when the measured consumption of each room/appliance was found to be higher with respect to the expected one (i.e., calculated energy consumption).

Finally, two new direct and indirect feedback strategies were adopted by the authors based on the following aspects: (i) the results of the survey; (ii) the personal interaction between the authors and the participants of the experimental campaign, which occurred during the informative meetings; (iii) the characteristics of the installed energy systems and measuring devices; and (iv) the analysis of existing scientific literature regarding the best practices in feedback about energy consumption.

## 4. Results and Discussions

Through the administration of specific designed surveys in the ATER buildings, the authors gathered useful information about users' attitudes to adopt energy saving strategies and to interact with monitoring and control systems. The response rate to the questionnaires provided was 100%. Figure 6 shows the list of the questions together with the overall analysis of the answers obtained. With regard to the installation of monitoring and control systems, the users, although they declare themselves satisfied (100%) and quite familiar with such systems (64%), were wary of the potential effectiveness in terms of savings (71%). As for indoor temperature perception, most users feel that they do not perceive too high (71%) or too low (78%) indoor temperatures.

The characteristics of the feedback designed within this work are summarized in Table 3.

### 4.1. Direct Feedback

The dashboard built for direct feedback (daily frequency) is made up of two sections. The first one for sub-metering shows the energy consumption (kWh and %) of each room/appliance, using a bar graph. In the second section (metering), through a multi-scale display, the user can simultaneously access the energy data consumed by the apartment (in kWh and in €) and the corresponding $CO_2$ emitted (in kg). In this way, the user receives, in real time, information about his own energy consumption, the related costs, and the environmental impact, as well as on their distribution among different environments, leading, at the same time, to adopting efficient behaviors at the energy, economic, and environmental levels. A daily frequency of this feedback was chosen by the authors owing to metering and sub-metering devices' characteristics and to the related costs of data transmission (e.g., battery consumption). Figure 7 shows the dashboard developed by the authors to display the daily energy consumption of a typical user both for heating and for electrical energy consumption.

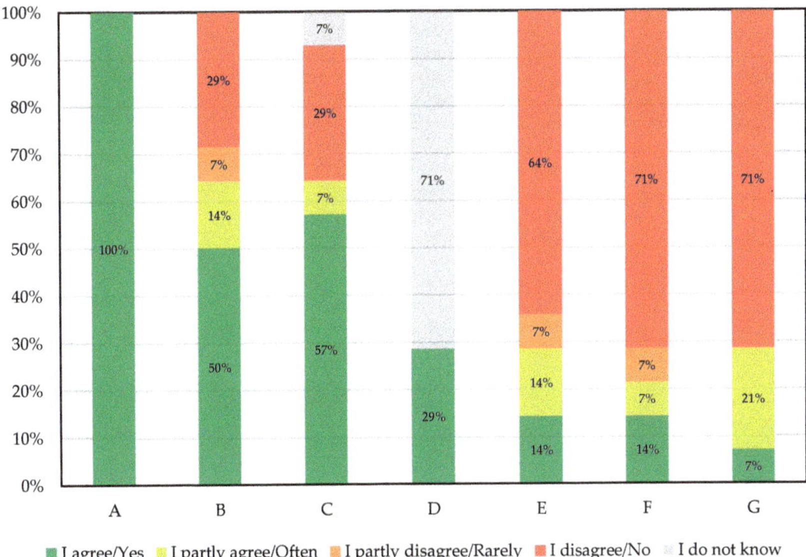


**Figure 6.** Results of the analysis of surveys.

**Table 3.** Technical specification of direct and indirect feedback.

| Characteristic | Direct Feedback | Indirect Feedback |
|---|---|---|
| Frequency | Daily | Monthly |
| Content | Consumed energy (kWh, %)<br>Cost (€)<br>$CO_2$ emitted (kg) | Consumed energy (kWh, %)<br>Cost (€)<br>$CO_2$ emitted (kg)<br>Consumption indexes |
| Aggregation | By room/appliance<br>By apartment | By room/appliance<br>By apartment<br>By building |
| Presentation | Energy dashboard | Bar charts<br>Ring diagrams<br>Histograms |
| Benchmark | - | Historical consumption<br>With other users (building)<br>Expected consumption (tailored rating)<br>Share of consumption for appliance |
| Further information | - | Useful tips for savings and efficiency |

## 4.2. Indirect Feedback

As above reported, through indirect feedback (monthly frequency), users also receive information on their consumption in terms of performance indices, personalized suggestions, comparison with historic consumption, and so on. In this way, the authors also tried to capture the users' attention by using emoticons, colors, and "user friendly" information.

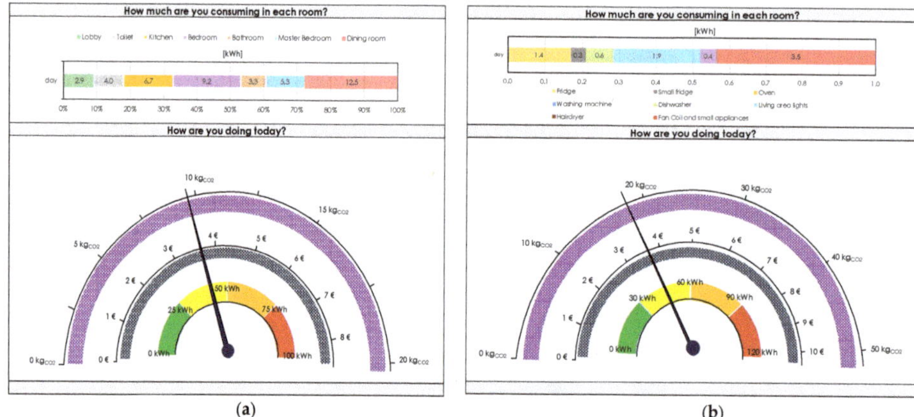

**Figure 7.** Direct feedback: (**a**) heating, (**b**) electricity.

To all users involved in the experimentation, the authors provided indirect feedback sheets, collecting impressions and suggestions. In particular, from the meetings carried out, the following emerged:

(i)    a high participation (about 95%) of the users;

(ii)    almost no user was able to understand the non-rated units provided by heat cost allocators;

(iii)    users who had a poor awareness and responsible behavior generally did not perceive the over consumption because previously, the experimentation heat accounting was carried out exclusively by floor area and costs due to building inefficiencies were divided into tenants accordingly;

(iv)    not all users were familiar with the control systems perceiving the intervention of thermostatic valves as a malfunction of the plant;

(v)    the cultural level of the users strongly affects the understanding of energy consumption data and, consequently, the adoption of retrofit actions;

(vi)    the information received during initial installation was not sufficient to understand operation and use.

Figures 8 and 9 show the form designed by the authors for indirect feedback of heating divided into six sections:

1.    aggregate and disaggregated monthly energy consumption for each room (energy consumed, related costs, and environmental impact);

2.    local consumption indexes in percentage and economic units;

3.    total consumption indexes for the user;

4.    personalized advice and tips aimed at saving energy;

5.    historical consumption and related average outdoor temperature;

6.    benchmark with the other apartments in the building or with average data of the building.

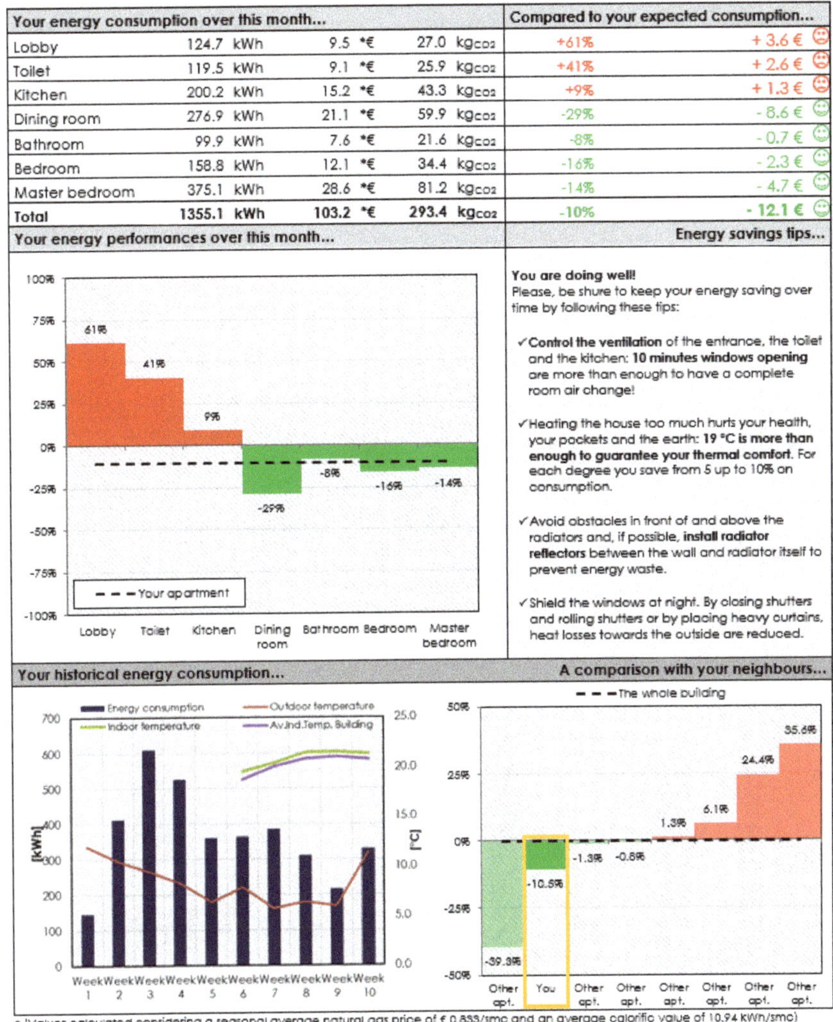

**Figure 8.** Indirect feedback for heating service.

Figure 9 shows the informative sheet for indirect feedback of electrical energy consumption.

In the electrical energy case study (Figure 9), no comparison with other user was possible, as the family lives in a detached house. A ring chart shows the share of energy usage for each monitored appliance in the reference period (month).

Tables 4 and 5 show the usage coefficients calculated as described in the methodology section for the available data [22] and the calculated expected energy consumption of each of the electrical appliances installed in the building #3 case study, respectively.

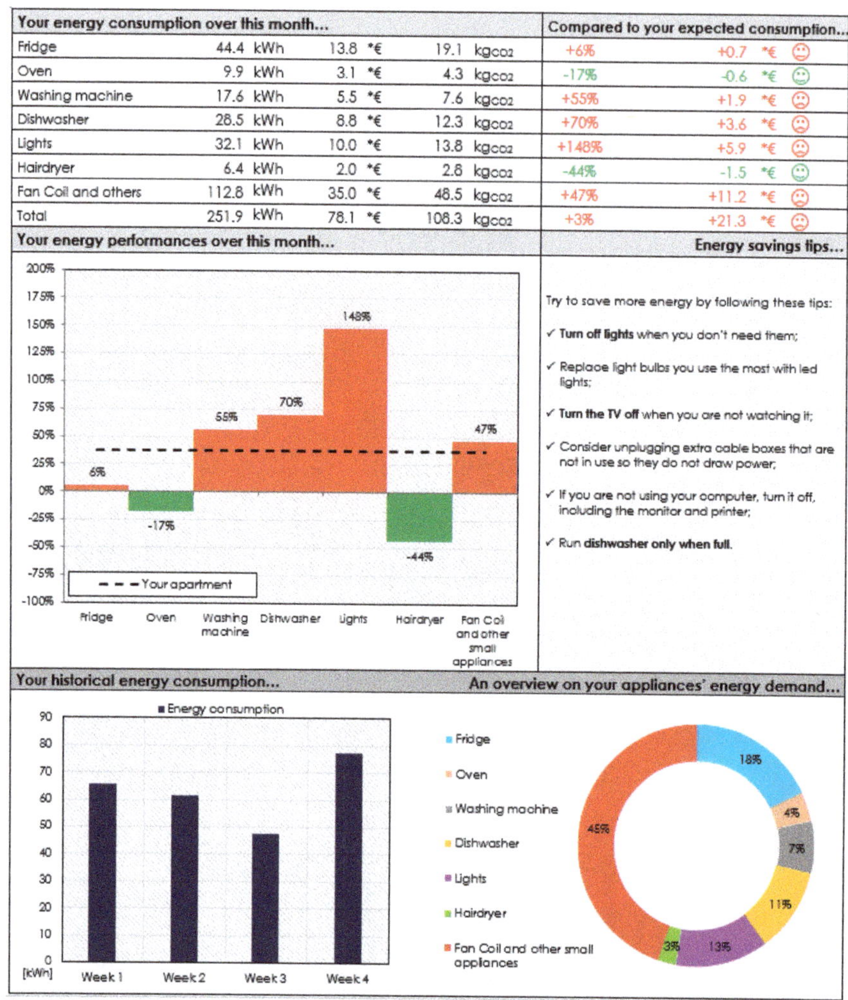

**Figure 9.** Indirect feedback for electrical energy.

**Table 4.** Calculated usage coefficients (Italian National Institute of Statistics (ISTAT), 2014).

| Family Components | Expenditure Coefficient | Washing Machine Coefficient | Dishwasher Coefficient |
| --- | --- | --- | --- |
| 1 | 0.80 | 0.54 | 0.66 |
| 2 | 0.95 | 0.83 | 0.91 |
| 2.4 | 1.00 | 1.00 | 1.00 |
| 3 | 1.07 | 1.26 | 1.14 |
| 4 | 1.12 | 1.60 | 1.36 |
| 5 | 1.23 | 1.94 | 1.46 |

**Table 5.** Expected energy consumption of main electrical appliances of the analyzed case-study (ISTAT, 2014).

| Device | Q.ty | Declared Consumption [kWh/cycle] | Power [kW] | Usage Coefficient | Cycles Per Week | Expected Time of Use [hours/year] | Energy Label [kWh/year] | Expected Consumption [kWh/year] |
|---|---|---|---|---|---|---|---|---|
| Refrigerator, Rex FI 22/10 H | 1 | n.a. | n.a. | n.a. | n.a. | n.a. | 511 | 511 [c] |
| Oven, Rex FMS 041 X | 1 | n.a. | 1.100 | 1.12 | 1.0 [a] | 133 [a,b] | 105 | 163 |
| Microwave, Panasonic NN-k-108 WM | 1 | n.a. | 0.390 | 1.12 | 13 [b] | 56 [b] | n.a. | 25 |
| Washing machine, Electrolux EWF 1286 DOW | 1 | 0.58 | n.a. | 1.60 | 3.5 [a] | n.a. | 134 | 169 |
| Dishwasher, Bosch SMV 46 KX 01 E | 1 | 0.92 | n.a. | 1.36 | 4.0 [a] | n.a. | 262 | 260 |
| Hairdryer, Bosch PHD9760/01 | 2 | n.a. | 2.000 | n.a. | 4.0 [b] | 35 [b] | n.a. | 139 |
| TV, Sony KD-49XF90 | 2 | n.a.<br>n.a. | 0.114 (on)<br>0.001 (standby) | 1.12 | n.a.<br>n.a. | 1456 [b]<br>7280 [b] | 158 | 386 |
| Energy saving fluorescent lamps | 20 | n.a. | 0.009 | | n.a. | 815 [a,b] | n.a. | |
| Fluorescent lamps | 5 | n.a. | 0.011 | 1.12 | n.a. | 408 [a,b] | n.a. | 290 |
| LED strips | 1 | n.a. | 0.013 | | n.a. | 815 [a,b] | n.a. | |
| LED lamps | 14 | n.a. | 0.007 | | n.a. | 815 [a,b] | n.a. | |
| Fan coil | 6 | n.a. | 0.050 | n.a. | n.a. | 1540 [b] | n.a. | 462 |

[a] calculated based on the works of [21,22]; [b] hypothesized based on user's declaration; [c] the expected consumption of the refrigerator is the declared annual energy consumption from the energy label.

It is underlined that the expected consumption in the present research has a high degree of tailoring, as it is assumed that the user has preliminary specified the technical characteristics (i.e., energy consumption per cycle and/or energy label) of each appliance installed. It is thus possible to state that the benchmark indices presented in this research are built with a tailored bottom-up approach based on the actual characteristics of the house analyzed and not with a top-down approach (i.e., a statistical approach), thus replicating the commonly applied methodology to determine the "tailored rating" for heating energy consumption applications.

## 5. Conclusions

In this work, the authors presented the first results of a study aimed at designing a proper feedback strategy for IoT energy consumption monitoring systems, so as to increase user awareness on energy consumption. In this framework, two case studies were presented and discussed related to heating and electrical energy consumption monitoring in residential buildings and specific tailored benchmarks indices were developed to present energy consumption data at the sub-metering level. To this aim, frequent informative meetings were carried out with a total of 18 end-users in three different buildings.

The direct interaction between the authors and the end users over the campaign, together with the provision of simple questionnaires, showed that the huge amount of measured data and the complexity of the monitored systems make analysis and feedback particularly complex for non-skilled users. Compatibly with the scientific literature, most of the users were not able to understand the measuring units expressed in kWh (or similar), while they showed particular interest in the consumption data expressed in euro and in percentage.

The design and the content of the designed feedback strategy was greatly appreciated by the final users, whose attention on the desired information was captured through simple information (e.g., different colors and emoticons). Owing to the characteristics of the sample of users (mostly elderly people with limited ability to manage technical information), these last favored the simplicity and immediacy of information, which has been given in the following form:

(i)    direct feedback, made up of a simple multi-scale energy dashboard giving information on both the total energy consumption of the house (metering level) and the share of consumption of each room/appliance of the house (sub-metering level);

(ii)   indirect feedback, represented by an informative sheet in which information was gathered in a more detailed manner by also showing the user tailored benchmark indicators on both metering and sub-metering levels.

All the users showed particular appreciation for the benchmarking indicators developed at the sub-metering level (generally not available neither in the energy bills nor in the energy monitoring systems available on the market), as this information allowed them to precisely locate the main energy consuming rooms/appliances of the house and to take specific actions in order to reduce their consumption.

Personalized suggestions were also given to specific participants whose consumption was found to be much lower than that expected, in order to make them aware that incorrect energy-saving practices, such as the closure of most radiators, can dangerously penalize the indoor air temperature, thus also affecting the perceived thermal comfort, which was not the purpose of the present campaign.

The adopted IoT technologies demonstrated high potential in terms of energy savings, as effective and frequent feedback contributes significantly to motivating and supporting change in occupant behavior. The analysis of the data showed some incorrect behavior that users were not aware of, such as excessive ventilation of some rooms (e.g., entrance, bathrooms, and kitchens), incorrect management of the thermostatic valves, and incorrect management of some domestic appliances.

It is highlighted that the benchmarking methodology developed by the authors for the purpose of the present campaign could be very useful for researchers and software developers looking for targeted strategies for energy benchmarking. Referring to benchmarking for electricity consumption, in fact, the authors proposed a simplified, detailed approach to estimate the energy consumption of a given appliance, which could be simply implemented in an IoT application for energy monitoring. In fact, this methodology requires the knowledge of the family composition and basic information about the appliance, retrievable by the technical booklet, by specific databases, or directly by the smart device.

Further progress of the present research is represented by the development, currently ongoing, of specific diagnostic routines for natural gas, electricity and heating systems that, based on the load profiles measured by the system, will allow the detection of possible faults or incorrect management, and will automatically recommend the most suitable personalized suggestion.

**Author Contributions:** Conceptualization, M.D. and G.F.; methodology, M.D., G.F., and L.C.; formal analysis, L.C. and B.I.P.; investigation, M.D., G.F., and L.C.; resources, M.D. and G.P.; data curation, L.C.; writing—original draft preparation, M.D., G.F., L.C., and B.I.P.; writing—review and editing, B.I.P. and G.P.; supervision, M.D., G.F., and G.P.; project administration, M.D. and G.P.; funding acquisition, M.D. and G.P.

**Funding:** This work was developed under the projects "Ricerca di Sistema Elettrico PAR 2016" funded by ENEA (grant number I12F16000180001) and "PRIN Riqualificazione del parco edilizio esistente in ottica NZEB" funded by MIUR (grant number 2015S7E247_002). The authors wish to thank ATER of Frosinone, the Territorial Agency for Social Housing, for the technical support during the on field experimental campaign.

**Conflicts of Interest:** The authors declare no conflict of interest.

## References

1.    Håkon, S.; Hege, W. A Multi-Method Evaluation of the Potential for Using the Electricity Bill to Encourage Energy Savings in Norwegian Households. *Energy Environ. Res.* **2013**, *3*. [CrossRef]

2.    Sonderegger, R.C. Movers and stayers: The resident's contribution to variation across houses in energy consumption for space heating. *Energy Build.* **1978**, *1*, 313–324. [CrossRef]

3.    Marinakis, V.; Doukas, H. An Advanced IoT-based System for Intelligent Energy Management in Buildings. *Sensors* **2018**, *18*, 610. [CrossRef] [PubMed]

4.    Abrahamse, W.; Steg, L.; Vlek, C.; Rothengatter, T. A review of intervention studies aimed at household energy conservation. *J. Environ. Psychol.* **2005**, *25*, 273–291. [CrossRef]

5.  Brandon, G.; Lewis, A. Reducing household energy consumption: A qualitative and quantitative field study. *J. Environ. Psychol.* **1999**, *19*, 75–85. [CrossRef]

6.  Guerassimoff, G.; Thomas, J. Enhancing energy efficiency and technical and marketing tools to change people's habits in the long-term. *Energy Build.* **2015**, *104*, 14–24. [CrossRef]

7.  Chen, V.L.; Delmas, M.A.; Kaiser, W.J. Real-time, appliance-level electricity use feedback system: How to engage users? *Energy Build.* **2014**, *70*, 455–462. [CrossRef]

8.  Anderson, W.; White, V. *The Smart Way to Display. Full Report: Exploring Consumer Preferences for Home Energy Display Functionality. A Report For the Energy Saving Trust by the Centre for Sustainable Energy*; Energy Saving Trust: London, UK, 2009.

9.  Gans, W.; Alberini, A.; Longo, A. Smart meter devices and the effect of feedback on residential electricity consumption: Evidence from a natural experiment in Northern Ireland. *Energy Econ.* **2013**, *36*, 729–743. [CrossRef]

10. Wilson, C.; Dowlatabadi, H. Models of decision making and residential energy use. *Annu. Rev. Environ. Resour.* **2007**, *32*, 169–203. [CrossRef]

11. Buchanan, K.; Russo, R.; Anderson, B. The question of energy reduction: The problem(s) with feedback. *Energy Policy* **2015**, *77*, 89–96. [CrossRef]

12. Nilsson, A.; Wester, M.; Lazarevic, D.A.B. Nils, Smart homes, home energy management systems and real-time feedback: Lessons for influencing household energy consumption from a Swedish field study. *Energy Build.* **2018**, *179*, 15–25. [CrossRef]

13. ELIH-Med Smart Metering. Analysis of Current Projects of Multi-Energies Smart Metering in Low-Income Housing in the MED Area. MED Programme Priority-Objective 2-2: Promotion and Renewable Energy and Improvement of Energy Efficiency Contract n. IS-MED10-029 Intelligent Energy Management Systems. 2014. Available online: http://www.elih-med.eu/uploads/image/ELIHMed_D611_Analysis_of_smart_metering_in_LIH.pdf (accessed on 25 November 2019).

14. Podgornik, A.; Sucic, B.; Blazic, B. Effects of customized consumption feedback on energy efficient behaviour in low-income households. *J. Clean. Prod.* **2016**, *130*, 25–34. [CrossRef]

15. Christine, B.; Rory, V.J.; Sabine, P.; Alba, F. Do psychological factors relate to energy saving behaviours in inefficient and damp homes? A study among English social housing residents. *Energy Res. Soc. Sci.* **2019**, *47*, 146–155.

16. Australian Energy Regulator. Guidance on Energy Consumption Benchmarks on Residential Customers' Bills. AER Reference: 61307–D17/171822; 2017. Available online: https://www.aer.gov.au/system/files/Bill%20benchmark%20guidance%202018_0.pdf (accessed on 1 November 2019).

17. Ghajarkhosravi, M.; Huang, Y.; Fung, A.S.; Kumar, R.; Straka, V. Energy Benchmarking Analysis of Multi-unit Residential Buildings (MURBs) in Toronto, Canada. *J. Build. Eng.* **2020**, *27*, 100981. [CrossRef]

18. Nikolaou, T.; Kolokotsa, D.; Stavrakakis, G. Review on methodologies for energy benchmarking, rating and classification of buildings. *Adv. Build. Energy Res.* **2011**, *5*, 53–70. [CrossRef]

19. Kavousian, A.; Rajagopal, R.; Fischer, M. Ranking appliance energy efficiency in households: Utilizing smart meter data and energy efficiency frontiers to estimate and identify the determinants of appliance energy efficiency in residential buildings. *Energy Build.* **2015**, *99*, 220–230. [CrossRef]

20. Zhengwei, L.; Yanmin, H.; Peng, X. Methods for benchmarking building energy consumption against its past or intended performance: An overview. *Appl. Energy* **2014**, *124*, 325–334.

21. ISTAT. Censimento Popolazione Abitazioni. 2011. Available online: http://dati-censimentopopolazione.istat.it/Index.aspx?lang=it (accessed on 1 September 2018).

22. ISTAT. I Consumi Energetici Delle Famiglie—Anno 2013. (in Italian Language). Available online: http://www.istat.it/it/archivio/142173 (accessed on 1 September 2018).

23. Celenza, L.; Dell'Isola, M.; Ficco, G.; Greco, M.; Grimaldi, M. Economic and technical feasibility of metering and sub-metering systems for heat accounting. *Int. J. Energy Econ. Policy* **2016**, *6*, 581–587.

24. Celenza, L.; Dell'Isola, M.; Ficco, G.; Palella, B.I.; Riccio, G. Heat accounting in historical buildings. *Energy Build.* **2015**, *95*, 47–56. [CrossRef]

25. Canale, L.; Dell'Isola, M.; Ficco, G.; Cholewa, T.; Siggelsten, S.; Balen, I. A comprehensive review on heat accounting and cost allocation in residential buildings in EU. *Energy Build.* **2019**, *202*, 109398. [CrossRef]

26. Canale, L.; Dell'Isola, M.; Ficco, G.; di Pietra, B.; Frattolillo, A. Estimating the impact of heat accounting on Italian residential energy consumption in different scenarios. *Energy Build.* **2018**, *168*, 385–398. [CrossRef]

27. Dell'Isola, M.; Ficco, G.; Arpino, F.; Cortellessa, G.; Canale, L. A novel model for the evaluation of heat accounting systemsreliability in residential buildings. *Energy Build.* **2017**, *150*, 281–293. [CrossRef]

28. Dell'Isola, M.; Ficco, G.; Canale, L.; Frattolillo, A.; Bertini, I. A new heat cost allocation method for social housing. *Energy Build.* **2018**, *172*, 67–77. [CrossRef]

29. D'Ambrosio Alfano, F.R.; Dell'Isola, M.; Ficco, G.; Palella, B.I.; Riccio, G. Experimental Air-Tightness Analysis in Mediterranean Buildings after Windows Retrofit. *Sustainability* **2016**, *8*, 991. [CrossRef]

30. Aghajan, H.; Augusto, J.C.; Lopez-Cozar Delgado, R. Human-Centric Interfaces for Ambient Intelligence. *J. Ambient Intell. Smart Environ.* **2010**, *2*, 345–346.

31. Fischer, C. Feedback on household electricity consumption: A tool for saving energy? *Energy Effic.* **2008**, *1*, 79–104. [CrossRef]

32. European Committee for Standardization. *EN ISO 13790 Energy Performance of Buildings—Calculation of Energy Use for Space Heating and Cooling*; European Committee for Standardization: Brussels, Belgium, 2008.

33. ENEA. Cambiamento Comportamentale ed Efficienza Energetica per Difendersi dal Caldo in Modo Intelligente. (In Italian Language). 2019. Available online: http://www.efficienzaenergetica.enea.it/Cittadino/risparmiare_energia (accessed on 1 November 2019).

*Article*

# Hybrid Ventilation System and Soft-Sensors for Maintaining Indoor Air Quality and Thermal Comfort in Buildings

**Nivetha Vadamalraj [1], Kishor Zingre [2],*, Subathra Seshadhri [1], Pandarasamy Arjunan [3] and Seshadhri Srinivasan [3]**

1    Department of Instrumentation and Control Engineering and Center for Research in Automatic Control Engineering (C-RACE), Kalasalingam University, Srivilliputhur 626125, India; nivetha@klu.ac.in (N.V.); b.subathra@klu.ac.in (S.S.)

2    Department of Architecture and Built Environment, University of Northumbria at Newcastle, Newcastle Upon Tyne NE1 8ST, UK

3    Berkeley Education Alliance for Research in Singapore, Singapore 138602, Singapore; samy@bears-berkeley.sg (P.A.); seshadhri.srinivasan@bears-berkeley.sg (S.S.)

*    Correspondence: kishor.zingre@northumbria.ac.uk

Received: 9 December 2019; Accepted: 31 December 2019; Published: 16 January 2020

**Abstract:** Maintaining both indoor air quality (IAQ) and thermal comfort in buildings along with optimized energy consumption is a challenging problem. This investigation presents a novel design for hybrid ventilation system enabled by predictive control and soft-sensors to achieve both IAQ and thermal comfort by combining predictive control with demand controlled ventilation (DCV). First, we show that the problem of maintaining IAQ, thermal comfort and optimal energy is a multi-objective optimization problem with competing objectives, and a predictive control approach is required to smartly control the system. This leads to many implementation challenges which are addressed by designing a hybrid ventilation scheme supported by predictive control and soft-sensors. The main idea of the hybrid ventilation system is to achieve thermal comfort by varying the ON/OFF times of the air conditioners to maintain the temperature within user-defined bands using a predictive control and IAQ is maintained using Healthbox 3.0, a DCV device. Furthermore, this study also designs soft-sensors by combining the Internet of Things (IoT)-based sensors with deep-learning tools. The hardware realization of the control and IoT prototype is also discussed. The proposed novel hybrid ventilation system and the soft-sensors are demonstrated in a real research laboratory, i.e., Center for Research in Automatic Control Engineering (C-RACE) located at Kalasalingam University, India. Our results show the perceived benefits of hybrid ventilation, predictive control, and soft-sensors.

**Keywords:** indoor air quality (IAQ); hybrid ventilation; demand controlled ventilation (DCV); internet of things (IoT); soft-sensor; convolution neural networks

---

## 1. Introduction

The building sector in India currently contributes to ~37% of the total energy consumption of the nation and predicted to further increase by 8% annually due to recently proposed construction of 40 billion m$^2$ by 2050 which are driven by rapidly growing population and urbanization [1]. Statistics suggest that space cooling alone contributes to about 40–45% of the total building energy consumption in India. Consequently, energy optimization for space cooling maintaining thermal comfort has attracted significant attention recently (see [2,3] and references therein). While the importance of energy consumption is often exacerbated, the indoor air quality (IAQ) is not usually discussed [4]. However, IAQ is an important parameter that determines the productivity and performance of occupants in the building. The IAQ refers to the air quality within and around buildings and structures, especially it relates to the health and comfort of building occupants (https://www.epa.gov/indoor-air-quality-iaq/introduction-indoor-air-quality). The American Society of Heating, Refrigerating and Air-Conditioning Engineers (ASHRAE) defines IAQ as Air in which there are no known contaminants at harmful concentrations as determined by cognizant authorities and with which a substantial majority (80% or more) of the people exposed do not express dissatisfaction (http://cms.ashrae.biz/iaqguide/pdf/IAQGuide.pdf?bcsi_scan_C17DAEAF2505A29E=0&bcsi_scan_filename=IAQGuide.pdf).

It is widely perceived that understanding the common pollutants and controlling them could reduce health concerns, which can either be short- or long-term. Common short-term effects include dizziness, headaches, and fatigue, whereas long-term effects could be respiratory diseases, heart diseases, and even cancer. There are many surveys which have shown that both IAQ and thermal comfort are not well maintained by current ventilation systems (see, for example, in [5–8]). This is mainly because IAQ, thermal comfort, and energy consumption are competing objectives. As an increase in IAQ means more fresh air induction which will increase the energy consumption and can lower down thermal comfort as well. Therefore, maintaining IAQ, thermal comfort and minimizing energy is a major challenge in buildings, but important for achieving energy goals as well as occupant performance in buildings.

There are two procedures commonly used for maintaining IAQ: (i) Ventilation Rate Procedure and (ii) Indoor Air Quality Procedure [9]. The later aims to strike a good balance between energy savings and IAQ. Furthermore, it provides direct reduction of indoor contaminants. Similarly, in practice, there are three approaches to improve IAQ: (i) source control, (ii) improved ventilation, and (iii) air-cleaners. In source control, we eliminate individual sources of pollution or reduce their emissions and is considered an effective method for maintaining IAQ. Ventilation improvements means forcing fresh air into the buildings by opening windows and doors when weather permits, but this may lead to energy loos. Air cleaners aim to remove the particle from indoor air and uses filtering blocks to remove pollutants. Among these methods, forcing fresh air is by far the cheapest method for maintaining IAQ. However, optimal fresh air should be infused to reduce energy consumption.

Over the years, several control approaches for maintaining both IAQ and thermal comfort have been studied but with certain limitations. Among those, few studies implemented model predictive control (MPC) to investigate the combined (i) thermal comfort and $CO_2$ optimization by regulating fresh air [10], (ii) IAQ (particle concentration) and energy optimization [11], (iii) multi-objective optimization of IAQ (particulate matter) and energy consumption in subway ventilation system [12], (iv) optimization energy and IAQ with focus on $CO_2$ [13], and (v) energy optimization and air-quality through fresh air induction [3]. More recently, the role of demand controlled ventilation(DCV) on IAQ and energy savings was studied [14]. Another study implemented MPC for energy optimization and monitoring carbon dioxide in commercial HVAC systems with an Internet of Things (IoT) based control [15]. A combined control for IAQ, energy and thermal comfort was proposed in [16] for variable air volume (VAV) controlled systems. Similarly, thermal comfort and IAQ in chilean schools with surveys was studied in [17]. The control of IAQ in direct expansion air-conditioning system was studied in [18]. However, most of these investigations have concentrated only on temperature, humidity, and carbon dioxide levels while discussing IAQ.

As for sensors, a WiFi-enabled IAQ monitoring and control system for buildings was proposed in [19]. A soft-sensor for measuring carbon dioxide content in the building by fusing carbon dioxide and PIR sensor was proposed in [20]. A soft-sensor for estimating cooling load using long short-term memory (LSTM) was proposed in [21]. A review of sensors for measuring Indoor Air Quality (IAQ) was presented in [22]. There are many studies on sensor design for IAQ or studying sensor installation/monitoring issues reviewing which are out of the scope of the paper. More recently, the use of soft-sensors for IAQ modeling was studied in [23] wherein sensor readings was used to train deep subspace network to model IAQ parameters. A soft-sensor for detecting urban IAQ was proposed in [24] using Bayesian networks. These results illustrated that by combining compact sensors with databased techniques capabilities of sensor could be overarched and they can be made more smarter. Albeit, such significant advantages, the role of soft-sensor has not been studied extensively.

A review of the literature reveals three gaps in control and monitoring in buildings. First, maintaining IAQ, thermal comfort and optimizing energy simultaneously is a challenge that has not been fully addressed in the literature. In particular, the ones concerning IAQ parameters such as Volatile Organic Compounds, Carbon-dioxide and others have not been combined with energy optimization based control and thermal comfort. Second, although low-cost sensors are available their reliability and performance is limited. To overcome this, soft-sensors which combine compact sensors with data analytics methods are gaining popularity. The advantage of soft-sensors is that they can also predict the variables and can also model variables which are not measured, e.g., cooling load. In addition, they provide smartness to the control. Third, the use of soft-sensors within control has not been reported. Our main objective is to address these research gaps in the literature by proposing a soft-sensor based ventilation method that could optimize energy consumption while maintaining IAQ and thermal comfort. The main contributions can be summarized as below.

1. Novel design of a hybrid ventilation system design which simultaneously optimizes energy consumption, maintains IAQ and meets thermal comfort as well.
2. Present the Internet of Things (IoT) architecture and sensor prototype for implementing the hybrid ventilation system.
3. Design soft-sensors for modeling and predicting IAQ parameters and cooling load.
4. Controller design for hybrid ventilation scheme that could use the soft-sensors and the IoT devices to implement a flexible control architecture.

The paper is organized into five sections. Section 2 presents the problem formulation and challenges. The hybrid ventilation system, soft-sensor design, and control techniques are presented in Section 3. Deployment results of the different components are presented in Section 4. Conclusions and future directions of the investigation are discussed in Section 5.

## 2. Problem Formulation

The objective is to design a novel hybrid ventilation system to maintain IAQ, thermal comfort in terms of set-point temperature and minimize energy consumption at the Center for Research in Automatic Control Engineering (C-RACE) at the International Research Center (IRC), Kalasalingam University, India. The lab houses faculty offices, startups in the building automation laboratory, research labs, researchers and visitors as shown in Figure 1. The current system consists of fans and variable refrigerant volume (VRV)-based air-conditioning system (package units) providing space cooling controlled by a thermostat which turns ON or OFF depending upon the user defined temperature limits. The lab has air quality issues due to the use of chemicals for making printed circuit boards in the startups, presence of human beings, particulate matter suspended due to tests conducted in electric vehicles laboratory and other contamination. Furthermore, some parts of the lab does not have air-conditioners rather fans for circulating air. There are exhaust fans for pushing the indoor air outside, but fresh air induction is currently only by opening doors.

The VRV systems are located in the electric vehicle lab (EV lab), building automation lab, research center on artificial pancreas, and software development lab. There is also the IAQ lab from where the hybrid ventilation system is to be implemented for maintaining IAQ, thermal comfort and energy optimization. Additionally, we plan not to make modifications to existing systems for example, the VRV system control or installing fresh air dampers in the space to avoid costly retrofits. In our analysis, the IAQ is modeled as

$$J_{IAQ} = f(PM, CO_2, TVOC, H, T) \tag{1}$$

where $PM, CO_2, TVOC, H,$ and $T$ denote the particulate matter, carbon dioxide(measured in ppm), total volatile organic compounds (measured in ppb), humidity, and temperature, respectively. These factors are time-varying and their dynamic is complex to model. The control variable that can influence the IAQ is the fresh air induction rate into the rooms $u_{FA}$, the fraction of total mass flow rate of fresh air that can be supplied to the room $m_{FA}$, and is given by

$$u_{FA} = (1 - d) \times m_{FA} \tag{2}$$

where $(1 - d)m_{FA}$ denotes the fresh air fraction that is supplied to the room. Similarly, to control the temperature and thermal comfort, we aim to change the control from being a simple thermostat to have additional degrees of freedom. This is required as there is no coarse control on the space possible with thermostat. To this extent, we could modify the time for which the air-conditioner is turned ON or OFF to modulate the average cooling power being supplied and this provides finer control over the temperature with pulse-width modulation sort of control. To this extent, we modify the control variable from being set-temperature to ON/OFF time. Assuming that the period for which the energy optimization is $T_p$ which depends on the time constant of the room, then the control variable $u_{AC}$ models the time for which the AC is turned ON and is given by

$$u_{AC} = \frac{t_{ON}}{T_p} \tag{3}$$

where $t_{ON}$ is the time-period for which the AC stays ON. However, a turn-off AC could cause the air circulation to go down, thereby making occupants to feel thermal discomfort. This could be avoided by turning ON fans during such times. This procedure of turning on AC and fans, we term it as toggling. The energy optimization problem then could be implemented as a toggling action between the AC and fan. The duty ratio for the fan is

$$u_{Fan} = \frac{T_p - t_{ON}}{T_p} \tag{4}$$

We used eight fans in our work. Their heat generation was not considered in our study as they were very minimal compared to the room size and cooling load.

The energy consumption in AC and fan is given by

$$J_{Power} = u_{AC} \times P_{AC} + u_{Fan} \times P_{Fan} \tag{5}$$

where $P_{AC}$ and $P_{Fan}$ are the power ratings of the fan and air-conditioner, respectively. The thermal comfort of the occupant is modeled using the temperature bounds on the room which is given by

$$T_{min} \leq T(k) \leq T_{max} \tag{6}$$

where $T(k)$ is the temperature at time-instant $k$ and $T_{min}, T_{max}$ denote, respectively, the minimum and maximum temperature supplied by the occupant based on their thermal comfort. Determining that the temperature is within the user defined comfort band requires model of the zone corresponding to the ON/OFF times of the air-conditioner and fan. However, fan does not vary the room temperature, but only the skin temperature of the occupant. Therefore, the zone temperature depends on the current room temperature $T(k)$, ambient temperature $w(k)$ and stray heating due to occupancy, lighting loads and others modeled as a disturbance term $v(k)$. Following [2], we model the room temperature dynamics to be

$$T(k+1) = aT(k) - bu_{AC}(k) + cw(k) + v(k) \tag{7}$$

where $a, b, c$ denote the thermal time constant of the room, the parameter that models the influence of air-conditioner cooling energy, and effect of weather on the room temperature, respectively. Similarly, there exists a minimum and maximum time for which the air-conditioner can be turned ON/OFF. This is modeled as

$$u_{AC_{min}} \leq u_{AC}(k) \leq u_{AC_{max}} \tag{8}$$

where $u_{AC_{min}}$ and $u_{AC_{max}}$ denote the minimum and maximum duration for which AC can be turned ON/OFF for the considered duration. Based on the model, we select $T_p$ to be 15 min as a change in control input gets reflected after 8–12 min of air-conditioner operation. The problem of maintaining IAQ, reducing energy consumption and maintaining user thermal comfort can be modeled as a multi-objective model predictive controller given by

$$J = \min_{u_{AC}, u_{FA}} \sum_{k=k+1}^{k+N_p} J_{IAQ}(k) + J_{Power}(k) \tag{9}$$

s. t.

Building thermal dynamics: $T(k+1) = aT(k) - bu_{AC}(k) + cw(k) + v(k)$,

Thermal comfort constraints: $T_{min} \leq T(k) \leq T_{max}$,

Control input limits: $u_{AC_{min}} \leq u_{AC}(k) \leq u_{AC_{max}}$,

Fresh air limits: $u_{FA_{min}} \leq u_{AC}(k) \leq u_{FA_{max}}$,

TVOC limits: $0 \leq TVOC(k) \leq \epsilon_{VOC}$,

$CO_2$ limits: $\epsilon_{CO_2}^{min} \leq CO_2(k) \leq \epsilon_{CO_2}^{max}$,

PM limits: $0 \leq PM(k) \leq \epsilon_{PM}$.

where $\epsilon_{TVOC}, \epsilon_{CO_2}$, and $\epsilon_{PM}$ are the bounds on the IAQ and could be computed as per the ASHRAE standard. The problem in (9) has the following challenges.

(C1) The function $J_{IAQ}$ and the variables (TVOC, $CO_2$, PM) are difficult to model with dynamical equations as they are highly time-varying with respect to the fresh air induction, temperature, humidity, and other factors influencing the thermal dynamics and IAQ in the room.
(C2) The problem in (9) is a multi-time step, multi-objective optimization problem and solving it in IoT devices is a challenging task.
(C3) Implementing the MPC requires both measurements on IAQ variables as well as forecasts. Obtaining such forecasts is complex task.
(C4) The proposed controller should coordinate the action of the fresh air damper and energy optimizer for the air-conditioner which is difficult due to absence of equations describing their interconnections.
(C5) Sensor and the IoT architecture for implementing the controller above should be implemented.

In what follows, we address the challenges above and modify the existing VRV system into an effective ventilation scheme.

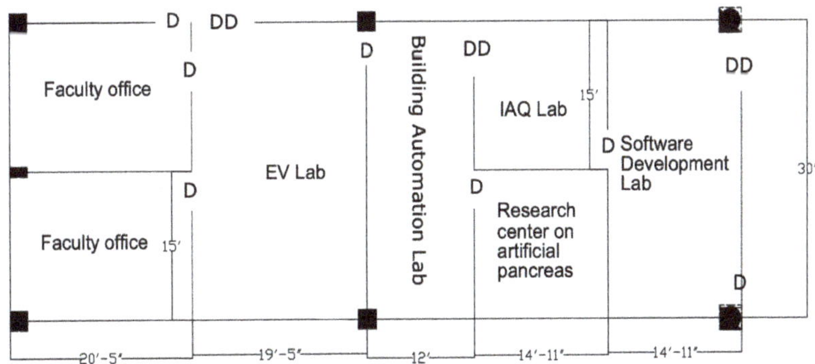


**Figure 1.** C-RACE laboratory layout located at Kalasalingam University, India (D–Door, DD–Double Door).

## 3. Novel Hybrid Ventilation System and Soft-Sensor

To implement the MPC with multiple objectives, we first decompose the problem into two parts: (i) Energy optimizer and (ii) IAQ optimizer. Both the optimizers are coupled through the fresh air induction, which influences the thermal dynamics of the zone thereby the energy consumption. Our idea of energy optimizer is to design a device that switches between air-conditioner and fan to maintain thermal comfort but reduce energy consumption using a predictive controller. To this extent, we first require sensor measurements on temperature, humidity and occupancy (to detect cooling loads). Then, the next step is to design the energy optimizer based on the fresh air flow which will be discussed later.

### 3.1. Sensor Module: Energy Optimizer

The sensor module for the energy optimizer is shown in Figure 2. It consists of a temperature and humidity sensor which we use a DHT11 sensor which can measure both temperature and humidity. To measure occupancy, we use carbon dioxide sensor and passive infra-red (PIR) sensor. The carbon dioxide sensor has delays to measure the occupancy and PIR sensor have errors in counting. To overcome this, we build a soft-sensor on top of the sensor to validate the $CO_2$ levels (see Section 3.3).

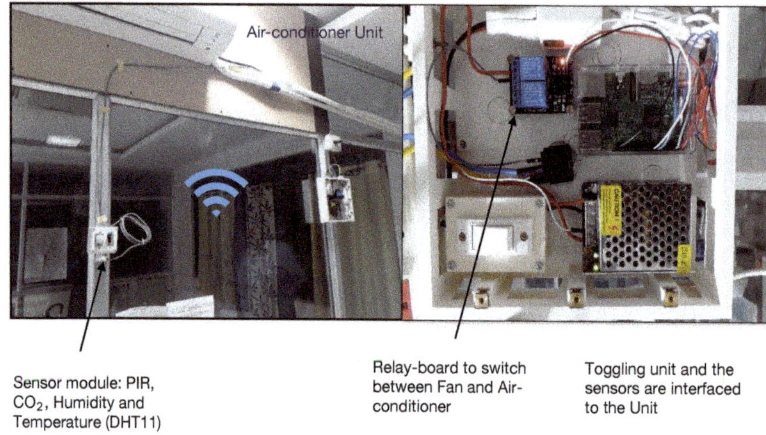


**Figure 2.** The sensor and controller units used for energy savings.

The PIR motion sensor, carbon dioxide sensor, temperature, and humidity sensor are interfaced to RPi. The RPi is connected to the Ethernet and it acts as a gateway to transmit the sensed information to the server. We use the Message Queuing Telemetry Transport protocol at the application layer to transmit the sensed information. The server uses SQL routines to store the information in the database.

### 3.2. Energy and IAQ Optimizer

The second step is to implement the control that saves energy, maintains thermal comfort and guarantees IAQ. However, as mentioned earlier these are competing objectives and also there are challenges (C1)–(C5) that needs to be addressed. To address challenges (C1) and (C2), we first decouple the problem of maintaining IAQ and energy optimization plus thermal comfort by designing a hybrid ventilation system. Next, we solve the problems independently but exploit the power of IoT to couple them. In what follows, we describe the different elements used in our implementation.

The hybrid ventilation system consists of two parts: energy optimizer and IAQ optimizer. To decouple the multiple objectives but coordinate the actions, we select the fresh air infusion flow rate as the variable coupling the energy and IAQ optimizers. One can observe that the problem of fresh air induction is quite challenging as the function to approximate IAQ is time-varying and depends on too many parameters. To overcome this, instead of designing a fresh air damper, we use a demand controlled ventilation system for maintaining the IAQ. In this paper, we select the RENSONs' Healthbox3.0 based on analysis performed with various scheme. The Healthbox3.0 is a DCV that controls fresh air induction based on TVOC. (measured in ppb), $CO_2$ (measured in ppm), humidity, and temperature. The installation of the RENSON Healthbox 3.0 in the C-RACE lab is shown in Figure 3. Consequently, the DCV will introduce fresh air depending on the IAQ in the particular room. Still particulate matter is not considered and we install PM2.5 sensor additionally to measure this variable. The Healthbox 3.0 is installed in the IAQ lab of C-RACE and it samples air from the different labs to change the fresh air induction. The system has inbuilt sensors to measure the IAQ variables which could be transmitted to database through simple SQL scripts by writing the variables into an excel or CSV files.

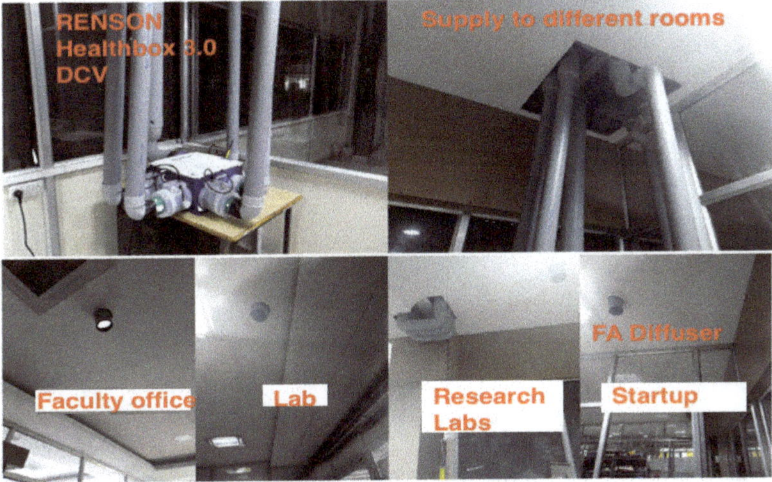


**Figure 3.** RENSON Healthbox 3.0 DCV Installation.

The next step is to connect the DCV to the IoT and other interfaces. However, the sensors need to be calibrated. To this extent, we first install the mobileApp provided by RENSON and then the calibration is performed. The flow information about the fresh air is then passed on to the Gateway (RPi) which uses this information to optimize the energy and also for the soft-sensor to compute carbon dioxide. The hybrid ventilation scheme is shown in Figure 4.

As the IAQ objective is now decoupled with the DCV, the energy optimizer solves the following optimization problem to implement the MPC.

$$J = \min_{u_{AC}} \sum_{k=k+1}^{k+N_p} P_{AC} u_{AC} \tag{10}$$

s. t.

Building thermal dynamics: $T(k+1) = aT(k) - bu_{AC}(k) + cw(k) + v(k)$

Thermal comfort constraints: $T_{min} \leq T(k) \leq T_{max}$

Control input limits: $u_{AC_{min}} \leq u_{AC}(k) \leq u_{AC_{max}} \qquad \forall k \in \{k+1, \ldots, k+N_p\}$

The problem in Equation (10) is a linear programming problem and we used Gnu Linear Programming Kit (GLPK) to implement the controller in RPi. The implemented controller obtained weather forecasts using webservices in Python, and the cooling load estimates were obtained based on the $CO_2$ predictions from soft-sensor. To address challenge (C4), i.e., absence of coupling between energy and IAQ optimization, we use the flow information and this coupling is solved inherently. Moreover, the hybrid ventilation system also implemented the IoT architecture for realizing the controller addressing the challenge (C5). The only problem to be tackled now is the development of soft-sensor for IAQ which is presented in the rest of the section.

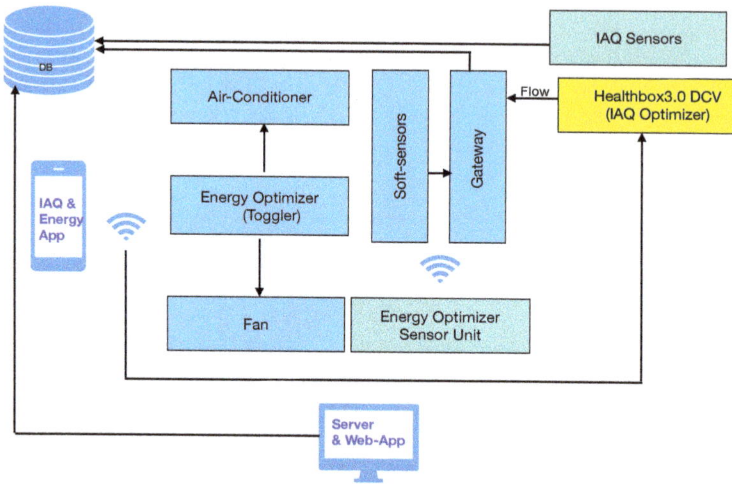


**Figure 4.** Schematic of the novel hybrid ventilation system.

### 3.3. Soft-Sensor for IAQ and Cooling Loads

As stated earlier, IAQ and cooling load due to fresh air induction due to DCV is a challenging problem due to complex dynamics of these variables. Finding dynamical equations for modeling it is a challenging task. Therefore, it makes perfect sense to develop models using databased methods to estimate the IAQ and cooling load. In this context, time-series data of IAQ variables is available from the hybrid ventilation system designed by us. Therefore, data required for building soft-sensors required for our application is available to us as shown in Figure 5. These are IAQ sensor readings (temperature, RH, $CO_2$ and TVOC) collected from the Helathbox 3.0 for 10 days with a sampling interval of 15 min. To study the dependence of the parameters, we studied the correlation among the parameters (see, Table 1). One can see that there is no strong correlation between the parameters. Therefore, the univariate forecasting model is used to simplify our analysis. Similarly, the cooling load should be estimated from coupling variable fresh air flow in individual zones. However, cooling loads are also affected by occupancy, stray heating loads such as lighting, computers etc. Therefore, building soft-sensors for cooling loads using data-based methods requires rethinking. In this paper, we use the zone thermal dynamics to estimate the cooling load as a function of fresh air flow and heat generated in the zone as will be illustrated in this section.

**Table 1.** Correlation matrix of indoor air quality (IAQ) parameters.

|        | TVOC  | RH    | $CO_2$ | Temp  |
|--------|-------|-------|--------|-------|
| TVOC   | 1     | −0.07 | 0.03   | −0.05 |
| RH     | −0.07 | 1     | −0.45  | 0.05  |
| $CO_2$ | 0.03  | −0.45 | 1      | 0.3   |
| temp   | −0.05 | 0.05  | 0.3    | 1     |

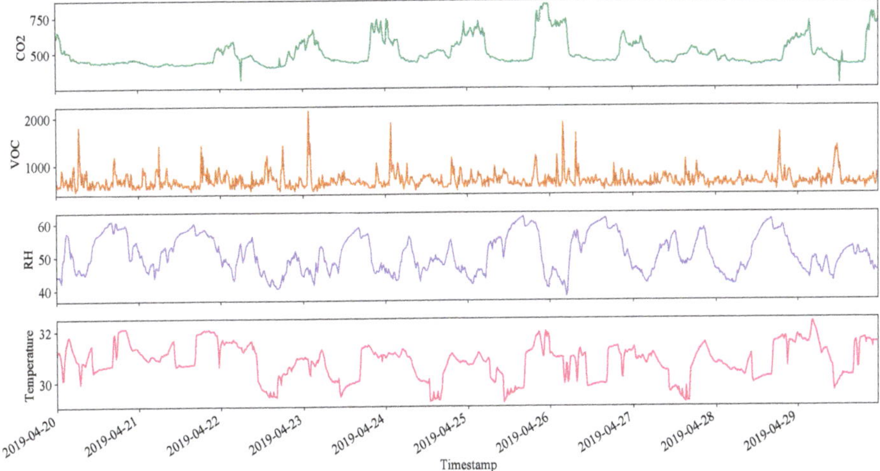

**Figure 5.** IAQ measurements for 10 days with 15 min interval.

### 3.3.1. Soft-Sensors for Cooling Load Measurement

The cooling load is also due to fresh air induction and the effect of cooling load is measured from the prediction equation using the zone thermal dynamic model as

$$Q(k) = T(k+1) - aT(k) + bu_{AC}(k) + cw(k) \tag{11}$$

The above equation gives an estimate of the cooling load as well as the effect of fresh air induction. The fresh air induction and the raise in temperature due to weather, current room temperature, and control input, i.e., the time for which the air-conditioner is kept ON/OFF is used to train the cooling load estimator. We use a simple radial basis function neural networks (RBFNN) to estimate the cooling load.

### 3.3.2. Soft-Sensors for IAQ

As our idea is to design a soft-sensor for IAQ with univariate analysis, we consider the use of deep neural networks as they tend to outperform other models used in computer vision. In recent years, deep neural networks have been shown to outperform previous state-of-the-art machine learning approaches in various application domains [25]. Convolutional Neural Networks (CNN) is a special type of deep neural networks which is mainly used to handle 2D input data such as images. A CNN architecture is built by stacking three main types of layers: *convolutional, pooling,* and *fully-connected.* The convolutional layer maps the input data to a feature map by performing the convolution operation (dot product between the input and a filter) by sliding over the input data. The output of convolution operation is passed through an activation function that controls neuron activation. Rectified Linear Unit (ReLU) is a commonly used activation function to introduce non-linearity into the neuron's output. The pooling layers are often included between the successive convolutional layers. They are useful in reducing the dimensionality of feature maps to avoid overfitting. Max pooling is commonly kept as it is the most dominant spatial relationship. Next, the pooling layer output is flattened and given as input to the fully connected dense layer that computes the final model output.

Our CNN model architecture for short-term forecasting is shown in Figure 6. It takes the past IAQ sensor readings as input and performs a one-step prediction. In our experiments, the past 24 sensor readings (6 h long) are used as input. Next, we added two 1-dimensional convolutional layers (kernel size 2) with ReLU activation function that learns the trend and seasonality in the sensor data.

This is followed by a 1-dimensional max pooling layer (pool size 2) that reduces the feature map dimension into half. Consequently, the pooling layer output is fed to the fully connected layer that yields the final prediction. We implemented our CNN architecture in Python 3 using the keras library (https://keras.io/). An early stopping method was also used to optimize the training process.

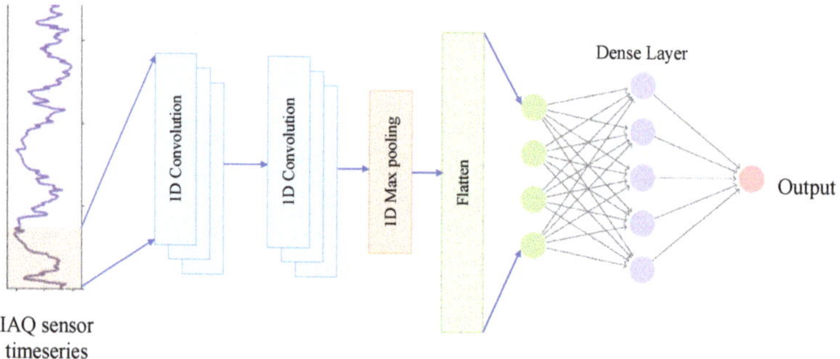


**Figure 6.** The convolutional neural network (CNN) model architecture for forecasting IAQ sensor data.

*3.4. Calibration of IAQ Sensors Using MobileApp*

To calibrate the sensors, a mobileApp provided by Renson was used. It is an Android application for Renson Healthbox 3.0 that allows the user to access and calibrate the device. Calibration is done by establishing connection between the device and mobile. The calibration tool performs auto-configuration and calibration is done for the actuators and sensors using the mobileApp automatically. Once installed, the Renson Healthbox 3.0 could connect to cloud and other platforms to port the data. There is also a portal provided by the Healthbox 3.0 for porting the data. The calibration is for control valves that open the vanes for supplying fresh air and takes up to 3–6 min. This will adjust the nominal air-flow rate to individual zones as well. The calibration procedure with the mobileApp is shown in Figure 7.

**Figure 7.** Calibration with MobileApp for Renson Healthbox 3.0.

## 4. Results

### 4.1. Soft-Sensors for IAQ

The soft sensor effectiveness was validated using the data collected from the hybrid ventilation system in Figure 4. We used 80% of our dataset to training our CNN models and the remaining 20% for testing. The descriptive statistics of the collected data is shown in Table 2. We summarize the performance of our short term forecasting model for each IAQ sensor using Mean Absolute Error (MAE) and Mean Absolute Percentage Error (MAPE) in Table 3 and visualization in Figure 8. The forecasting error of TVOC is high compared to other sensors due to no seasonality in the raw data. This requires further investigation, possibly extending our CNN model with Long Short-Term Memory [26]. The forecasting error of all other sensors is less than 5%.

**Table 2.** Descriptive statistics of the collected IAQ sensor data.

| Statistics | $CO_2$ | TVOC | RH | Temperature |
|---|---|---|---|---|
| **Mean** | 492.90 | 656.81 | 50.70 | 30.79 |
| **Std** | 89.23 | 172.37 | 5.62 | 0.68 |
| **Min** | 279.67 | 450.00 | 37.90 | 29.20 |
| **25%** | 433.33 | 567.89 | 45.93 | 30.27 |
| **50%** | 453.83 | 611.24 | 49.89 | 30.89 |
| **75%** | 525.92 | 700.16 | 55.76 | 31.28 |
| **Max** | 839.00 | 2,140.36 | 62.11 | 32.32 |

**Table 3.** Comparison of short-term forecasting results using mean absolute percentage error (MAPE) and mean absolute error (MAE).

| Error Metrics | $CO_2$ | TVOC | RH | Temperature |
|---|---|---|---|---|
| **MAPE (%)** | 4.1 | 13.5 | 1.9 | 0.6 |
| **MAE** | 19.1 | 96.9 | 1.0 | 0.2 |

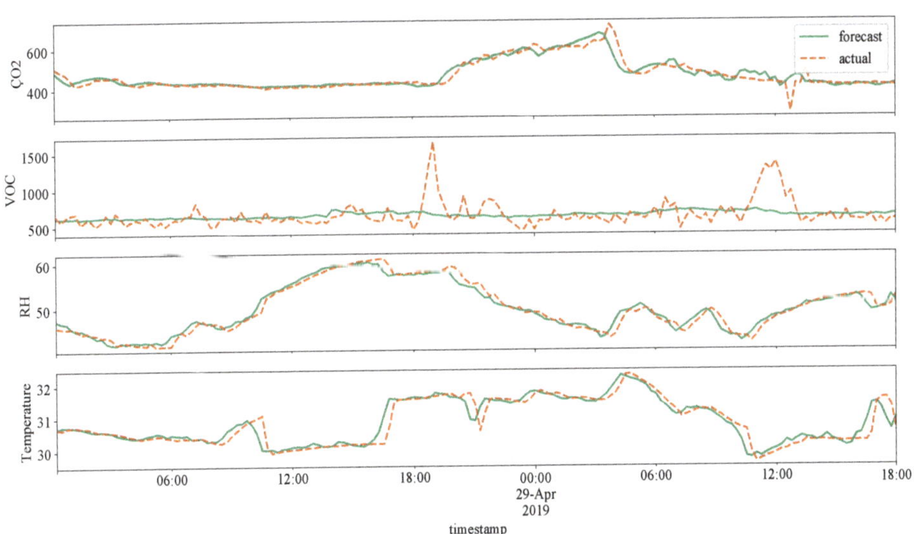


**Figure 8.** Visualization of IAQ sensor forecasting using CNN model.

As an example, consider the scenario wherein the objective is to maintain both IAQ and minimize power consumption. Suppose, if the IAQ is bad, then the Healthbox 3.0 which is a DCV introduces more fresh air which increases the cooling load. However, the DCV controls the IAQ parameters efficiently. On the other hand, the increase in cooling load is forecasted as increase in cooling loads and fed to the toggler which is a model based controller. It fuses this information into the model and predicts future temperature evolution in a particular zone and then tries to optimize the energy supplied to the individual zones based on temperature set-points provided by the user by adjusting the duty cycle of the toggler. This way both IAQ and energy is optimized by the hybrid ventilation scheme.

### 4.2. Indoor Air Quality

The performance of the DCV with the proposed energy optimizer is evaluated. For this study, we first collect information on the ventilated space over different periods. Then, we compare the results with and without hybrid ventilation system. It was seen that without the hybrid ventilation system the TVOC, carbon dioxide, temperature, and humidity levels were not maintained, as there was no control on IAQ and space cooling.

The variations in humidity with the hybrid ventilation scheme and the energy optimizer is shown in Figure 9. One can see that the humidity is maintained within acceptable values of 60 even during the nights when the ventilation system is shut out. This is due to the pre-conditioning created by the fresh air introduced by the proposed hybrid ventilation system. Contrary to this before installing the Healthbox 3.0, the humidity cannot be maintained and the behavior is almost random.

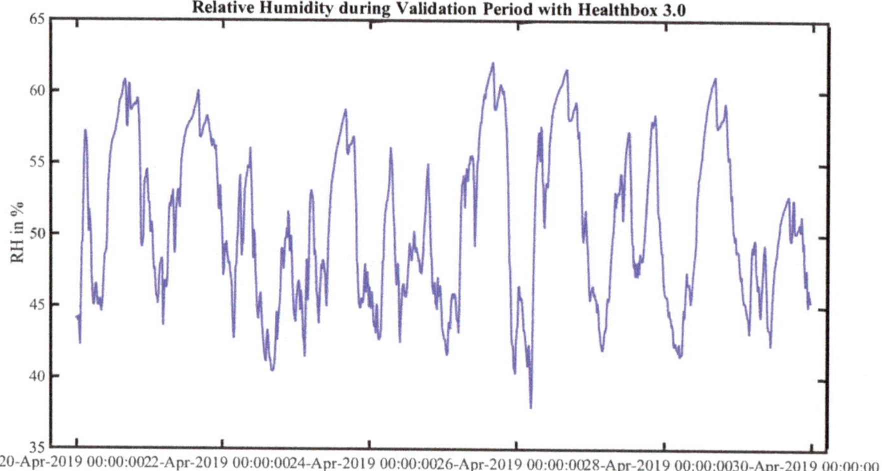


**Figure 9.** Humidity variations with the hybrid ventilation system.

The variation in $CO_2$ levels with the hybrid ventilation system installed is shown in Figure 10. The $CO_2$ levels are well maintained with the proposed energy and IAQ optimizer. A clear pattern of the variation is shown and the maximum value reaches 850 ppm though there are approximately 5–6 occupants in the room. This results shows the capability of the proposed system to maintain $CO_2$ levels. While the Helathbox maintains the IAQ, the energy optimizer also coordinates its action by using the fresh-air flow as the variable.

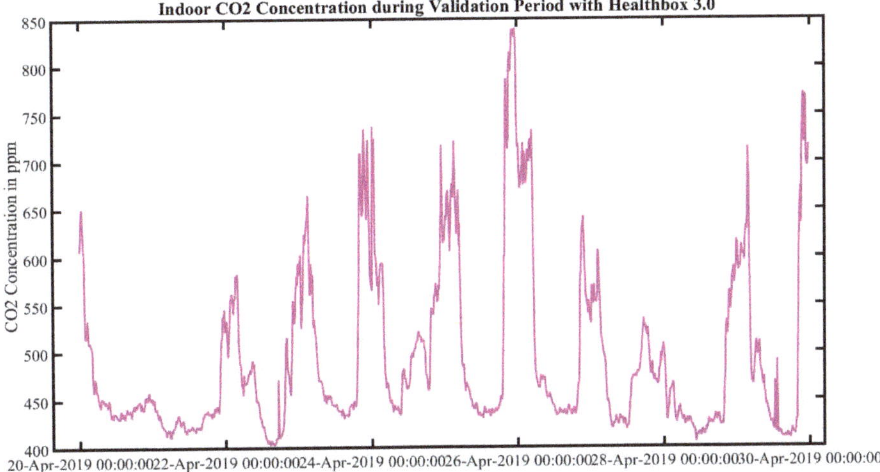

**Figure 10.** The carbon dioxide levels with Hybrid Ventilation system.

The variations in TVOC with the proposed hybrid ventilation system is shown in Figure 11. One can see that the TVOC is maintained within a ppb of 1000 during normal days and exceeds to 1400 only when the system is turned OFF. This results shows the ability of the system to maintain IAQ by limiting the TVOC.

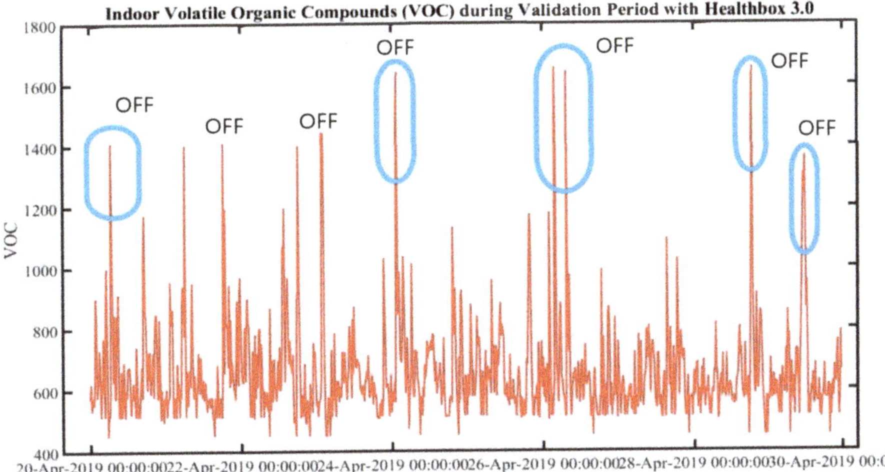


**Figure 11.** The Total Volatile Organic Compound levels with Hybrid Ventilation system.

### 4.3. IAQ and Thermal Comfort

As seen in the results, one can see that the IAQ variables are maintained within acceptable values throughout the period, and the temperature is also maintained within user-defined comfort values by the system. The occasional spurts represent the night times when the system is kept off, but the temperature is maintained well within the user defined comfort margin of 20–25 degree °Celsius.

### 4.4. Energy Optimizer

The performance of the energy optimizer for a 12 h period with a sampling time of 15 min is shown in Figure 12. Here, the control input is the duty cycle, i.e., $u_{AC} = \frac{t_{ON}}{T_p}$. A value of $u_{AC} = 0.5$ means that the air-conditioner should be operated for 7.5 min within 15 min. However, this gives rise to frequent switching and transient energy gets wasted. Therefore, to overcome this, we use an averaging approach, i.e., the control inputs computed at two time-instants are implemented for a time-frame of 30 min. For example, $u_{AC}(k) = 0.5$ and $u_{AC}(k-1) = 0.3$, where $k-1$ was not implemented to avoid transient energy loss, then a total of 12.5 min is turned ON during the $k$th time-instant. Such an averaging may sometimes result in the zone temperature to raise above the comfort band of 28 °C. However, considering that the time constant of the zone is very high in the order of 12–13 min, this is negligible period for which the thermal comfort could be breached by a small margin, but the savings in transient energy is quite high. Therefore, we implement the wait and apply strategy to reduce the transient energy consumption.

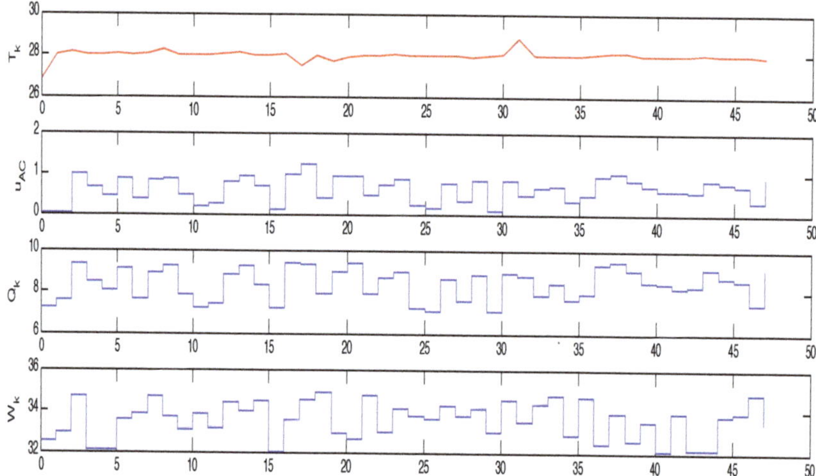

**Figure 12.** Variations in zone temperature, control input, heating due to cooling load, and ambient temperature (*x*-axis: 1 unit = 15 min).

The energy consumption was recorded with and without the proposed control strategy. We found that an average saving of 19.2–21.4% was observed with the proposed control strategy over conventional thermostat control while maintaining the IAQ. These savings are for 12 h period in a working day. The savings are mainly due to switching between fan and air conditioner. This result illustrates the energy savings potential of the proposed hybrid ventilation and control scheme.

Our results showed that the proposed hybrid ventilation scheme achieved good IAQ and energy savings without breaching the comfort band proposed by the occupant. We also showed that an energy savings of up to 19–21% can be achieved, and the IAQ parameters were maintained within bounds specified. We also demonstrated the role of soft-sensors and implementation aspect of the proposed hybrid ventilation scheme, i.e., demand controlled ventilation, sensors for measuring IAQ parameters and energy relevant data, communication architecture and other aspects were also presented. The results demonstrated the value of soft-sensor to measure parameters like cooling loads, TVOC, carbon dioxide concentration and others which were not possible before. An additional benefit of soft-sensor is the ability to generate short-term predictions which could be used to plan ventilation schedules.

## 5. Conclusions

This paper presented a novel design of a hybrid ventilation system for optimizing energy consumption in variable air volume (VRV) systems while maintaining indoor air quality (IAQ) and the thermal comfort of the building occupants. We showed that the problem is multi-objective, multi-time step optimization problem as the objective to optimize both IAQ and energy requires a model predictive controller (MPC). As such implementing MPC schemes on low-cost hardware is a challenge and the absence of sensed information/predictions on IAQ and cooling loads posed additional challenges. To overcome this, we proposed IoT based sensing that was extended with deep learning tools to design soft-sensors that provided measurements on parameters, which could not be measured with physical sensors. Then, the multiple objectives was decoupled into IAQ and energy savings. To achieve IAQ we proposed a novel hybrid ventilation system which used RENSONs' Healthbox 3.0, which is an off-the-shelf demand controlled ventilation. On the other hand, we designed an MPC in Rasperry Pi using the Gnu Linear Programming Kit (GLPK) which provided optimal solution to the multi-time step optimization problem to realize the energy optimizer. We provided results which showed the performance of soft-sensors, IAQ and energy optimization capabilities of the hybrid ventilation scheme while meeting user defined comfort bands. Extending the hybrid ventilation scheme to other IAQ measures and including thermal comfort with predictive mean vote or other standards is the future course of this investigation.

**Author Contributions:** Conceptualization, K.Z. and S.S. (Seshadhri Srinivasan); methodology, S.S. (Seshadhri Srinivasan) and P.A.; software, N.V. and S.S. (Subathra Seshadhri); validation, S.S. (Subathra Seshadhri), N.V. and K.Z.; writing—original draft preparation, S.S. (Seshadhri Srinivasan); writing—review and editing, P.A., K.Z. and S.S. (Subathra Seshadhri); supervision, S.S. (Seshadhri Srinivasan); project administration, S.S. (Seshadhri Srinivasan) and S.S. (Subathra Seshadhri); funding acquisition, S.S. (Seshadhri Srinivasan) and K.Z. All authors have read and agreed to the published version of the manuscript.

**Funding:** This project is supported by the project Resilient and Optimal-micro-energy grid (ROME): Funded by Department of Science and Technology and Research Council of Norway, Norway through Resilient and Optimal Micro-Energy grid INT/NOR/RCN/ICT/P-05/2018.

**Acknowledgments:** The authors thank RENSON Belgium for providing support with Healthbox 3.0. In particular, we would like to thank Makarand Kendre from RENSON India for his support in the initiative. The authors also thank Northumbria University, UK for providing support to do the open access publication. The authors also thank the Center for Research in Automatic Control Engineering, International Research Center, Kalasalingam University for its support to conduct the research.

**Conflicts of Interest:** The authors declare no conflict of interest.

## Abbreviations

The following abbreviations are used in this manuscript.

| | |
|---|---|
| HVAC | Heating, Ventilation and Air-Conditioning |
| VRV | Variable Refrigerant Volume |
| IoT | Internet of Things |
| IAQ | Indoor Air Quality |
| TVOC | Total Volatile Organic Compound |
| PM | Particulate Matter |
| MPC | Model Predictive Control |
| ASHRAE | American Society of Heating, Refrigerating and Air-Conditioning Engineers |
| DCV | Demand Controlled Ventilation |
| CNN | Convolution Neural Network |
| EPA | Environmental Protection Agency |
| C-RACE | Center for Research in Automatic Control Engineering |

## References

1. Chaturvedi, V.; Eom, J.; Clarke, L.E.; Shukla, P.R. Long term building energy demand for India: Disaggregating end use energy services in an integrated assessment modeling framework. *Energy Policy* **2014**, *64*, 226–242. [CrossRef]
2. Soudari, M.; Srinivasan, S.; Balasubramanian, S.; Vain, J.; Kotta, U. Learning based personalized energy management systems for residential buildings. *Energy Build.* **2016**, *127*, 953–968. [CrossRef]
3. Radhakrishnan, N.; Srinivasan, S.; Su, R.; Poolla, K. Learning-based hierarchical distributed HVAC scheduling with operational constraints. *IEEE Trans. Control Syst. Technol.* **2017**, *26*, 1892–1900. [CrossRef]
4. Lube, G.; Knopp, T.; Rapin, G.; Gritzki, R.; Rösler, M. Stabilized finite element methods to predict ventilation efficiency and thermal comfort in buildings. *Int. J. Numer. Methods Fluids* **2008**, *57*, 1269–1290. [CrossRef]
5. Mendes, A.; Pereira, C.; Mendes, D.; Aguiar, L.; Neves, P.; Silva, S.; Batterman, S.; Teixeira, J. Indoor air quality and thermal comfort—Results of a pilot study in elderly care centers in Portugal. *J. Toxicol. Environ. Health Part A* **2013**, *76*, 333–344. [CrossRef] [PubMed]
6. Baniassadi, A.; Sailor, D.J.; Olenick, C.R. Indoor air quality and thermal comfort for elderly residents in Houston TX—A case study. In Proceedings of the 7th International Building Physics Conference, Syracuse, NY, USA, 23–26 September 2018; pp. 787–792.
7. Heracleous, C.; Michael, A. Experimental assessment of the impact of natural ventilation on indoor air quality and thermal comfort conditions of educational buildings in the Eastern Mediterranean region during the heating period. *J. Build. Eng.* **2019**, *26*, 100917. [CrossRef]
8. Huizenga, C.; Abbaszadeh, S.; Zagreus, L.; Arens, E.A. Air quality and thermal comfort in office buildings: Results of a large indoor environmental quality survey. In Proceedings of the Healthy Buildings, Lisboa, Portugal, 4–8 June 2006; Volume III, pp. 393–397.
9. Zhang, X.; Ding, X.; Wang, X.; Talifu, D.; Wang, G.; Zhang, Y.; Abulizi, A. Volatile Organic Compounds in a Petrochemical Region in Arid of NW China: Chemical Reactivity and Source Apportionment. *Atmosphere* **2019**, *10*, 641. [CrossRef]
10. Wu, Z.; Rajamani, M.R.; Rawlings, J.B.; Stoustrup, J. Model predictive control of thermal comfort and indoor air quality in livestock stable. In Proceedings of the 2007 European Control Conference (ECC), Kos, Greece, 2–5 July 2007; pp. 4746–4751.
11. Kolokotsa, D.; Pouliezos, A.; Stavrakakis, G.; Lazos, C. Predictive control techniques for energy and indoor environmental quality management in buildings. *Build. Environ.* **2009**, *44*, 1850–1863. [CrossRef]
12. Liu, H.; Lee, S.; Kim, M.; Shi, H.; Kim, J.T.; Wasewar, K.L.; Yoo, C. Multi-objective optimization of indoor air quality control and energy consumption minimization in a subway ventilation system. *Energy Build.* **2013**, *66*, 553–561. [CrossRef]
13. Becker, R.; Goldberger, I.; Paciuk, M. Improving energy performance of school buildings while ensuring indoor air quality ventilation. *Build. Environ.* **2007**, *42*, 3261–3276. [CrossRef]
14. Hesaraki, A.; Holmberg, S. Demand-controlled ventilation in new residential buildings: Consequences on indoor air quality and energy savings. *Indoor Built Environ.* **2015**, *24*, 162–173. [CrossRef]
15. Png, E.; Srinivasan, S.; Bekiroglu, K.; Chaoyang, J.; Su, R.; Poolla, K. An internet of things upgrade for smart and scalable heating, ventilation and air-conditioning control in commercial buildings. *Appl. Energy* **2019**, *239*, 408–424. [CrossRef]
16. Anand, P.; Sekhar, C.; Cheong, D.; Santamouris, M.; Kondepudi, S. Occupancy-based zone-level VAV system control implications on thermal comfort, ventilation, indoor air quality and building energy efficiency. *Energy Build.* **2019**, *204*, 109473. [CrossRef]
17. Rivera, M.I.; Kwok, A.G. Thermal comfort and air quality in Chilean schools, perceptions of students and teachers. In Proceedings of the ARCC Conference Repository, Toronto, ON, Canada, 29 May–1 June 2019.
18. Mei, J.; Xia, X. Energy-efficient predictive control of indoor thermal comfort and air quality in a direct expansion air conditioning system. *Appl. Energy* **2017**, *195*, 439–452. [CrossRef]
19. Yang, X.; Yang, L.; Zhang, J. A WiFi-enabled indoor air quality monitoring and control system: The design and control experiments. In Proceedings of the 2017 13th IEEE International Conference on Control & Automation (ICCA), Ohrid, Macedonia, 3–6 July 2017; pp. 927–932.

20. Jiang, C.; Chen, Z.; Png, L.C.; Bekiroglu, K.; Srinivasan, S.; Su, R. Building Occupancy Detection from Carbon-dioxide and Motion Sensors. In Proceedings of the 2018 15th International Conference on Control, Automation, Robotics and Vision (ICARCV), Singapore, 18–21 November 2018; pp. 931–936.

21. Zingre, K.; Srinivasan, S.; Marzband, M. Cooling load estimation using machine learning techniques. In Proceedings of the 32nd International Conference on Efficiency, Cost, Optimization, Simulation and Environmental Impact of Energy Systems, Warsaw, Poland, 23–28 June 2019.

22. Karagulian, F.; Barbiere, M.; Kotsev, A.; Spinelle, L.; Gerboles, M.; Lagler, F.; Redon, N.; Crunaire, S.; Borowiak, A. Review of the Performance of Low-Cost Sensors for Air Quality Monitoring. *Atmosphere* **2019**, *10*, 506. [CrossRef]

23. Sun, X.; Xu, W. Deep Random Subspace Learning: A Spatial-Temporal Modeling Approach for Air Quality Prediction. *Atmosphere* **2019**, *10*, 560. [CrossRef]

24. Misra, P.; Imasu, R.; Takeuchi, W. Impact of Urban Growth on Air Quality in Indian Cities Using Hierarchical Bayesian Approach. *Atmosphere* **2019**, *10*, 517. [CrossRef]

25. LeCun, Y.; Bengio, Y.; Hinton, G. Deep learning. *Nature* **2015**, *521*, 436–444. [CrossRef] [PubMed]

26. Hochreiter, S.; Schmidhuber, J. Long short-term memory. *Neural Comput.* **1997**, *9*, 1735–1780. [CrossRef] [PubMed]

*Article*

# Adaption of an Evaporative Desert Cooler into a Liquid Desiccant Air Conditioner: Experimental and Numerical Analysis

**Mustafa Jaradat *, Mohammad Al-Addous and Aiman Albatayneh**

Department of Energy Engineering; German Jordanian University, Amman Madaba Street, P.O. Box 35247, Amman 11180, Jordan; Mohammad.Addous@gju.edu.jo (M.A.-A.); aiman.albatayneh@gju.edu.jo (A.A.)
* Correspondence: Mustafa.Jaradat@gju.edu.jo

Received: 25 November 2019; Accepted: 25 December 2019; Published: 28 December 2019

**Abstract:** Desert coolers have attracted much attention as an alternative to mechanical air conditioning systems, as they are proving to be of lower initial cost and significantly lower operating cost. However, the uncontrolled increase in the moisture content of the supply air is still a great issue for indoor air quality and human thermal comfort concerns. This paper represents an experimental and numerical investigation of a modified desert air cooler into a liquid desiccant air conditioner (LDAC). An experimental setup was established to explore the supply air properties for an adapted commercial desert cooler. Several experiments were performed for air–water and air–desiccant as flow media, at several solutions to air mass ratios. Furthermore, the experimental results were compared with the result of a numerical simplified effectiveness model. The outcomes indicate a sharp reduction in the air humidity ratio by applying the desiccant solutions up to 5.57 g/kg and up to 4.15 g/kg, corresponding to dew point temperatures of 9.5 °C and 12.4 °C for LiCl and $CaCl_2$, respectively. Additionally, the experimental and the numerical results concurred having shown the same pattern, with a maximal deviation of about 18% within the experimental uncertainties.

**Keywords:** desert cooler; evaporative cooling; indoor air quality; liquid desiccant; effectiveness model; moisture removal

---

## 1. Introduction

Air conditioning represents a considerable part of current electrical power consumption due to the significant growth in populations, in addition to rapid developments in lifestyle and comfort standards [1,2]. In certain countries of the Middle East and North Africa (MENA), for example, Jordan, Lebanon, Syria, and Egypt, having air conditioners was considered as a luxury in the last few decades. However, the climate changes and hot summers in the last years have made it a must. The recent heatwaves have sharply increased the demand for air conditioners as simple ventilators were not enough to moderate the heat.

The total global electrical energy consumption for cooling in buildings using air conditioners accounts for about 20% nowadays [3]. Additionally, the share of energy consumption for space cooling is higher for regions with hot climates. In some of Middle East countries, such as Arab Gulf countries, the domestic ownership of air conditioning units is close to 100%.

There are various techniques for space cooling. The most utilized is the conventional vapor compression air conditioners. Although conventional air conditioners have many advantages, such as their reliability, stability, effective heat transfer rates, and compact sizes, they still have the problem of high overall electricity demand and peak electricity loads [4]. Besides, handling the latent load in vapor compression systems requires cooling the supply air beneath its dew point temperature. Significant energy is required to reach the apparatus dew point (ADP) and further cooling below the ADP is required to compensate the bypass effect of the evaporator coils [5]. The dehumidified air must then be reheated to the required comfort level and this leads to extra energy waste [6]. Numerous issues have been associated with conventional coolers, starting with their relatively high costs, their operational costs, and maintenance costs; it has likewise been understood that recently the demand on using air conditioning for cooling has been increasing due to climate change and due to increase in the urban heat island impact, particularly in populated territories.

In some countries, because of the unaffordable conventional air conditioners (ACs) being accompanied by high operating expenses, people are now looking for alternatives of conventional power supply systems. Evaporative coolers have attracted much attention as an alternative to conventional air conditioning systems, as they are considered an economic solution to cool the supply air temperature [7]. Direct evaporative coolers use the latent heat of water to absorb heat from the supply air as it directly contacts the water and evaporates it. In this way, the system loses sensible heat with a lower dry bulb temperature while maintaining a constant wet bulb temperature. Simultaneously, the water vapor that evaporated is also added to the air, which increases the water content in the supply air leading to extreme indoor air quality (IAQ) issues and can increase the risk of respiratory diseases [8,9].

Regarding the most suitable climates where evaporative coolers are widely used, it has been noticed that they operate more efficiently in hot, dry regions; as the temperature of ambient air increases and its humidity decreases, the better will be the performance of the evaporative cooler at providing indoor thermal comfort. Those regions are the Middle East, Far East, some regions in North and West America, most regions in Australia, in addition to a number of European countries [10]. Figure 1 shows the regions that have hot and dry climates (red colored) that are preferable for desert coolers.

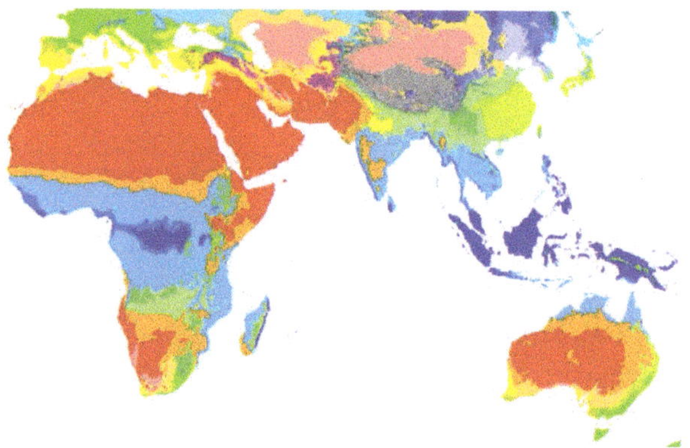

**Figure 1.** Köppen–Geiger climate classification map (1980–2016) [10].

Although evaporative cooling systems have been used as a more progressive and environmentally friendly replacement for conventional ACs, as it only requires electricity to drive a pump unlike vapor compression air conditioning systems which draw more electric energy [11–13], the issue remains with the fact that these systems rely on the ambient conditions. The cooling effectiveness of desert coolers is low comparing it to conventional vapor compression systems; they can also have relatively larger sizes and most importantly the potential for affecting human comfort is higher as it can increase the humidity level or even temperature may not be reduced enough. The increment in the water vapor level in the supplied air can lead to a fertile environment to sweat in, and for mold and bacterial growth. Furthermore, increased moisture levels can affect the stability of household hygroscopic material, such as woody furniture. Figure 2 shows the humidity levels been linked to comfort, mold growth, and the incidence of respiratory illness [14,15].

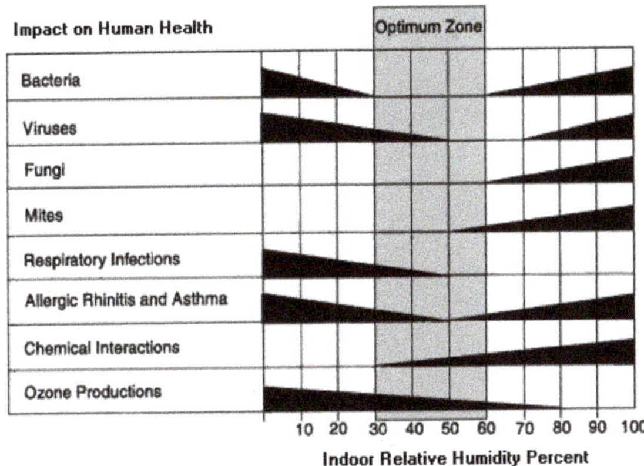



**Figure 2.** Recommended space humidity levels linked to its impact on human health [15].

Despite that evaporative cooling is known as an old system, there are few modifications in these systems, and very little regarding the humidity of the supply air leaving the cooler. Most of the modifications in the literature were coordinated toward the determination of air–water contact surfaces. Different cooling pads were evaluated regarding cooling efficiency, such as jute, luffa fibers, and palm fibers [16]; coarse and fine polyvinyl carbonate [17]; rice straw and palm leaf fibers [18]; and CELdek®, straw, and slice wood [19]. Additionally, several researchers have discussed the main factors when testing a cooling pad, such as surface area, thickness, and type and size of its perforations [20–22]. Some researchers also further discussed the decrease in electricity consumption by natural draught without the use of a blower [23].

One of the promising and energy efficient strategies for handling the latent load of air conditioning systems is to apply desiccant systems. By applying desiccants, air can be dehumidified without the need to cool the air below its dew point temperature [24]. The most used type of desiccant systems is solid desiccants and namely the desiccant wheels [25]. However, liquid desiccant systems possess the advantages of a separation between the absorption and the regeneration processes [26], regeneration at lower temperatures [27], and the possibility for energy storage in a thermochemical energy form [28–30]. Liquid desiccants are sorbents that have a high affinity to sorbates. In open cycle systems (under atmospheric pressure), the sorbate is the water vapor in the air.

The desiccant solution's effectiveness normally depends on the temperature and the concentration giving it a better performance at high concentrations and lower temperatures, so the controlling method for a solution can be attained by varying its temperature, concentration, or both according to the air specifications required for the output.

Due the appealing properties of desiccants with their affinity for more than water vapor, such as microorganisms, they are widely used in the applications in which air borne microorganisms could cause major issues, for example hospitals and medical facilities [31–33].

There are several common liquid desiccants that were investigated in many researches. Some examples of the investigated solutions are hygroscopic salt solutions, such as lithium bromide (LiBr), $CaCl_2$, and LiCl, or organic compounds like mono/triethylene glycol [34–36]. In the literature, LiCl and $CaCl_2$ solutions are the most investigated.

This paper represents an adaption of liquid desiccant solutions into a commercial desert cooler. $LiCl\text{-}H_2O$ with a mass fraction of 0.43 and $CaCl_2\text{-}H_2O$ with a mass fraction of 0.45 were used as desiccants. The desert cooler was firstly tested as a normal evaporative cooler and then it was tested with externally cooled desiccants with temperatures of 20 °C and 24 °C. The experiments for both desiccants were also performed by varying the solution mass flow rates at eight values, starting with

50 kg/h with an increment of 50 kg/h. Furthermore, the experimental results were compared with the numerical results of a basic effectiveness model.

## 2. Description of the Modified Desert Cooler

An experimental teat rig was constructed to experimentally study the air outlet characteristics. The analyses were performed on a typical commercial desert cooler that cost about US $80. The desert cooler represented a drip-type direct evaporative cooler with corrugated packing at the air outlet. Water streams flow down by gravity over the structured packing material and are recycled from the basin driven by a pump. The water sprinkled onto the top edges of the pad is distributed further by gravity and capillarity. Figure 3 shows the inner structure of the desert evaporative cooler and the setup as modified with desiccant solutions.

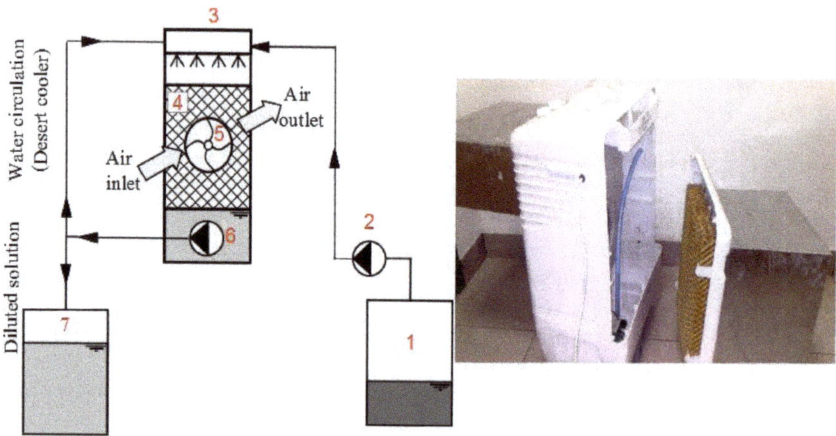


**Figure 3.** Inner configuration of the evaporative cooler and the setup with a desiccant solution (**left**), and the modified desert cooler (**right**). Numerical labels are (1) concentrated desiccant tank, (2) desiccant solution pump, (3) water/desiccant distributor, (4) packing material, (5) fan, (6) water circulation/desiccant-outlet pump, and (7) diluted desiccant tank.

The commercial desert cooler consisted of a corrugated packing material (4) made of cellulose with a volumetric surface area of 77 $m^2/m^3$. The cooler had a built-in pump (6) and a fan (5). The pump was originally used to circulate the water in the conventional evaporative cooler measurements. The built-in pump was used to extract the collected diluted solution to its tank (7).

The desert cooler inlet and outlet were modified by applying two rectangular ducts in order to connect it to the laboratory air ductwork, as shown in Figure 3 (right). The desiccant solution is introduced from above and it flows through a distributor above the cellulose packing material. The distributor is used originally to distribute the water in the desert evaporative cooler.

## 3. Instrumentation and Experimental Setup

Ambient air was supplied to an air handling unit accompanied by an air heater/cooler and humidifier/dehumidifier to create the target conditions required for each experiment. Air volume flow rate ($\dot{V}_a$) was measured using a volume flow meter. Air temperature ($\vartheta_a$) and relative humidity ($\varphi$) were measured using temperature and humidity sensors. Two temperature and relative humidity sensors were embedded at the inlet and outlet of the absorber. The three autonomous properties, namely temperature, relative humidity, and pressure, were used to derive all other air properties, such as the air humidity ratio and enthalpy, using humid air state equations from the American Society of Heating, Refrigerating and Air-Conditioning Engineers (ASHRAE) [37].

The desiccant solution circuit consisted of plastic tanks for the concentrated and diluted desiccant solutions. A pump was used to circulate the liquid desiccant. The volume flow rate of the desiccant solution was measured by a magnetic inductive flowmeter. The temperature of the desiccant solution was measured with Pt100 resistance thermometer sensors connected to the absorber inlet and outlet. The density of the solution ($\rho$) was measured by taking samples of the solution at the absorber inlet and outlet. The desiccant concentration was derived by the measured desiccant density and temperature by applying the correlations given by Conde [38]. Figure 4 shows the instrumentation setup of the absorber air circuit.

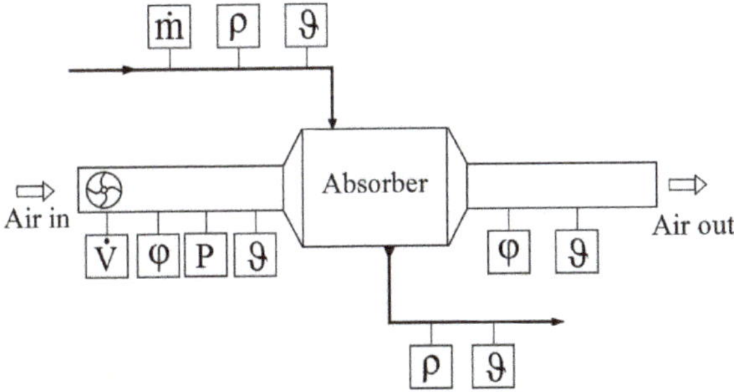


**Figure 4.** The experimental setup of the device tested with desiccant solutions. For the air circuit: volume flow rate ($\dot{V}$), relative humidity ($\varphi$), and temperature ($\vartheta$) were measured. For the desiccant solutions: mass flow rate ($\dot{m}$), density ($\rho$), and temperature ($\vartheta$) were measured.

The relevant parameters of the desiccant solution and air were measured at the inlet and outlet of the modified desert cooler. Table 1 shows the uncertainties of the connected instruments.

**Table 1.** Applied instrumentation and uncertainties as given by the manufacturer.

| Instrument | Model | Uncertainty | Range of Operation |
|---|---|---|---|
| Air temperature | Hygroflex 420 | 0.5 K | −50–100 °C |
| Relative humidity | Hygroflex 420 | 2% | 0%–100% |
| Air volume flow rate | Optiswirl 4070 DN 150 | 3% | 370–4800 m³/h |
| Solution temperature | Pt100 | 0.5 K | |
| Solution volume flow rate | Optiflux 1050 | 1% | 20–3900 L/h |
| Solution density | L-Dens 323 | 1 g/cm³ | 0.5–2 g/cm³ |

The modified desert cooler was tested by varying the desiccant solution mass flow rate in the range of 50–400 kg/h in increments of 50. The air mass flow rate, temperature, and humidity ratio were kept steady as 400 kg/h, 32 °C, and 13.12 g/kg for the entire experiment. The mass fractions of the desiccant solutions were 0.43 for the LiCl-H$_2$O solution and 0.45 kg/kg for the CaCl$_2$-H$_2$O solution. The solution temperature varied at values 20 °C and 24 °C. Table 2 shows the experimental inlet parameters.

**Table 2.** Experimental setup of the modified desert cooler: inlet conditions.

| $\dot{m}_{sol}$ kg/h | $\dot{m}_a$ kg/h | $\vartheta_a$ °C | $\omega_a$ g/kg | $\vartheta_{sol}$ °C | $\xi_{LiCl}$ kg/kg | $\xi_{CaCl2}$ kg/kg |
|---|---|---|---|---|---|---|
| 50–400 | 400 | 32 | 13.12 | 20–24 | 0.43 | 0.45 |

A total of 32 experiments were carried out by varying one of the inlet parameters, denoted by the blue shaded cells in Table 2. The change in air temperature $\Delta\vartheta_a$, air humidity ratio $\Delta\omega$, and the rate of moisture removal $\dot{m}_v$ were assessed. The rate of moisture removal is given as shown in Equations (1) and (2) [31]:

$$\dot{m}_{v,AS} = \dot{m}_{da}(\omega_i - \omega_o), \tag{1}$$

$$\dot{m}_{v,SS} = \dot{m}_{salt}(X_o - X_i), \tag{2}$$

where $\dot{m}_{v,AS}$ represents the moisture removal rate calculated from the air side, which represents the change in the air humidity ratio ($\Delta\omega$) multiplied by the dry air mass flow rate $\left(\dot{m}_{da}\right)$; and $\dot{m}_{v,SS}$ represents the change in water content in the salt ($\Delta X$) multiplied by the salt mass flow rate $\left(\dot{m}_{salt}\right)$.

The water content per mass of salt ($X$) is given in Equation (3):

$$X = \frac{1-\xi}{\xi} \tag{3}$$

where $\xi$ is the mass fraction of the salt solution, which represents the mass of the salt divided by the mass of the solution.

## 4. Numerical Model

A basic effectiveness model was modified for a quick estimation of the outlet conditions for the heat and mass exchanger from the entry conditions and the exchange surface. In addition, the model helps to determine the deviations between the measured data and simulation. This paper demonstrates a mathematical model for a cross-flow heat and mass exchanger for determining $\varepsilon$-NTU relations. The present model is based on the work of Stevens [39]. Effectiveness, $\varepsilon$, is defined as the ratio of the actual heat and mass transfer rate to the maximum possible rate, namely,

$$\varepsilon = \frac{\omega_{a,i} - \omega_{a,o}}{\omega_{a,i} - \omega_{eq}}, \tag{4}$$

where $\omega_{eq}$ is the humidity ratio of air at equilibrium with the desiccant solution and is given as

$$\omega_e = 0.62198 \frac{P_s\left(\vartheta_{sol,i}, \xi_{sol,i}\right)}{P - P_s\left(\vartheta_{sol,i}, \xi_{sol,i}\right)}. \tag{5}$$

The saturated pressure $p_s$ is realized as a nonlinear correlation of temperature and mass fraction according to [38] for aqueous solutions of LiCl and $CaCl_2$.

By applying the mass transfer coefficient ($\beta$) and the area of exchanger surface ($A$), the number of transfer units is derived:

$$NTU = \frac{\beta A}{\dot{m}_a}. \tag{6}$$

Applying the Lewis relationship, the rate of mass transfer coefficient to heat transfer coefficient is given in Equation (7) [40]:

$$\frac{\beta}{\alpha} = \frac{DLe^{1/3}}{\lambda}. \tag{7}$$

The $\varepsilon$-NTU relation for cross-flow configuration depends on the number of rows. For the corrugated packing material, with an infinite number of rows (air passages), the use of the approximate relation is described by Equation (8) and represents the infinite series solution taken from [39].

$$\varepsilon = \frac{1}{C^*NTU} \sum_{n=0}^{\infty} \left\{ \left[ 1 - e^{-NTU} \sum_{m=0}^{n} \frac{(NTU)^m}{m!} \right] \left[ 1 - e^{-C^*NTU} \sum_{m=0}^{n} \frac{(C^*NTU)^m}{m!} \right] \right\}. \tag{8}$$

## 5. Results and Discussion

### 5.1. Conventional Direct Evaporative Cooling

Prior to the desiccant investigations, the desert cooler was tested as a conventional evaporative cooler using water for the same air properties. Tap water was used with a temperature of 26 °C to fill the desert cooler tank and the water was circulated internally. Table 3 shows the inlet parameters and outlet results of the desert evaporative cooler.

**Table 3.** Inlet and outlet conditions of the desert cooler tested in evaporative cooling mode.

| $\dot{m}_a$ kg/h | $\vartheta_{a,i}$ °C | $\omega_{a,i}$ g/kg | $\vartheta_{a,o}$ °C | $\omega_{a,o}$ g/kg | $\varepsilon$ – |
|---|---|---|---|---|---|
| 400 | 32.0 | 13.12 | 24.7 | 16.07 | 0.68 |

As shown in Table 3, a reduction in the air dry bulb ($\Delta\vartheta_a = 7.3\ K$) and an increment in the air humidity ratio ($\Delta\omega = 2.95\ g/kg$) were observed as a result of the heat and mass transfer between the air and water.

The performance of the desert cooler was assessed by determining the saturation effectiveness. The effectiveness of the desert cooler was $\varepsilon = 0.68$. The saturation effectiveness represents the extent to which the air exiting the desert cooler approaches the wet bulb temperature, as shown in Equation (9):

$$\varepsilon = \frac{\vartheta_{db,o} - \vartheta_{db,i}}{\vartheta_{db,o} - \vartheta_{wb}}. \tag{9}$$

*5.2. Modified Desert Cooler: Adiabatic Liquid Desiccant Dehumidification*

Two test sequences were performed to study the effect of solution flow rate and temperature on the absorption process. The test sequences were performed for the desiccant solutions of LiCl and CaCl$_2$. Even though that LiCl is more effective as a desiccant [32], CaCl$_2$ was also applied as it is more accessible and has a low cost compared to LiCl salt. The solution temperatures were set to 20 °C and 24 °C. The duration of each experiment was set to about 75 min. The shown experiments represent the averaged last 30 min of each experiment with a time-step of 10 s. This time represent a quasi steady-state density of the desiccant solutions at the absorber outlet.

The aim of the experiments was to study the supply air conditions represented by the rate of moisture removal and outlet temperature as shown in Figure 5, Figure 6, Figure 7, Figure 8, Figure 9, and Figure 10. Figures 5 and 6 show the rate of moisture removal rate from the airstream as a function of the solution mass flow rate using externally cooled desiccant solutions at 20 °C (Figure 5) and 24 °C (Figure 6). The experimental results of each of the solutions were compared; LiCl-H$_2$O (solid red line) and CaCl$_2$-H$_2$O (solid blue line). Moreover, the experimental results were compared with the results from the simplified effectiveness model (the dashed red and blue lines for LiCl and CaCl$_2$ solutions, respectively).

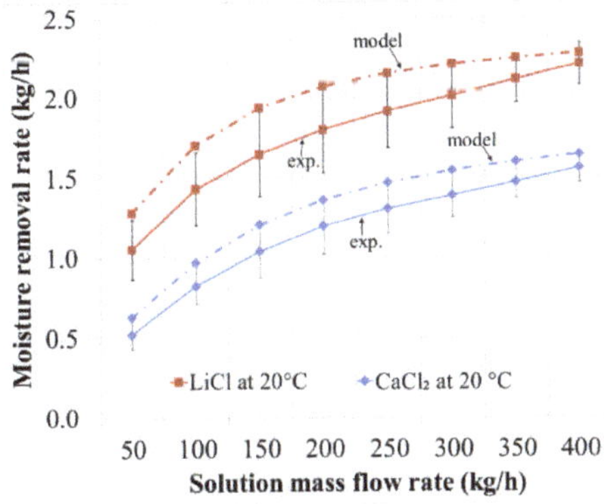


**Figure 5.** Absorbed water vapor versus the solution flow rate for the desiccants with an inlet temperature of 20 °C.

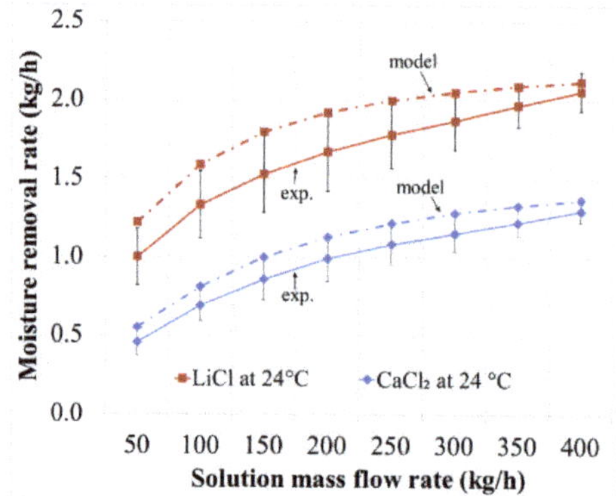


**Figure 6.** Absorbed water vapor versus the solution flow rate for the desiccants with an inlet temperature of 24 °C.

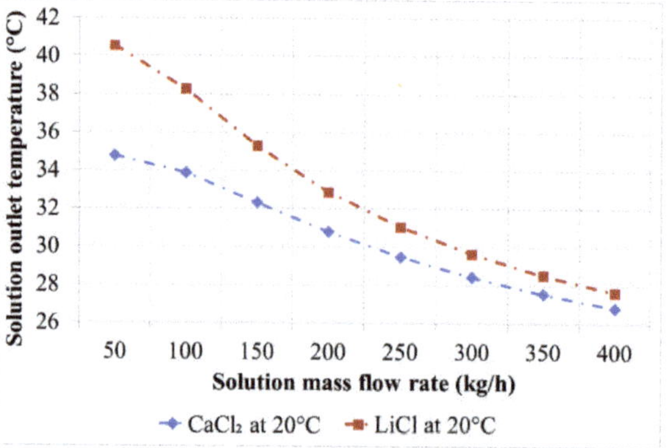


**Figure 7.** Solutions outlet temperature as a function of the solution flow rate for the desiccants with an inlet temperature of 20 °C.

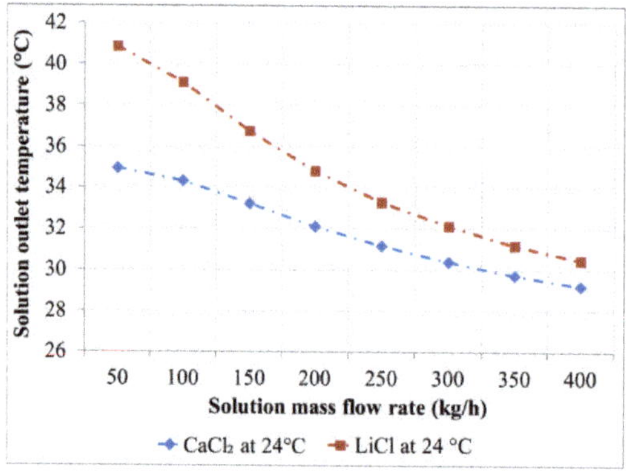


**Figure 8.** Solutions outlet temperature as a function of the solution flow rate for the desiccants with an inlet temperature of 24 °C.

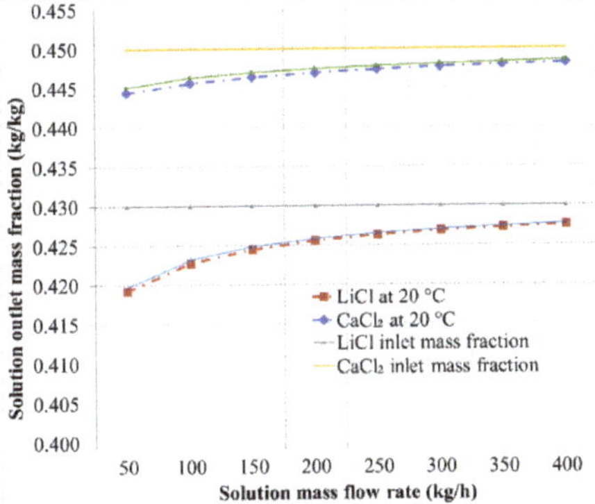

**Figure 9.** Mass fraction spread for the LiCl and CaCl$_2$ desiccant solutions at 20 °C and 24 °C as a function of solution flow rate.

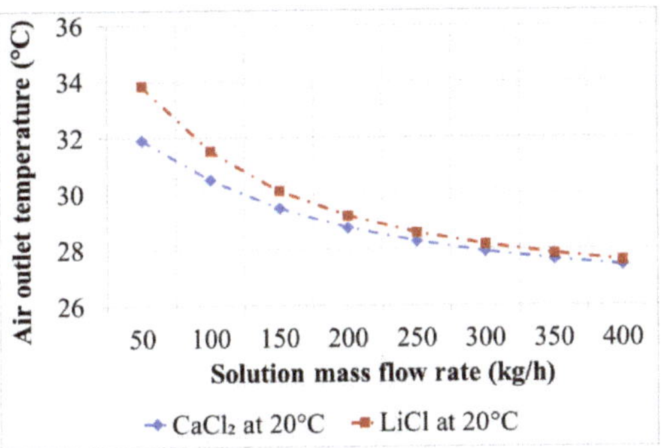

**Figure 10.** Air outlet temperature as a function of the solution flow rate for the desiccants with an inlet temperature of 20 °C.

As shown in Figures 5 and 6, the moisture removal rate increases by increasing the solution mass flow rate. For the LiCl solution there was an increase in the moisture removal rate of 1.16 kg/h (110% increase) for the given range of solution flow rate. This trend is expected as increasing the solution flow rate will lead to lower increments in the solution temperature, as shown in Figures 7 and 8. Increasing the solution flow rate also leads to lower decrement in the solution concentration, as shown in Figure 9. As a result, this will keep the solution at a low vapor pressure and thus improve the mass transfer of water vapor from air to the concentrated solution. Furthermore, an increased solution flow rate will enhance the wetted transfer area of the corrugated packing material.

The deviation between the experimental and numerical results was in the range of 3%–18% and decreased by increasing the solution flow rate. An explanation could be that increasing the solution flow rate will increase the surface wetting of the corrugated material and thus coming closer to the assumption of complete surface wetting assumed in the model. The deviation was within the experimental uncertainties of 20%.

The rate of moisture removal using the LiCl solution is higher than for the CaCl$_2$ solution due to its higher absorption capacity. The rate has also been noted to be higher for solutions with lower temperatures, hence at 20 °C. For the LiCl solution with an inlet temperature of 20 °C, the moisture removal rate was in the range of 1.05–2.21 kg/h compared to 0.52–1.56 kg/h for CaCl$_2$ for the same inlet temperature.

An increment in the solutions outlet temperature was observed as a result of the enthalpy of condensation and enthalpy of dilution. This increment is inversely proportional to the solution flow rate and is lower for the $CaCl_2$ solution compared to LiCl solution, as shown in Figures 7 and 8. An increment in the solution temperature of $\Delta\vartheta_{sol} = 20.6\ K$ was observed for the LiCl solution with an inlet temperature 20 °C and a mass flow rate of 50 kg/h compared to $\Delta\vartheta_{sol} = 14.8\ K$ for the $CaCl_2$ solution at the same inlet conditions. The increase in the solution temperature decreases with increasing the solution flow rate as a result of the low heat capacity of the desiccant solutions and the reduced exposure time. An increment of the solution temperature of about 7 K was observed for a solution flow rate of 400 kg/h for both desiccants.

Figure 9 shows the outlet mass fraction of both desiccants as a function of solution flow rate. The spread of desiccant solutions is inversely proportional to the solution mass flow rate, as shown in Figure 9. The spread of the mass fraction was higher for the LiCl solution with a maximal spread of 2.5% compared to the $CaCl_2$ solution with a maximal spread of 1.2%; both for a solution mass flow rate of 50 kg/h. While, the mass fraction spread was in the range of 0.6% and 0.4% for a solutions mass flow of 400 kg/h for LiCl and $CaCl_2$, respectively. Furthermore, the mass fraction spread was slightly affected by the solution inlet temperature for the given temperature range.

Figures 10 and 11 show the air outlet temperature as a function of solution flow rate for a solution inlet temperature of 20 °C and 24 °C. As an adiabatic process, the air outlet temperature decreased by increasing the solution flow rate. An increment in the air outlet temperature was observed with LiCl as the desiccant for a solution flow rate of 50 kg/h.

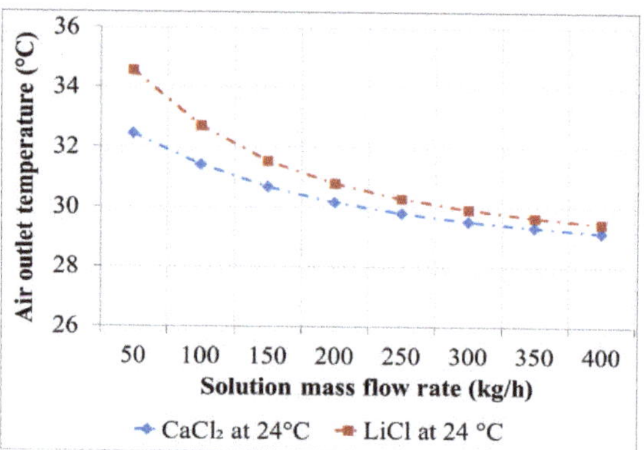


**Figure 11.** Air outlet temperature as a function of the solution flow rate for the desiccants with an inlet temperature of 24 °C.

A summary of the performed experiments is presented in Figure 12. The prototype was tested in three modes: in the conventional direct evaporative cooling mode (the red line), and adiabatic dehumidification using LiCl (blue points) and $CaCl_2$ (green points) solutions.

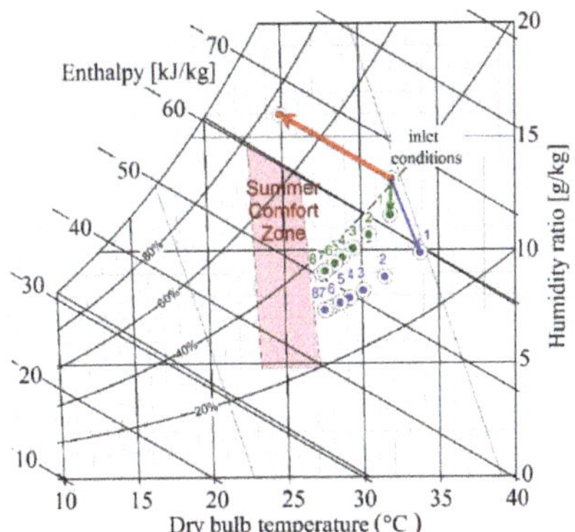


**Figure 12.** Illustration of the performed experiments on the psychrometric chart (Mollier diagram): The red line represents the conventional direct evaporative desert cooler, the green dots represents the performed experiments with $CaCl_2$ as a desiccant, and the blue dots represents the performed experiments with LiCl as a desiccant.

Figure 12 represents a graphical user interface for thermal comfort prediction based on ASHRAE Standard 55 [41]. Summer comfort zone specifies boundaries of air temperature and humidity for people in typical summer clothing during primarily sedentary activity. Since people feel comfort subjectively and each person has their own preferences for room climate, there is a comfort range of room air temperature and relative humidity in which the vast majority of room users feel comfortable. The limits of summer comfort zone, of ASHRAE Standard 55, were developed theoretically. As shown in Figure 12, this zone extends from around 23 °C to 27 °C and 35% to 75% relative humidity; i.e., the absolute humidity should be between 5 and 12 g/kg air for comfort.

The air that left the conventional desert cooler had a significant drop in the dry bulb temperature of 7.3 K. However, the air humidity ratio is significantly increased to 16.1 g/kg; this can be noticed while observing the red line path on the psychrometric chart (Mollier diagram) above. The air outlet conditions for the liquid desiccant system are shown in Figure 12, and the points were determined based on the experimental evaluation carried out in the paper. The results of absorption using LiCl and $CaCl_2$ solutions at 20 °C were chosen to be represented in the psychrometric chart. The blue line shows the path of the experimental results using the LiCl solution for the experiments 1 to 8 (by increasing the solution flow rate) while the green dots represent the experimental results using $CaCl_2$. As shown, the humidity ratio decreases significantly by applying the desiccant solutions and it is higher for the LiCl solution compared to the $CaCl_2$ solution. Additionally, the temperature does not decrease like it does in a conventional desert cooler, the air outlet conditions hit the comfort zone margins for a solution to air mass ratio of $r_m = 1$.

## 5.3. Dehumidification Load

For the dehumidification and cooling by applying liquid desiccants in the modified desert cooler, the cooling load is calculated by multiplying the air mass flow rate by the change in specific enthalpy as given in Equation (10) [5]:

$$\dot{Q}_{deh} = \dot{m}_{da}(h_i - h_o). \tag{10}$$

The specific enthalpy ($h$) of the air entering the modified evaporative cooler is $\left(h_i = 65.7\frac{kJ}{kg}\right)$. For the performed experiments, presented in Figure 12, the dehumidification/cooling load ($Q_{deh}$) and the air outlet conditions are given in Table 4.

**Table 4.** Air outlet conditions and the dehumidification load for both desiccants at a temperature of 20 °C.

| Exp. | Air Outlet Conditions Using LiCl-H$_2$O as Desiccant | | | | Air Outlet Conditions Using CaCl$_2$-H$_2$O as Desiccant | | | |
|---|---|---|---|---|---|---|---|---|
| | $\vartheta_o$, °C | $\omega_o$, $\frac{g}{kg}$ | $h_o$, $\frac{kJ}{kg}$ | $\dot{Q}_{deh}$, kW | $\vartheta_o$, °C | $\omega_o$, $\frac{g}{kg}$ | $h_o$, $\frac{kJ}{kg}$ | $\dot{Q}_{deh}$, kW |
| 1 | 33.8 | 9.87 | 59.4 | 0.71 | 31.9 | 11.53 | 61.6 | 0.46 |
| 2 | 31.5 | 8.81 | 54.3 | 1.27 | 30.5 | 10.67 | 58.0 | 0.86 |
| 3 | 30.1 | 8.22 | 51.3 | 1.60 | 29.5 | 10.07 | 55.4 | 1.15 |
| 4 | 29.2 | 7.89 | 49.5 | 1.80 | 28.8 | 9.67 | 53.7 | 1.34 |
| 5 | 28.6 | 7.68 | 48.4 | 1.93 | 28.3 | 9.41 | 52.5 | 1.47 |
| 6 | 28.2 | 7.54 | 47.6 | 2.02 | 27.9 | 9.22 | 51.6 | 1.57 |
| 7 | 27.9 | 7.43 | 47.0 | 2.08 | 27.6 | 9.07 | 51.0 | 1.64 |
| 8 | 27.6 | 7.36 | 46.6 | 2.13 | 27.4 | 8.97 | 50.5 | 1.70 |

As shown in Table 4, the dehumidification load increases by increasing the solution flow rate up to 2.13 kW and 1.7 kW for solution mass flow rates of 400 kg/h of LiCl-H$_2$O and CaCl$_2$-H$_2$O, respectively. The table also shows that the LiCl-H$_2$O is more efficient in dehumidification compared to CaCl$_2$-H$_2$O, as expected. The percentage change in the dehumidification load was up to 25.3% for an air to solution mass ratio of $r_m = 1$.

However, an additional load is required for the regeneration of the diluted liquid desiccant solution that can be regenerated by solar energy.

## 6. Conclusions

In this paper, a modification of a direct evaporative (desert) cooler was performed by applying LiCl-H$_2$O and CaCl$_2$-H$_2$O as desiccants. The prototype was tested by running it in a direct evaporative cooler and in an externally cooled (adiabatic) desiccant system. Several experiments were performed by varying the solution flow rate at the two inlet temperatures. The experimental results were also compared with a single-node effectiveness model.

The results show more beneficial air outlet conditions by using liquid desiccants regarding the air humidity ratio: A reduction in the air humidity ratio of up to 5.76 g/kg for the LiCl solution and up to 4.15 g/kg for the CaCl$_2$ solution compared to an increase of 2.95 g/kg in the humidity ratio for the conventional desert cooler. Thus, the indoor air quality is significantly improved in the modified desert cooler.

The moisture removal rate calculated from the measured data was compared with the numerical results from the $\epsilon - NTU$ effectiveness model. The results show a systematic pattern and the results were in good agreement. The maximal deviation in the rate of moisture removal between the experimental and the modeled results was in the range of 18%.

The results show that the LiCl solution is more effective in moisture removal than the CaCl$_2$ solution. However, the difference in moisture removal between LiCl and CaCl$_2$ solutions decreased by increasing the solution flow rate. The percentage difference was 51% for $r_m = 0.125$ and it was reduced to 28% for $r_m = 1$. Considering the huge price difference between LiCl and CaCl$_2$, it is highly recommended to apply CaCl$_2$ to the conventional desert coolers, which will make the price competitive for the target group of users.

**Author Contributions:** Conceptualization, M.J.; methodology, M.J. and A.A.; software, M.J.; formal analysis, M.A.-A. and A.A.; investigation, M.A.-A.; resources, A.A.; data curation, M.A.-A.; writing—original draft, M.J.; writing—review and editing, A.A.; supervision, A.A. All authors have read and agreed to the published version of the manuscript.

**Funding:** This research received no external funding.

**Acknowledgments:** The authors are grateful for the support of the Deanship of Graduate Studies and Research at the German Jordanian University.

**Conflicts of Interest:** The authors declare no conflict of interest.

## Nomenclature

| Symbol | Meaning | Unit |
| --- | --- | --- |
| $A$ | Area | $m^2$ |
| $\alpha$ | heat transfer coefficient | $W/m^2.K$ |
| $\beta$ | mass transfer coefficient | $m/s$ |
| $\varepsilon$ | effectiveness | |
| $h_{fg}$ | enthalpy of vaporization | $kJ/kg$ |
| $\dot{m}$ | mass flow rate | $kg/h$ |
| $p$ | pressure | $Pa$ |
| $r_m$ | air to solution mass flow ratio | |
| $X$ | mass of water per mass of salt | $kg_{H2O}/kg_{salt}$ |
| $V$ | volume | $m^3$ |
| $\Delta$ | difference | |
| $\lambda$ | thermal conductivity | $W/m.K$ |
| $\omega$ | humidity ratio | $g_w/kg_{da}$ |
| $\varphi$ | relative humidity | |
| $\rho$ | density | $kg/m^3$ |
| $\vartheta$ | temperature | $^\circ C$ |
| $\xi$ | mass fraction of the desiccant | $kg_{salt}/kg_{sol}$ |
| **Subscripts** | | |
| a | air | |
| $CaCl_2$ | calcium chloride | |
| i | inlet conditions | |
| da | dry air | |
| deh | dehumidification | |
| LiCl | lithium chloride as salt | |
| NTU | number of transfer units | |
| o | outlet conditions | |
| s | saturated | |
| sol | solution | |
| v | water vapor | |
| w | water | |

## References

1. Damiano, L.; Dougan, D. ANSI/ASHRAE Standard 62.1-2004. In *Encyclopedia of Energy Engineering and Technology—3 Volume Set (Print Version)*; Capehart, B., Ed.; CRC Press: Boca Raton, FL, USA, 2007; pp. 50–62. [CrossRef]
2. Abdel-Salam, A.H.; Simonson, C.J. State-of-the-art in liquid desiccant air conditioning equipment and systems. *Renew. Sustain. Energy Rev.* **2016**, *58*, 1152–1183. [CrossRef]
3. International Energy Agency (IEA). World Energy Outlook 2018. ISBN 9789264306776 (PDF). Available online: https://doi.org/10.1787/weo-2018-en (accessed on 27 August 2019).
4. Zegenhagen, M.T.; Ricart, C.; Meyer, T.; Kühn, R.; Ziegler, F. Experimental Investigation of A Liquid Desiccant System for Air Dehumidification Working With Ionic Liquids. *Energy Procedia* **2015**, *70*, 544–551. [CrossRef]
5. Wang, S.K. *Handbook of Air Conditioning and Refrigeration*, 2nd ed.; McGraw-Hill Handbooks; McGraw-Hill: New York, NY, USA, 2001.
6. Tu, M.; Ren, C.-Q.; Tang, G.-F.; Zhao, Z.-S. Performance comparison between two novel configurations of liquid desiccant air-conditioning system. *Build. Environ.* **2010**, *45*, 2808–2816. [CrossRef]

7.  Bishoyi, D.; Sudhakar, K. Experimental performance of a direct evaporative cooler in composite climate of India. *Energy Build.* **2017**, *153*, 190–200. [CrossRef]
8.  Lemons, A.R.; Hogan, M.B.; Gault, R.A.; Holland, K.J.; Sobek, E.; Olsen-Wilson, K.A.; Green, B.J. Fungal Metagenomic Analysis of Indoor Evaporative Cooler Environments in the Great Basin Desert Region. *J. Allergy Clin. Immunol.* **2016**, *137*, AB181. [CrossRef]
9.  Kim, W.; Dong, H.-W.; Park, J.; Sung, M.; Jeong, J.-W. Impact of an Ultraviolet Reactor on the Improvement of Air Quality Leaving a Direct Evaporative Cooler. *Sustainability* **2018**, *10*, 1123. [CrossRef]
10. Feddema, J.J. A Revised Thornthwaite-Type Global Climate Classification. *Phys. Geogr.* **2005**, *26*, 442–466. [CrossRef]
11. Xuan, Y.M.; Xiao, F.; Niu, X.F.; Huang, X.; Wang, S.W. Research and application of evaporative cooling in China: A Review (I)—Research. *Renew. Sustain. Energy Rev.* **2012**, *16*, 3535–3546. [CrossRef]
12. Kang, D.; Strand, R.K. Analysis of the system response of a spray passive downdraft evaporative cooling system. *Build. Environ.* **2019**, *157*, 101–111. [CrossRef]
13. Yang, Y.; Cui, G.; Lan, C.Q. Developments in evaporative cooling and enhanced evaporative cooling—A review. *Renew. Sustain. Energy Rev.* **2019**, *113*, 109230. [CrossRef]
14. ASHRAE. *Ventilation for Acceptable Indoor Air Quality*; ASHRAE Standard 62-1999; Thermal Comfort; American Society of Heating, Refrigerating and Air-Conditioning Engineers: Atlanta, GA, USA, 1999.
15. Arundel, A.; Sterling, E.; Biggin, J.; Sterling, T. Indirect Health Effects of Relative Humidity in Indoor Environments. *Environ. Health Perspect.* **1986**, *65*, 351. [CrossRef] [PubMed]
16. Al-Sulaiman, F. Evaluation of the performance of local fibers in evaporative cooling. *Energy Convers. Manag.* **2002**, *43*, 2267–2273. [CrossRef]
17. Liao, C.-M.; Chiu, K.-H. Wind tunnel modeling the system performance of alternative evaporative cooling pads in Taiwan region. *Build. Environ.* **2002**, *37*, 177–187. [CrossRef]
18. Darwesh, M.; Abouzaher, S.; Fouda, T.; Helmy, M. Effect of Using Pad Manufactured from Agricultural Residues on the Performance of Evaporative Cooling System. *Jordan J. Agric. Sci.* **2009**, *5*, 17.
19. Ahmed, E.M.; Abaas, O.; Ahmed, M.; Ismail, M.R. Performance evaluation of three different types of local evaporative cooling pads in greenhouses in Sudan. *Saudi J. Biol. Sci.* **2011**, *18*, 45–51. [CrossRef]
20. Malli, A.; Seyf, H.R.; Layeghi, M.; Sharifian, S.; Behravesh, H. Investigating the performance of cellulosic evaporative cooling pads. *Energy Convers. Manag.* **2011**, *52*, 2598–2603. [CrossRef]
21. Harby, K.; Al-Amri, F. An investigation on energy savings of a split air-conditioning using different commercial cooling pad thicknesses and climatic conditions. *Energy* **2019**, *182*, 321–336. [CrossRef]
22. Wang, J.; Meng, Q.; Zhang, L.; Zhang, Y.; He, B.-J.; Zheng, S.; Santamouris, M. Impacts of the water absorption capability on the evaporative cooling effect of pervious paving materials. *Build. Environ.* **2019**, *151*, 187–197. [CrossRef]
23. Sharma, A.K.; Bishnoi, P. Development and testing of natural draught desert cooler. *Int. J. Sci. Eng. Appl.* **2013**, *2*, 19–22. [CrossRef]
24. Elhelw, M. Performance evaluation for solar liquid desiccant air dehumidification system. *Alex. Eng. J.* **2016**, *55*, 933–940. [CrossRef]
25. Jaradat, M.; Fleig, D.; Vajen, K.; Jordan, U. Investigations of a Dehumidifier in a Solar-Assisted Liquid Desiccant Demonstration Plant. *J. Sol. Energy Eng.* **2018**, *141*, 031001. [CrossRef]
26. Öberg, V.; Goswami, D.Y. Experimental Study of the Heat and Mass Transfer in a Packed Bed Liquid Desiccant Air Dehumidifier. *J. Sol. Energy Eng.* **1998**, *120*, 289. [CrossRef]
27. Lowenstein, A. Review of Liquid Desiccant Technology for HVAC Applications. *HVAC&R Res.* **2008**, *14*, 819–839. [CrossRef]
28. Kessling, W.; Laevemann, E.; Peltzer, M. Energy storage in open cycle liquid desiccant cooling systems. *Int. J. Refrig.* **1998**, *21*, 150–156. [CrossRef]
29. Yu, N.; Wang, R.Z.; Wang, L.W. Sorption thermal storage for solar energy. *Prog. Energy Combust. Sci.* **2013**, *39*, 489–514. [CrossRef]
30. Kabeel, A.E.; Khalil, A.; Elsayed, S.S.; Alatyar, A.M. Theoretical investigation on energy storage characteristics of a solar liquid desiccant air conditioning system in Egypt. *Energy* **2018**, *158*, 164–180. [CrossRef]
31. Jaradat, M. Construction and Analysis of Heat-and Mass Exchangers for Liquid Desiccant Systems. Ph.D. Thesis, Universität Kassel, Kassel, Germany, 2016.

32.  Rafique, M.M.; Gandhidasan, P.; Bahaidarah, H.M.S. Liquid desiccant materials and dehumidifiers—A review. *Renew. Sustain. Energy Rev.* **2016**, *56*, 179–195. [CrossRef]

33.  Park, J.-Y.; Yoon, D.-S.; Li, S.; Park, J.; Bang, J.-I.; Sung, M.; Jeong, J.-W. Empirical analysis of indoor air quality enhancement potential in a liquid-desiccant assisted air conditioning system. *Build. Environ.* **2017**, *121*, 11–25. [CrossRef]

34.  Al-Farayedhi, A.A.; Gandhidasan, P.; Al-Mutairi, M.A. Evaluation of heat and mass transfer coefficients in a gauze-type structured packing air dehumidifier operating with liquid desiccant. *Int. J. Refrig.* **2002**, *25*, 330–339. [CrossRef]

35.  Abdul-Wahab, S.A.; Zurigat, Y.H.; Abu-Arabi, M.K. Predictions of moisture removal rate and dehumidification effectiveness for structured liquid desiccant air dehumidifier. *Energy* **2004**, *29*, 19–34. [CrossRef]

36.  Elsarrag, E. Performance study on a structured packed liquid desiccant regenerator. *Sol. Energy* **2006**, *80*, 1624–1631. [CrossRef]

37.  ASHRAE. *Handbook—Fundamentals*; Chapter 6: Psychrometrics; American Society of Heating, Refrigerating and Air-Conditioning Engineers: Atlanta, GA, USA, 1997.

38.  Conde, M.R. Properties of aqueous solutions of lithium and calcium chlorides: Formulations for use in air conditioning equipment design. *Int. J. Therm. Sci.* **2004**, *43*, 367–382. [CrossRef]

39.  Stevens, D.I.; Braun, J.E.; Klein, S.A. An effectiveness model of liquid-desiccant system heat/mass exchangers. *Sol. Energy* **1989**, *42*, 449–455. [CrossRef]

40.  Baehr, H.D.; Stephan, K. *Heat and Mass Transfer*, 3rd ed.; Springer-Verlag: Berlin, Germany, 2011.

41.  ASHRAE. *Handbook—Fundamentals*; Chapter 8: Thermal Comfort; American Society of Heating, Refrigerating and Air-Conditioning Engineers: Atlanta, GA, USA, 1997.

 *atmosphere*

*Article*

# Assessment of a Real-Time Prediction Method for High Clothing Thermal Insulation Using a Thermoregulation Model and an Infrared Camera

**Kyungsoo Lee [1], Haneul Choi [2], Hyungkeun Kim [2], Daeung Danny Kim [3] and Taeyeon Kim [2,***

[1]   Energy & Environment Business Division, KCL (Korea Conformity Laboratories), Jincheon 27872, Korea; kslee@kcl.re.kr
[2]   Department of Architecture & Architectural Engineering, Yonsei University, 50 Yonsei-ro, Seodaemun-gu, Seoul 03722, Korea; chn7960@yonsei.ac.kr (H.C.); hang0621@hanmail.net (H.K.)
[3]   Architectural Engineering Department, KFUPM, Dhahran 31261, Saudi Arabia; dkim@kfupm.edu.sa
*   Correspondence: tkim@yonsei.ac.kr; Tel.: +82-2-2123-5783

Received: 4 November 2019; Accepted: 12 January 2020; Published: 15 January 2020

**Abstract:** For evaluating the thermal comfort of occupants, human factors such as clothing thermal insulation (clo level) and metabolic rate (Met) are one of the important parameters as well as environmental factors such as air temperature (Ta) and humidity. In general, a fixed clo level is commonly used for controlling heating, ventilation, and air conditioning using the thermal comfort index. However, a fixed clo level can lead to errors for estimating the thermal comfort of occupants, because clo levels of occupants can vary with time and by season. The present study assesses a method for predicting the clo level of occupants using a thermoregulation model and an infrared (IR) camera. The Tanabe model and the Fanger model were used as the thermoregulation models, and the predicted performance for high clo level (winter clothing) was compared. The skin and clothing temperatures of eight subjects using a non-contact IR camera were measured in a climate chamber. In addition, the measured values were used for the thermoregulation models to predict the clo levels. As a result, the Tanabe model showed a better performance than the Fanger model for predicting clo levels. In addition, all models tended to predict a clo level higher than the traditional method.

**Keywords:** clothing thermal insulation; thermoregulation model; Tanabe model; infrared camera; thermal comfort

---

## 1. Introduction

Maintaining a pleasant thermal environment is one of the major goals of heating, ventilation, and air conditioning (HVAC) systems in buildings [1]. According to American [2] and international standards [3], six factors are needed to assess the thermal comfort of occupants. Four environmental factors—temperature, relative humidity (RH), mean radiant temperature (MRT), and air velocity (V)—and two personal factors—clothing thermal insulation (clo level) and metabolic rate (Met)—are required. While environmental factors can be measured using various types of equipment, personal factors, especially the clo level, are difficult to predict accurately since they can vary by season. In general, a fixed clo level (0.5 clo in summer, 1.0 clo in winter) is generally employed for thermal comfort index-based HVAC control, in accordance with the ASHRAE Standard 55 [2] or ISO 7730 [3].

The clo level of an occupant differs with respect to the time and season. Therefore, a fixed clo level may cause errors in predicting the thermal comfort. Newsham [4] reported that increasing the clo-level flexibility in an office can enhance the comfort of occupants and significantly reduce the building energy consumption. Lee and Schiavon [5] reported that fixing the clo level according to a predicted mean vote (PMV), a representative thermal comfort index, in a PMV-based HVAC control in summer and winter caused poor predictability of the room temperature and energy consumption, and they suggested the need for a dynamic clothing model.

The clo level is typically evaluated using thermal manikins or human subjects under laboratory conditions. Several studies have been performed on predicting the clo levels dynamically. Konarska et al. [6] measured skin temperature by attaching a thermocouple to the body of a subject and applied the measured temperature to the heat-balance equation for calculating the clo level. Although this method can provide accurate clo levels, it is difficult to attach thermocouples to the skin of occupants in real-life situations. Olesen and Nielsen [7] and McCullough et al. [8] analyzed the relationship between the weight of garments and the clo level through regression analysis. De Carli et al. [9] and Schiavon and Lee [10] developed models that consider the selection of clothing based on the weather condition in the morning. Although these methods can predict the clo level more flexibly than the conventional methods, they have limitations for predicting the changing clo level in real time.

Recently, with the development of image-recognition techniques, several attempts have been made to predict clo levels in real time using cameras. Matsumoto et al. [11] proposed a method for estimating the clo level by recognizing the images of clothes after constructing a simple database of clo levels according to the weight of the clothes. However, the experiment was performed only with images in the database, not on an actual person, and it was difficult to predict the clo level without knowing the weights of the clothes in advance. Lee et al. [12] used an infrared (IR) camera and evaluated the clo level according to the measured temperatures of the skin (forehead) and top and bottom clothing surfaces (chest and legs, respectively). In this study, the sensible heat loss from the skin was calculated using the heat-balance equation proposed by Fanger [13]. However, the sensible heat loss from the skin and the skin temperatures of each part of the human body cannot be calculated, owing to the limitation of Fanger model, which considers the human body as one node. The Tanabe model [14] is a multi-node thermoregulation model that overcomes the disadvantages of the Fanger model. It models the temperature of the human body in greater detail. However, there have been no attempts to predict the clo level in a non-contact manner using the Tanabe model and an IR camera. Additionally, the difference in the prediction results between the Fanger and Tanabe thermoregulation models remains unknown.

The objective of this study was to evaluate the applicability of the Tanabe thermoregulation model to predict clo level and to compare the prediction accuracy of two thermoregulation models (Fanger model and Tanabe model). Focusing on the objectives, experiments were performed to measure the temperatures of the skin (forehead) and the top and bottom winter clothing surfaces (chest and thigh, respectively) of subjects using an IR camera since thermoregulation models are better suited to predict higher clo levels (winter clothing) than lower clo levels (summer clothing) [6,12]. The measured skin and clothing surface temperatures were applied to two types of human thermoregulation models to calculate the sensible heat loss from the skin and to predict the clo levels.

## 2. Methods for Evaluating Clo Level

### 2.1. Standard Method Using a Thermal Manikin

The clo level is generally evaluated according to the procedures and standards provided by ISO 7730 [15] and ASTM F1291 [16]. This evaluation method uses a human-shaped thermal manikin. The dressed thermal manikin in a standing posture is heated, and the clo level is calculated using the measured skin surface temperature and operative temperature [17]. The clo level is calculated using Equations (1)–(3). The unit of the clo level (clo) is 0.155 m$^2$°C/W.

$$I_T = \frac{t_{sk} - t_o}{H}, \tag{1}$$

where $I_T$ is the total thermal insulation, including the clo level and boundary air layer (m²°C/W); $t_{sk}$ is the skin surface temperature (°C); $t_o$ is the operative temperature (°C); and $H$ is the sensible heat loss from the skin (W/m²).

$$I_a = \frac{1}{h_c + h_r},\tag{2}$$

where $I_a$ is the thermal insulation of the boundary air layer (m²°C/W), $h_c$ is the convective heat-transfer coefficient (W/m²°C), and $h_r$ is the radiative heat-transfer coefficient (W/m²°C).

$$I_{cl} = I_T - \frac{I_a}{f_{cl}},\tag{3}$$

Here, $I_{cl}$ is the intrinsic clo level (m²°C/W), and $f_{cl}$ is the clothing area factor (dimensionless).

### 2.2. Evaluation Method Using a Human Thermoregulation Model

Among the many types of thermoregulation models, the model proposed by Fanger [13,18] is most commonly employed. It was constructed according to experimental results for standardized clothing and activities under steady-state laboratory conditions. Figure 1 shows the process of heat transfer between the human body and the surrounding environment. The heat exchanged between the body and the surrounding environment passes through the clothing [19]. If the sensible heat loss from the skin is identified, the clo level can be evaluated using Equation (4).

$$R_{cl} = (t_{sk} - t_{cl})/(C + R),\tag{4}$$

where $C$ is the convective heat loss from the outer surface of the clothed body (W/m²), $R$ is the radiative heat loss from the outer surface of the clothed body (W/m²), $(C + R)$ is the sensible heat loss from the outer surface of the clothed body (W/m²), $t_{cl}$ is the clothing surface temperature (°C), and $R_{cl}$ is the clo level (m²K/W).

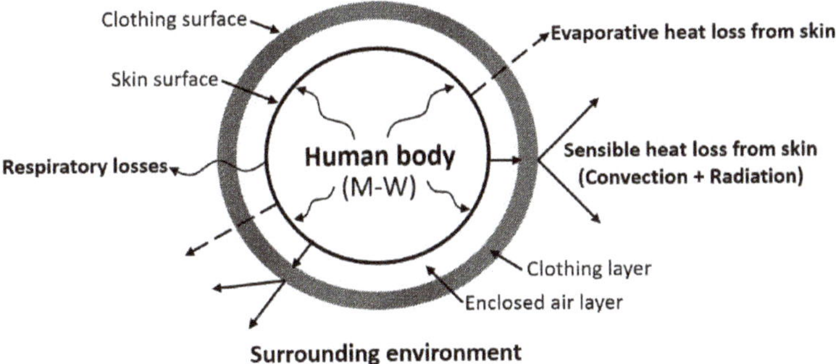


**Figure 1.** Heat balance between the human body and the surrounding environment, where M is the metabolic rate and W is the external work [19].

Equation (4) differs from Equation (3) because the clo level is directly calculated using the skin and clothing surface temperatures. When a thermal manikin is used, the sensible heat loss from the skin is clearly identified.

The human thermoregulation model involves a series of mathematical processes that remove heat from the body through the skin and respiration to maintain a constant core temperature. Using the thermoregulation model, it is possible to estimate the heat loss from the skin according to the air temperature (Ta), MRT, RH, air velocity (V), clo level, and metabolic rate (Met). The clo level can be calculated by substituting the estimated heat loss from the skin into Equation (4).

### 2.3. Multi-Node Human Thermoregulation Model

In general, thermoregulation models are classified as one-node, two-node, or multi-node according to the number of nodes representing the human body. As previously shown, the Fanger model has a simple design that only considers the heat characteristic of a human body. Givoni and Goldman [20] proposed another one-node model. Gagge [21,22] studied a two-node model, in which the human body is divided into a core layer and a skin layer. Jones [23] and Takada et al. [24] used a two-node model for the investigation of the physiological response. However, the human body adapts to various thermal environments through physiological responses, such as vasoconstriction, vasodilatation, shivering, and sweating, and heat is transferred to the surrounding environment via these processes [25]. Moreover, in one- and two-node models, it is difficult to describe the physiological phenomena that occur in different parts of the human body.

Considering the responses from various parts of the human body, a multi-node human thermoregulation model was proposed, and numerous thermoregulation models [14,26–28] have been developed to quantify the complex physiological phenomena of the human body in a thermal environment. In the multi-node model developed by Stolwijk [29], the human body is divided into six parts: head, trunk, arms, hands, legs, and feet. Each part consists of four layers: core, muscle, fat, and skin. Each layer is connected to the central blood compartment via the bloodstream. Huizenga et al. [26] divided the human body into 16 parts and added a clothing layer to allow for the transfer of heat moisture through clothing. Tanabe et al. [14] proposed a model based on the Stolwijk model and the results of thermal-manikin experiments in which the manikin was divided into 16 parts comprising 65 nodes with 4 layers and a central blood compartment. In other works [26,28,30], multi-node models based on the Stolwijk model were proposed. Owing to the advantages of the multi-node thermoregulation model, the model proposed by Tanabe et al. [14] was employed in the present study.

In the case of the multi-node thermoregulation model, the sensible heat loss from the skin of each part of the body is calculated using Equation (5). The clo level can then be calculated using that equation.

$$Q_t(i) = h_t(i) \times (t_{sk}(i) - t_o(i)) \times A_{Du}(i), \tag{5}$$

$$\frac{1}{h_t(i)} = 0.155 \times I_{cl}(i) + \frac{1}{(h_c(i) + h_r(i)) \times f_{cl}(i)}, \tag{6}$$

where $i$ is the body-segment number, $Q_t(i)$ is the sensible heat loss from the skin (W/m$^2$), $t_o(i)$ is the operative temperature (°C), $A_{Du}(i)$ is the surface area of each segment (m$^2$), $h_t(i)$ is the heat-transfer coefficient from the skin to the environment (W/m$^2$K), $h_c(i)$ is the convective heat-transfer coefficient (W/m$^2$K), and $h_r(i)$ is the radiative heat-transfer coefficient (W/m$^2$K).

### 2.4. Calculation of Clothing Insulation Using a Human Thermoregulation Model and an IR Camera

Recently, many studies [31–34] have been conducted to measure the skin temperatures of occupants using non-contact sensors. Using a non-contact sensor such as an IR camera, the skin temperature can be measured in real space without harming the occupants. Additionally, the thermal sensation vote of an occupant can be predicted directly.

Figure 2 shows the process of clo-level calculation using the human thermoregulation model and an IR camera. In this study, the Fanger model was used as the one-node model, and the Tanabe model was used as the multi-node model. The calculation of the clo level using the human thermoregulation model is outlined as follows:

a.  To determine the temperatures of the skin and clothing surface, the temperature of the human body was measured using an IR camera. Theoretically, to calculate the thermoregulation model, we only need to know the temperatures of several parts of the human body. Therefore, we measured the skin temperature at the forehead, the top clothing temperature at the chest, and the bottom clothing temperature at the thigh. These measurements were easily performed using the IR camera, and it was easy to extract stable values from the experiment. The skin temperature inside the clothing was predicted using the human thermoregulation model.

b.     The human thermoregulation model was simulated using Ta, MRT, RH, V, the assumed clo, and Met.

c.     The human thermoregulation model was used to calculate the sensible heat loss from the skin. In the Fanger model, the sensible heat loss from the skin ($C + R$ in Equation (4)) was calculated using the method specified in Annex D of ISO 7730. In the Tanabe model, the sensible heat loss from the skin ($Q_t$ in Equation (5)) of each body part was calculated.

d.     In each prediction model, the skin and clothing temperatures of each part were calculated using the sensible heat loss from the skin.

e.     The calculated skin and clothing temperatures of each part were compared with the measured temperature from step a. If a difference was found between the two temperatures, the clo level from step b was modified, and the calculation was performed again.

f.     The calculations of steps b–e were repeated to determine the clo level at which the predicted skin temperature and measured temperature of each part were equal. The identified clo levels were those evaluated using the Fanger and Tanabe models.

    Because this study focused on the method for evaluating the clo level using thermoregulation models, differences in the characteristics of the human thermoregulation models were not considered.

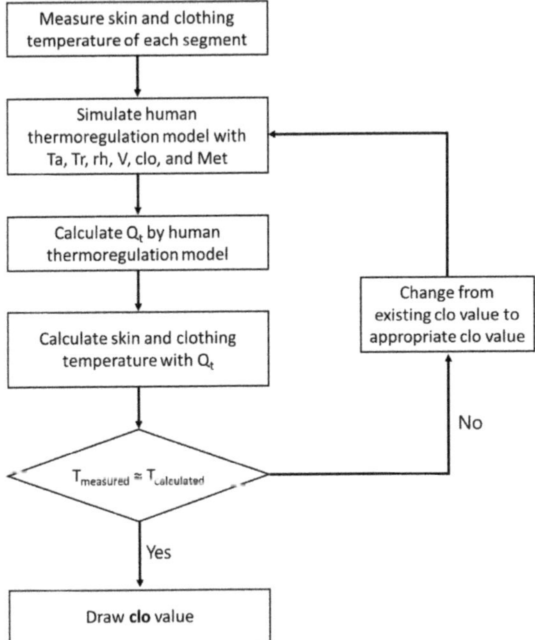


**Figure 2.** Flowchart of the evaluation of the clothing thermal insulation (clo level) using the human thermoregulation model.

## 3. Experiments for Evaluating Clothing Insulation

### 3.1. Outline

#### 3.1.1. Climate Chamber

Experiments were performed in a climate chamber located at Yonsei University, Seoul, Korea to evaluate the prediction models. As shown in Figure 3, the dimensions of the climate chamber were 4.2 (length) × 2.3 (width) × 2.1 m (height). The temperature could be controlled from 0 to 60 °C (±1 °C), and the RH could be controlled from 1% to 99% (±10%). The wind speed inside the chamber was maintained below 0.1 m/s, and the difference in the air temperature between the heights of the head (1.7 m) and feet (0.1 m) of the subject did not exceed 3 °C. The location of the subject and the measuring device inside the climate chamber are shown in Figure 3.

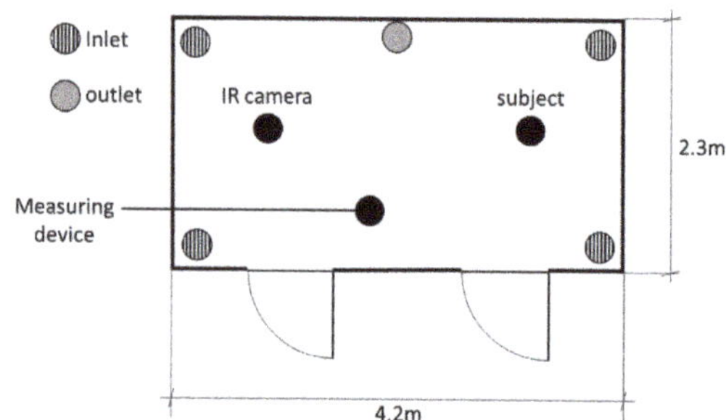


**Figure 3.** Floor plan of the climate chamber.

#### 3.1.2. Experimental Equipment

The environmental conditions of the climate chamber (air temperature, MRT, RH, and air velocity) were measured using a Testo 480 (TESTO, Inc., West Chester, PA, USA). The temperatures of the skin (forehead) and the top and bottom clothing surfaces (chest and thigh, respectively) were measured using an IR camera (Thermal Expert TE-Q1, i3system, Inc., Daejeon, Korea). The IR camera used in the experiments is shown in Figure 4. Table 1 presents details regarding the equipment used in the experiments.

**Table 1.** Equipment used in the experiments.

| Parameter | Measuring Device | Specifications |
|---|---|---|
| Air temperature | Testo 480 (thermal-flow probe) | ±0.5 °C |
| MRT | Testo 480 (globe probe) | ±1.5 °C |
| RH | Testo 480 (thermal-flow probe) | ±(1.8% RH + 0.7% of measured value) |
| Wind speed | Testo 480 (thermal-flow probe) | ±(0.03 m/s + 4% of measured value) |
| Skin and clothing temperatures | TE-Q1 (IR camera) | Scene range temperature: −10 to 150 °C<br>Accuracy: ±3% at ambient temperature<br>Thermal sensitivity <0.08 °C<br>Resolution: 384 × 288 (17 μm pitch)<br>Emissivity factor: 0.98 |

**Figure 4.** Infrared (IR) camera used in the experiments.

### 3.1.3. Evaluation of Clo Level for Clothes Used in Experiment According to ASTM F1291

The clo level used in the experiments was precisely determined for verifying the accuracy of the evaluation method using the IR camera and the human thermoregulation model. The clothing used in the experiments was evaluated according to the procedure prescribed in ASTM F1291. In this study, a 20-zone movable sweating thermal manikin (MTNW, Seattle, WA, USA) was used. In addition, the clothing used in the experiments were suitable for winter (briefs, undershirts, thick sweatshirts, and thick sweatpants, 22 °C and 50% RH).

As shown in Figure 5, the clo levels of the experimental clothing ensembles were assessed four times using the thermal manikin. The total clo value was 0.94 clo in winter. The standard deviations of the results of the thermal-manikin experiments for clothing ensembles were very small: 0.003 clo. In previous studies [35,36], occupants wore clothing of <1.0 clo during the winter. Therefore, in the present study, experiments were performed with a clothing ensemble having a relatively low clo level (<1.0 clo).

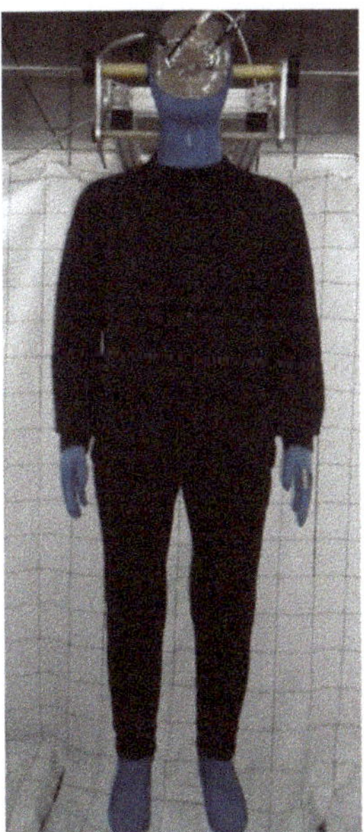

**Figure 5.** Evaluation of the clo level using a manikin with winter clothing.

### 3.1.4. Experimental Procedure

The IR camera and the human thermoregulation model were used to evaluate the clo levels of the subjects. All the subjects in the experiment were healthy people in the age range of 20–29 years and participated in the experiment voluntarily. Information regarding the subjects is presented in Table 2. The experiments involving the subjects were approved by the Institutional Review Board of Yonsei University. The procedures related to the experiment are presented below:

a.    The subject wore winter clothing in the climate chamber and was given 10 min to adapt to the winter conditions.

b.    After the 10 min of adaptation, the subject assumed a standing posture and relaxed for 20 min while looking at the front of the IR camera.

c.    The temperatures of the skin (forehead) and clothing surface (chest and thigh) were measured using the IR camera.

During the experiments, the PMV in the chamber was controlled within ±0.5, so that the surface temperatures of the skin and clothing were not affected by the unpleasant hot or cold environments.

The skin temperatures are affected by the ambient thermal condition with a certain level of time lag, which seems naturally occurred based on the physiological thermoregulation principle. Therefore, it is important to keep a sufficient time length per designed thermal condition. Choi and Yeom [37,38] and Lee et al. [39] recently determined 10–20 min as the time required to adapt subjects to thermal conditions in thermal comfort experiments. In addition, to measure the stable skin temperature of a person at rest using thermographic images, 10 min of acclimatization is needed [40]. Bach et al. [41] and Buono et al. [42] performed acclimation for 10–20 min prior to measuring the skin temperatures using IR cameras. In the present study, 10 min of acclimatization was employed to measure the stable skin temperature at rest using the thermographic images.

The IR camera measured the skin and clothing surface temperatures of a subject every 10 s at an angle of 90° and a distance of 1.8 m from the front of the subject. The temperature and RH of the chamber were measured at heights of 0.1, 1.1, and 1.7 m every 10 s. Figure 6 shows the images of a subject obtained using the IR camera.

**Table 2.** Information regarding subjects (mean ± standard deviation).

| Sex | Number of Subjects | Age | Height (cm) | Weight (kg) |
|---|---|---|---|---|
| Male | 8 | 24.3 ± 1.58 | 171.6 ± 4.27 | 64.7 ± 7.94 |

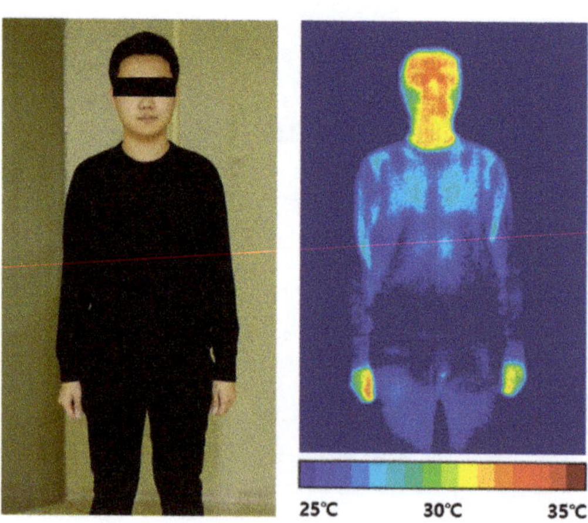


**Figure 6.** Skin and clothing temperatures measured using the IR camera.

## 3.2. Prediction Models for Evaluating Clothing Insulation

Figure 7 shows the measured skin and clothing surface temperatures used in the prediction models. The clothing surface temperature showed a maximum difference of >1 °C at the measurement points. Therefore, the top and bottom clothing temperatures at nine adjacent points were measured. The median of the nine measured temperatures was applied to the prediction model to reduce the errors at the measurement points.

Among the various human thermoregulation models, four prediction models were developed. Models 1 and 2 were based on the Fanger model, and Models 3 were 4 are based on the Tanabe model. Models 1 and 3 used only the skin and top clothing temperatures; the bottom clothing temperatures were not measured owing to situations where the occupants were seated or an obstacle, such as a table, was located between the occupant and the IR camera. Models 2 and 4 considered the bottom clothing temperatures. In Model 2, the area-weighted average temperature of the top and bottom clothing was used. In Model 3, the clo levels were evaluated using the skin and top clothing temperatures.

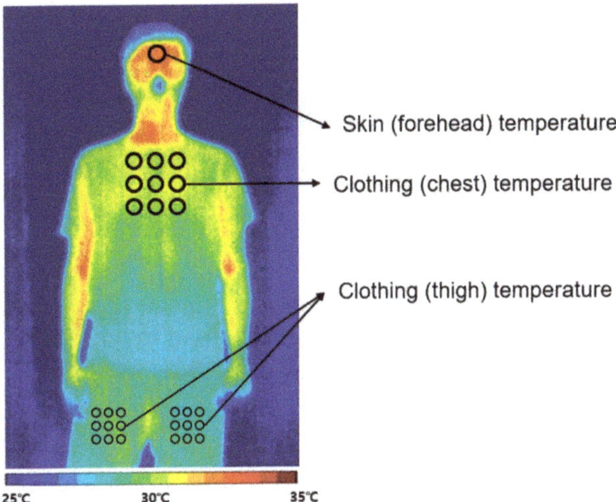


**Figure 7.** Skin and clothing temperatures used in the prediction models.

## 4. Results and Discussion

### 4.1. Results of Experiments

Tables 3 and 4 show the environmental conditions and results of the experiments. In all the experiments, the RH inside the climate chamber was maintained at 50%, and the air velocity was also kept lower than 0.1 m/s. As shown in the results, the temperatures of the skin (forehead), top clothing (chest), and bottom clothing (thigh) were similar in the winter. In addition, the top clothing temperatures were higher than the bottom clothing temperatures.

**Table 3.** Results of winter season experiments (mean ± standard deviation).

| Subject No. | Air Temp. (°C) | MRT (°C) | RH (%) | Air Velocity (m/s) | Skin Temp. (Forehead) (°C) | Top Clothing Temp. (Chest) (°C) | Bottom Clothing Temp. (Thigh) (°C) |
|---|---|---|---|---|---|---|---|
| 1 | 22.1 | 22.0 | 44.4 | 0.08 | 33.6 | 27.4 | 25.8 |
| 2 | 22.3 | 22.7 | 44.1 | 0.08 | 32.6 | 27.3 | 26.7 |
| 3 | 21.9 | 22.8 | 46.0 | 0.08 | 33.4 | 26.7 | 24.7 |
| 4 | 22.0 | 22.7 | 46.3 | 0.08 | 34.1 | 26.6 | 25.7 |
| 5 | 21.9 | 22.9 | 47.1 | 0.08 | 34.1 | 27.4 | 25.7 |
| 6 | 21.9 | 22.9 | 46.7 | 0.08 | 34.1 | 27.9 | 25.1 |
| 7 | 21.9 | 22.3 | 46.3 | 0.10 | 33.3 | 27.9 | 25.1 |
| 8 | 21.9 | 22.6 | 45.2 | 0.08 | 33.4 | 27.7 | 25.3 |
| Mean | 22.0 | 22.4 | 45.8 | 0.08 | 33.6 | 27.4 | 25.5 |
| SD | 0.13 | 0.19 | 1.07 | 0.01 | 0.54 | 0.52 | 0.62 |

### 4.2. Four Clo Prediction Models

The clo level in the winter was calculated by applying the measured data to each prediction model. As shown in Figure 8, the result was 0.89 clo for Model 1 and this was similar to the thermal-manikin measurement (0.94 clo). For Model 2, 1.20 clo was achieved. Even though the same Fanger model was applied, the result of Model 2 exhibited a larger difference from the manikin measurements than those of Model 1. This may be caused by the temperature of the bottom clothing. Because the temperature difference of each body part is not considered in the Fanger model, the area-weighted average temperature of the top and bottom clothing measured using the IR camera was used. The experimental results indicate that the bottom clothing temperature was slightly lower than the top clothing temperature. Therefore, the average temperature was lower than the top clothing temperature. This lower estimated temperature led to a higher clo level, in accordance with Equation (4). According to the calculation result of Model 3, the predicted clo value was similar to the measurement. Similar results were observed for Model 4. Therefore, the predicted values from the models based on the Tanabe model showed a good agreement with the manikin measurements.

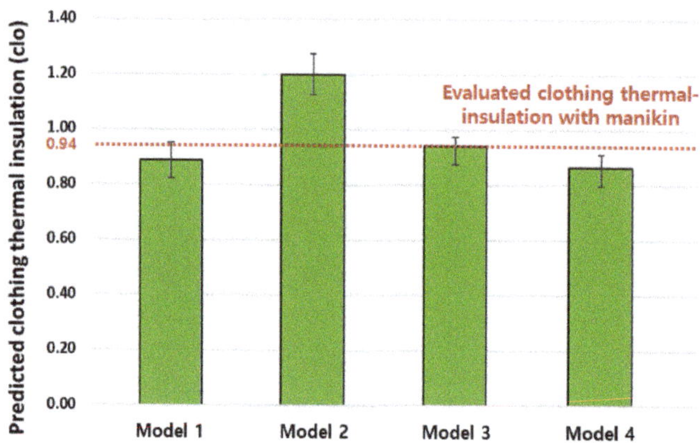


**Figure 8.** Results of the prediction models.

**Table 4.** Errors of the prediction models.

| Prediction Model | Model 1 | Model 2 | Model 3 | Model 4 |
|---|---|---|---|---|
| Error | 0.05 clo | 0.26 clo | 0.00 clo | 0.08 clo |

## 5. Discussion

The prediction models presented in the study exhibited accurate clo levels for the winter. In addition, the predicted results tended to be higher than the results of the manikin experiment. Similar results were obtained in previous studies. In the study of Konarska et al. [6], the clo level obtained from an experiment with a human body was higher than that obtained via a thermal-manikin measurement. The measurement with a human body conducted by Lee et al. [12] yielded similar results. The clo levels obtained through measurements with a human body in previous studies and the present study are shown in Figure 9. The similar trends of the present study and the previous studies indicate that the air layers between clothes significantly affect the clo level. Lee et al. [12] performed an experiment with air layers of three different thicknesses. Moreover, Mert et al. [43] demonstrated that the clo level can be affected by the presence and shape of the air layer between clothes. The air layer between the clothing layers depends on the characteristics of the clothing [44], the physical characteristics of the human body, and the posture of the individual [45]. In addition, demographic differences, gender, and ethnicity can also lead to clo level differences.

While the thickness and shape of the air layer between clothes is assumed to be distributed uniformly (left part of Figure 10), the air layer is formed irregularly in reality (right part of Figure 10). The results of the present study indicate that the shape and thickness of the air layer depended on the body shape and clothing conditions. Considering the obtained results, the thermal insulation from the enclosed air layer should be considered in evaluating the thermal insulation of the clothing level.

In addition, the emissivity factor of the IR camera was 0.98 in this study. This value is the emissivity factor for human skin [46], slightly different from 0.90 to 0.98 for typical fabrics [47,48]. Furthermore, the emissivity for clothing may vary depending on the color. However, it is difficult to apply different emissivity factors in real time to skin and clothing in one IR camera, therefore we fixed the emissivity as stated above. To improve the prediction accuracy of the method in the future, it is necessary to distinguish between skin and clothing and clothing color and apply different emissivity factors.

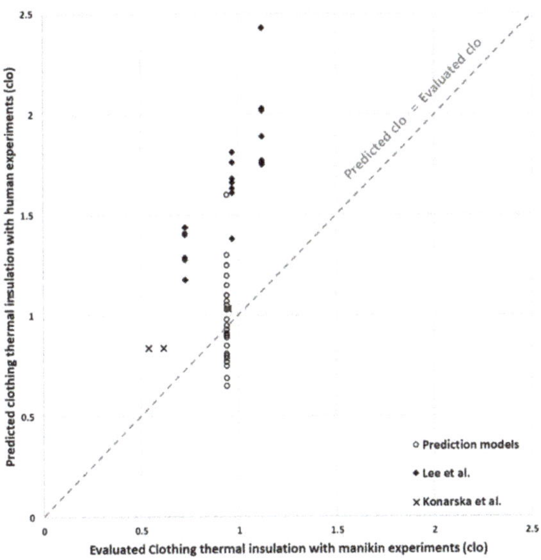


**Figure 9.** Predicted clothing insulation (four prediction models of the present study, Konarska et al. [6], and Lee et al. [12]).

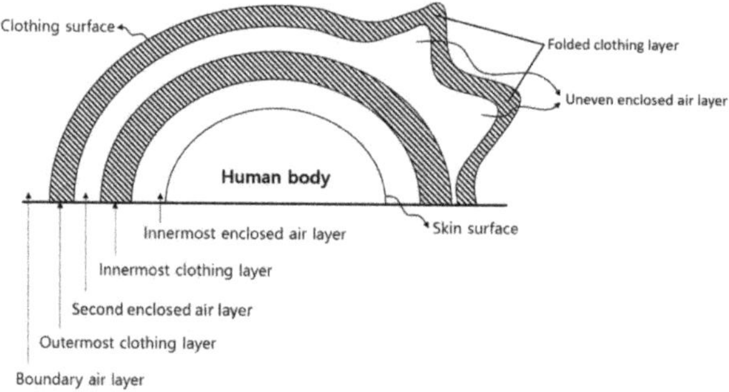


**Figure 10.** Clothing and air layers of the human body.

## 6. Conclusions

This study performed clo level measurements using an IR camera and the measured data were applied to two thermoregulation models to compare the accuracy of these models. The skin and clothing temperatures using an IR camera under specific conditions from eight subjects were measured. With the measurement data, the clo level was predicted using four models (Models 1 and 2 were based on the Fanger model, and Models 3 and 4 were based on the Tanabe model) and the results were compared. On the basis of our findings, the following conclusions are drawn:

(1) When skin temperature and top clothing temperature were used as input data, Model 3 predicted clo level better than Model 1. Model 4 also predicted clo level better than Model 2 when skin temperature and top and bottom clothing temperatures were used. As shown in the comparison results of two models, the clo levels predicted by the Tanabe model were closer to the manikin measurements than the Fanger models. Thus, the Tanabe model exhibited better prediction results than the Fanger model. The multi-node thermoregulation model (Tanabe model) was superior for predicting the sensible heat loss from the skin of each body part.

(2) Regardless of the thermoregulation model used, the high clo level for winter clothing was well predicted. In particular, the predicted values of Model 3 using the Tanabe model were similar to measurement values obtained using a mannequin. In addition, prediction models yielded somewhat higher clo levels than traditional methods.

By measuring the clo level of occupants in a non-contact manner, the predicted clo level may be less accurate than the existing clo-level evaluation method using the mannequin. However, it is meaningful in that it enables real-time prediction, which is impossible with the conventional method. Moreover, the outcome of the present study can be adopted for maintaining thermal comfort using an HVAC automatic control.

Further research will examine the clo levels by varying the thickness and surface colors of clothing with respect to the enclosed air layer in thin clothing and the emissivity effect, respectively. In addition, the experiment will be carried out with male and female participants. In this preliminary study, women were excluded from the experiment to eliminate the complexity and uncertainty caused by bras. In the future, it will be necessary to conduct the experiment with women.

**Author Contributions:** Conceptualization, H.K. and T.K.; methodology, K.L.; writing—original draft preparation, K.L.; writing—review and editing, H.C. and D.D.K.; supervision, T.K. All authors have read and agreed to the published version of the manuscript.

**Funding:** This research was supported by the Basic Science Research Program through the National Research Foundation of Korea (NRF) funded by the Ministry of Science and ICT (NRF-2017R1A2B3012914). It was also supported by the Korea Institute of Energy Technology Evaluation and Planning (KETEP) and the Ministry of Trade, Industry & Energy (MOTIE) of the Republic of Korea (No. 20174010201320).

**Conflicts of Interest:** The authors declare no conflict of interest.

## References


1. Dounis, A.I.; Caraiscos, C. Advanced control systems engineering for energy and comfort management in a building environment—A review. *Renew. Sustain. Energy Rev.* **2009**, *13*, 1246–1261. [CrossRef]
2. Ashrae Standard. *Standard 55-2017—Thermal Environmental Conditions for Human Occupancy*; ASHRAE Inc.: Tullie Circle, NE, USA, 2017; pp. 9–11.
3. Iso, E. Ergonomics of the thermal Environment-Analytical determination and interpretation of thermal comfort using calculation of the PMV and PPD indices and local thermal comfort criteria. *Management* **2005**, *3*, e615.
4. Newsham, G.R. Clothing as a thermal comfort moderator and the effect on energy consumption. *Energy Build.* **1997**, *26*, 283–291. [CrossRef]
5. Lee, K.; Schiavon, S. Influence of three dynamic predictive clothing insulation models on building energy use, HVAC sizing and thermal comfort. *Energies* **2014**, *7*, 1917–1934. [CrossRef]
6. Konarska, M.; Sołtynski, K.; Sudoł-Szopińska, I.; Chojnacka, A. Comparative evaluation of clothing thermal insulation measured in a thermal manikin and on volunteers. *Fibres Text. East. Eur.* **2007**, *15*, 73–79.
7. Olesen, B.; Nielsen, R. Thermal insulation of clothing measured on a movable thermal manikin and on human subjects. *ECSC Programme Res.* **1983**, *7206*, 914.
8. McCullough, E.A.; Jones, B.W.; Huck, J. A comprehensive data base for estimating clothing insulation. *ASHRAE Trans.* **1985**, *91*, 29–47.
9. De Carli, M.; Olesen, B.W.; Zarrella, A.; Zecchin, R. People's clothing behaviour according to external weather and indoor environment. *Build. Environ.* **2007**, *42*, 3965–3973. [CrossRef]
10. Schiavon, S.; Lee, K.H. Dynamic predictive clothing insulation models based on outdoor air and indoor operative temperatures. *Build. Environ.* **2013**, *59*, 250–260. [CrossRef]
11. Matsumoto, H.; Iwai, Y.; Ishiguro, H. Estimation of Thermal Comfort by Measuring Clo Value without Contact. In Proceedings of the MVA2011 IAPR Conference on Machine Vision Applications, Nara, Japan, 13–15 June 2011; pp. 491–494.
12. Lee, J.H.; Kim, Y.K.; Kim, K.S.; Kim, S. Estimating clothing thermal insulation using an infrared camera. *Sensors* **2016**, *16*, 341. [CrossRef]
13. Fanger, P.O. *Thermal Comfort: Analysis and Applications in Environmental Engineering*; Danish Technical Press: Copenhagen, Denmark, 1970; p. 244.
14. Tanabe, S.I.; Kobayashi, K.; Nakano, J.; Ozeki, Y.; Konishi, M. Evaluation of thermal comfort using combined multi-node thermoregulation (65MN) and radiation models and computational fluid dynamics (CFD). *Energy Build.* **2002**, *34*, 637–646. [CrossRef]
15. International Organization for Standardization. *ISO15831:2004 Clothing—Physiological Effects—Measurement of Thermal Insulation by Means of a Thermal Manikin*; ISO. Vernier: Geneva, Switzerland, 2004.
16. American Society of Testing and Materials International (ASTM). *Standard Test Method for Measuring the Thermal Insulation of Clothing Using a Heated Manikin*; Standard, F1291-10; ASTM International: West Conshohocken, PA, USA, 2010.
17. Havenith, G.; Holmér, I.; Parsons, K. Personal factors in thermal comfort assessment: Clothing properties and metabolic heat production. *Energy Build.* **2002**, *34*, 581–591. [CrossRef]
18. Fanger, P. Calculation of thermal comfort, Introduction of a basic comfort equation. *ASHRAE Trans.* **1967**, *73*, III4.1–III4.20.
19. ASHRAE. *Chapter 9 Thermal Comfort. ASHRAE Handbook-Fundamentals*; ASHRAE Inc.: Atlanta, GA, USA, 2017.
20. Givoni, B.; Goldman, R.F. Predicting metabolic energy cost. *J. Appl. Physiol.* **1971**, *30*, 429–433. [CrossRef]
21. Gagge, A.P.; Stolwijk, J.A.J.; Nishi, Y. Effective temperature scale based on a simple model of human physiological regulatory response. *ASHRAE Trans.* **2017**, *77*, 247–263.
22. Gagge, A.P. A two node model of human temperature regulation in FORTRAN. In *Bioastronautics Data*, 2nd ed.; Parker, J.F., Jr., West, V.R., Eds.; NASA Special Publication: Washington, DC, USA, 1973; pp. 142–148.
23. Jones, B. Transient interaction between the human and the thermal environment. *ASHRAE Trans.* **1992**, *98*, 189–195.
24. Takada, S.; Kobayashi, H.; Matsushita, T. Thermal model of human body fitted with individual characteristics of body temperature regulation. *Build. Environ.* **2009**, *44*, 463–470. [CrossRef]
25. Cheng, Y.; Niu, J.; Gao, N. Thermal comfort models: A review and numerical investigation. *Build. Environ.* **2012**, *47*, 13–22. [CrossRef]

26. Huizenga, C.; Hui, Z.; Arens, E. A model of human physiology and comfort for assessing complex thermal environments. *Build. Environ.* **2001**, *36*, 691–699. [CrossRef]

27. Fiala, D.; Lomas, K.J.; Stohrer, M. A computer model of human thermoregulation for a wide range of environmental conditions: The passive system. *J. Appl. Physiol.* **1999**, *87*, 1957–1972. [CrossRef]

28. Fiala, D.; Lomas, K.J.; Stohrer, M. Computer prediction of human thermoregulatory and temperature responses to a wide range of environmental conditions. *Int. J. Biometeorol.* **2001**, *45*, 143–159. [CrossRef] [PubMed]

29. Stolwijk, J.A. *A Mathematical Model of Physiological Temperature Regulation in Man*; NASA: Washington, DC, USA, 1971.

30. Salloum, M.; Ghaddar, N.; Ghali, K. A new transient bioheat model of the human body and its integration to clothing models. *Int. J. Therm. Sci.* **2007**, *46*, 371–384. [CrossRef]

31. Li, D.; Menassa, C.C.; Kamat, V.R. Non-intrusive interpretation of human thermal comfort through analysis of facial infrared thermography. *Energy Build.* **2018**, *176*, 246–261. [CrossRef]

32. Cosma, A.C.; Simha, R. Thermal comfort modeling in transient conditions using real-time local body temperature extraction with a thermographic camera. *Build. Environ.* **2018**, *143*, 36–47. [CrossRef]

33. Metzmacher, H.; Wölki, D.; Schmidt, C.; Frisch, J.; van Treeck, C. Real-time human skin temperature analysis using thermal image recognition for thermal comfort assessment. *Energy Build.* **2018**, *158*, 1063–1078. [CrossRef]

34. Cosma, A.C.; Simha, R. Machine learning method for real-time non-invasive prediction of individual thermal preference in transient conditions. *Build. Environ.* **2019**, *148*, 372–383. [CrossRef]

35. Ning, H.; Wang, Z.; Ji, Y. Thermal history and adaptation: Does a long-term indoor thermal exposure impact human thermal adaptability? *Appl. Energy* **2016**, *183*, 22–30. [CrossRef]

36. Bae, C.; Chun, C. Research on seasonal indoor thermal environment and residents' control behavior of cooling and heating systems in Korea. *Build. Environ.* **2009**, *44*, 2300–2307. [CrossRef]

37. Choi, J.H.; Yeom, D. Development of the data-driven thermal satisfaction prediction model as a function of human physiological responses in a built environment. *Build. Environ.* **2019**, *150*, 206–218. [CrossRef]

38. Choi, J.H.; Yeom, D. Investigation of the relationships between thermal sensations of local body areas and the whole body in an indoor built environment. *Energy Build.* **2017**, *149*, 204–215. [CrossRef]

39. Lee, K.; Choi, H.; Choi, J.H.; Kim, T. Development of a Data-Driven Predictive Model of Clothing Thermal Insulation Estimation by Using Advanced Computational Approaches. *Sustainability* **2019**, *11*, 5702. [CrossRef]

40. Marins, J.C.B.; Moreira, D.G.; Cano, S.P.; Quintana, M.S.; Soares, D.D.; de Andrade Fernandes, A.; da Silva, F.S.; Costa, C.M.A.; dos Santos Amorim, P.R. Time required to stabilize thermographic images at rest. *Infrared Phys. Technol.* **2014**, *65*, 30–35. [CrossRef]

41. Bach, A.J.E.; Stewart, I.B.; Disher, A.E.; Costello, J.T. A comparison between conductive and infrared devices for measuring mean skin temperature at rest, during exercise in the heat, and recovery. *PLoS ONE* **2015**, *10*, e0117907. [CrossRef]

42. Buono, M.J.; Jechort, A.; Marques, R.; Smith, C.; Welch, J. Comparison of infrared versus contact thermometry for measuring skin temperature during exercise in the heat. *Physiol. Meas.* **2007**, *28*, 855–859. [CrossRef] [PubMed]

43. Mert, E.; Psikuta, A.; Bueno, M.-A.; Rossi, R.M. Effect of heterogenous and homogenous air gaps on dry heat loss through the garment. *Int. J. Biometeorol.* **2015**, *59*, 1701–1710. [CrossRef] [PubMed]

44. Frackiewicz-Kaczmarek, J. Determination of the Air Gap Thickness and the Contact Area Under Wearing Conditions. Ph.D. Thesis, Université de Haute Alsace-Mulhouse, Mulhouse, France, 2013.

45. Li, X.; Wang, Y.; Lu, Y. Effects of body postures on clothing air gap in protective clothing. *J. Fiber Bioeng. Informat.* **2011**, *4*, 277–283.

46. Steketee, J. Spectral emissivity of skin and pericardium. *Phys. Med. Biol.* **1973**, *18*, 686. [CrossRef]

47. Anand, S.; Horrocks, A.R. *Handbook of Technical Textiles*; CRC Press/Woodhead Pub: Boca Raton, FL, USA, 2000.

48. Schleimann-Jensen, A.; Forsberg, K. New test method for determination of emissivity and reflection properties of protective materials exposed to radiant heat. In *Performance of Protective Clothing*; ASTM International: West Conshohocken, PA, USA, 1986; pp. 376–386.

*Article*

# Influence of Internal Structure and Composition on Head's Local Thermal Sensation and Temperature Distribution

**Shuai He [1], Yinghua Zhang [1], Zhian Huang [1,2,*], Ge Zhang [1] and Yukun Gao [1]**

[1]    State Key Laboratory of High-Efficient Mining and Safety of Metal Mines Ministry of Education, University of Science and Technology Beijing, Beijing 100083, China; He.Shuai@hotmail.com (S.H.); zyhustb@163.com (Y.Z.); uh-60a@163.com (G.Z.); gaoyukunustb@126.com (Y.G.)

[2]    Key Laboratory of Gas and Fire Control for Coal Mines (China University of Mining and Technology), Ministry of Education, Xuzhou 221116, China

*    Correspondence: huang_za@ustb.edu.cn

Received: 10 January 2020; Accepted: 17 February 2020; Published: 21 February 2020

**Abstract:** A personalized thermal environment is an effective way to ensure a good thermal sensation for individuals. Since local thermal sensation and temperature distribution are affected by individual physiological differences, it is necessary to study the effects of physiological parameters. The purpose of this study was to investigate the effects of internal structures and tissue composition on head temperature distribution and thermal sensation. A new mathematical model based on fuzzy logic control was established, the internal structure and tissue composition of the head were obtained by magnetic resonance imaging (MRI), and the local thermal sensation (LTS) index was used to evaluate the thermal sensation. Based on the mathematical model and the real physiological data, the head temperature and local sensation changes under different parameters were investigated, and the sensitivity of thermal sensation relative to the differences in tissue thickness was analyzed. The results show that skin tissue had the highest influence ($C_{skin} = 0.0180$) on head temperature, followed by muscle tissue ($C_{muscle} = 0.0127$), and the influence of adipose tissue ($C_{fat} = 0.0097$) was the lowest. LTS was most sensitive to skin thickness variation, with an average sensitivity coefficient of 1.58, while the muscle tissue had an average sensitivity coefficient of 0.2, and the sensitivity coefficient of fat was relatively small, at a value of 0.04.

**Keywords:** thermal sensation; biological structure and composition; tissue temperature; bioheat model; MRI analysis; sensitivity analysis

## 1. Introduction

The helmet is a basic and necessary protective piece of equipment, which is widely used in aviation, industrial, medical and military fields, etc. The variety of application, within both indoor and outdoor conditions, results in diverse thermal environments. In the design, production and use of these types of helmets, thermal comfort is one of the important factors that must be considered [1–7]. Good thermal comfort is a necessary condition for the wearer to work properly. However, due to differences in individual physiological attributes [3,5,7–9], the thermal sensation for the wearer from the equipment varies from person to person [10–15]. To ensure the thermal comfort of each individual, the thermal environment of the equipment needs to be personalized.

The thermal sensation of the human body is influenced by the external environment and individual physiological differences. In order to provide a personalized thermal environment of equipment, it is necessary to study the effects of individual physiological differences on human thermal sensation and temperature distribution. Study [16–18] shows that the thermal sensation and thermal comfort of the human body are strongly related to the internal heat balance, the bioheat as a by-product of metabolism is produced in living tissue at all times, and transferred to the skin from the internal organization to dissipate to the environment. In order to keep the body in a comfortable state, the relative balance of heat production and dissipation must be guaranteed [16,19–22]. A lot of studies [23–25] have been carried out on heat generation, heat dissipation and heat exchange with the environment of the body. Some scholars carried out the experiment with subjects, to study the heat generation and dissipation [26–30], and obtained the thermal response of the human body to the environment, which includes the temperature change and skin temperature distribution, as well as the subjective thermal sensation and thermal comfort state of subjects [7,11,20,29,31–34]. However, although these experiments can obtain the human body thermal response and the subjective thermal state of the human body, few of them can obtain the heat transfer law of internal tissues, which results in the incomplete explanation of the heat transfer process in the human body.

At the same time, mathematical models of the human body's thermoregulation were developed by researchers and used for physiological response predicting and thermal comfort assessments [9–12,14,16,35–37]. These models simplify the body into an active system and a passive system to simulate the body's thermoregulation process. The active system regulates heat production and heat loss by simulating the physiological processes of human sweating, muscle shivering, vasoconstriction and vasodilatation, and the passive system simulates the heat transfer in tissue and the heat exchange with the environment [9,16,35–39]. Based on these models, the thermal response of the human body to the environment can be calculated. Numerical methods can avoid deficiency in experimental studies, but the reliability of the results is affected by the applicability of the model, the degree of simplification of the model, and the similarity with human physiological processes and parameters.

Previous studies have paid more attention to the influence of external environmental factors on the human thermal state [7,12,35,40,41]. However, in fact, internal physiological difference will make the thermal state differ significantly, and hence needs to be considered in providing the personalized thermal environment or in the thermal management of personnel protective equipment [42–44], yet, it has not been deeply explored.

Therefore, the aim of this paper is to investigate the influence of the internal structure and tissue composition of the head on temperature distribution and local thermal sensation. Based on fuzzy logic control, a new mathematical model which is closer to the human thermal regulation process was established to calculate the temperature of the inner and outer layers of the head. To ensure the reliability of the results, we obtained the real internal data of the head through the MRI [45–48], and obtained the real internal head structure and tissue physiological parameters through the MRI images. Based on these real data and combined with the established mathematical model, the temperature of the inner and outer layers of the head were simulated. The local thermal sensation value was also calculated according to the local thermal sensation model. Through comparative analysis of the calculation results, the influence of internal structure and tissue composition on temperature and local thermal sensation will be clarified.

## 2. Materials and Methods

### 2.1. Mathematical Model

#### 2.1.1. Governing Equation and Boundary Conditions

The bioheat was generated inside the human body and transferred to the outer layers because of the heat gradient. Pennes [41] proposed the widely used bioheat model which considers the blood perfusion effect, and the partial differential equation in a one-dimensional cylindrical coordinate system is described in Equation (1). On the left side of Equation (1) is the heat storage term. The total heat storage equals the summation of the heat conduction, heat generation, and heat exchange with blood perfusion which correspond to the three terms on the right side of the Equation (1), respectively.

$$\rho c \frac{\partial T}{\partial t} = k \left( \frac{\partial^2 T}{\partial r^2} + \frac{\omega}{r} \frac{\partial T}{\partial r} \right) + \rho_{bl} w_{bl} c_{bl} (T_{bl} - T) + q_m \tag{1}$$

where, $\rho$ is the specific mass (kg/m$^3$), $c$ is the specific heat (J/kg·K), $T$ is tissue temperature (°C), $t$ is the time (s), $k$ is the conductivity (W/m·K), $r$ is the body element radius (m), $\omega$ is a dimensionless geometry factor, $q_m$ is the metabolic heat production (W/m$^3$), $\rho_{bl}$ is the specific mass of blood (kg/m$^3$), $w_{bl}$ is the blood perfusion rate (m$^3$/m$^3$·s), $c_{bl}$ is the specific heat of blood (J/kg·K), $T_{bl}$ is the blood temperature (°C). In this study, the blood perfusion term and the heat generation term were considered as constant variables according to Fiala et al. [35].

The heat transfers from the inner tissue to the skin and dissipates to environment. As the interface of heat exchange with environment, the heat flux of the skin's surface consists of convection, radiation and evaporation [16,35], as shown below:

$$-k \frac{\partial T_{skin}}{\partial r} = Q_c + Q_r + Q_e \tag{2}$$

The left side of Equation (2) is the heat flux on the skin's surface, the term $Q_c$ represents the convective heat exchange between the skin's surface and the ambient air, which consists of natural and forced convection, expressed as [35]:

$$Q_c = h_{c,mix} \left( T_{sf} - T_{air} \right) \tag{3}$$

The combined convection coefficient $h_{c,mix}$ is the function of the location of the body, the temperature difference between the surface and the air, and the effective air speed $v_{a,eff}$:

$$h_{c,mix} = \sqrt{a_{nat} \sqrt{|T_{sf} - T_{air}|} + a_{frc} v_{a,eff} + a_{mix}} \tag{4}$$

According to [35], the coefficients $a_{nat}$, $a_{frc}$, and $a_{mix}$ have been derived from experiments and provide different values for each of the body's elements—for the head, the values are 3.0, 113, 5.7, respectively.

The term $Q_r$ represents the heat exchange by radiation between head and the environment and can be expressed by [10]:

$$Q_r = \sigma \varepsilon \left( T_{sf}^4 - T_{air}^4 \right) \tag{5}$$

where, $\sigma$ is the Stefan–Bolzmann constant which is equal to $5.67 \times 10^{-8}$, and $\varepsilon$ is the emission coefficient of the skin surface, in this study, $\varepsilon = 0.9$.

The term $Q_e$ represents the heat loss by evaporation [16]:

$$Q_e = (3.054 + 16.7 h_{c,mix} W_{rsw}) \left( 0.256 T_{sf} - 3.37 - P_a \right) \tag{6}$$

where, $W_{rsw}$ is the wetness of the skin, generally, the wetness is between 0 and 1, corresponding to the skin being totally dry and entirely wet, respectively. $P_a$ stands for the vapor pressure of ambient air, and can be expressed as:

$$P_a = \phi_a P_a^* \tag{7}$$

where $\phi_a$ is the relative humidity of the surrounding air, $P_a^*$ is the saturated vapor pressure of the surrounding air that is affected by the air temperature, the $P_a^*$ is calculated by:

$$\lg P_a^* = 7.07406 - (1657.46 / (T_{air} + 227.02)) \tag{8}$$

### 2.1.2. Dynamic Response of the Model

The body temperature regulation mechanism is an important means with which to keep the body temperature relatively constant. The body regulates its temperature through vasodilation, vasoconstriction, shivering and sweating. These physiological activities are activated in response to different external environments. Generally, in a hot environment, the body will undergo vasodilation, increase sweating to increase heat dissipation, while in a cold environment, vasoconstriction and shivering will occur, increasing heat generation from reduced heat dissipation.

The dynamic response part of the model simulates the regulation of body temperature by simulating vasodilation, vasoconstriction, shivering, and sweating. Vasodilation and vasoconstriction directly affect blood flow, which corresponds to $w_{bl}$ in the governing equation. Shivering corresponds to the $q_m$ variable in the equation; and sweating activity affects the wetness of the skin's surface, which is the $W_{rsw}$ variable in boundary conditions.

The simulation of thermoregulation is based on the definition of the body's thermal signals. The signals were calculated by the temperature difference between each layer and its set point, which is $Error = T - T_{set}$. If the $Error > 0$, then the $Sig_{warm} = Error$ and the $Sig_{cold} = 0$, in this situation vasoconstriction and sweating occur. If the $Error < 0$, then the $Sig_{cold} = Error$ and the $Sig_{warm} = 0$, and the vasodilation and shivering occur. Here, $Sig_{warm}$ and $Sig_{cold}$ are the warm and cold signals, respectively.

Based on the above, a dynamic temperature regulation model based on fuzzy control is established. The fuzzy control model includes one input variable and three output variables. The input variable is the cold and warm sensation signal of the human body, and its calculation process is described above. The output variables are the blood flow control coefficient, metabolic heat production control coefficient and skin humidity control coefficient. The basic structure of the control model is shown in Figure 1a.

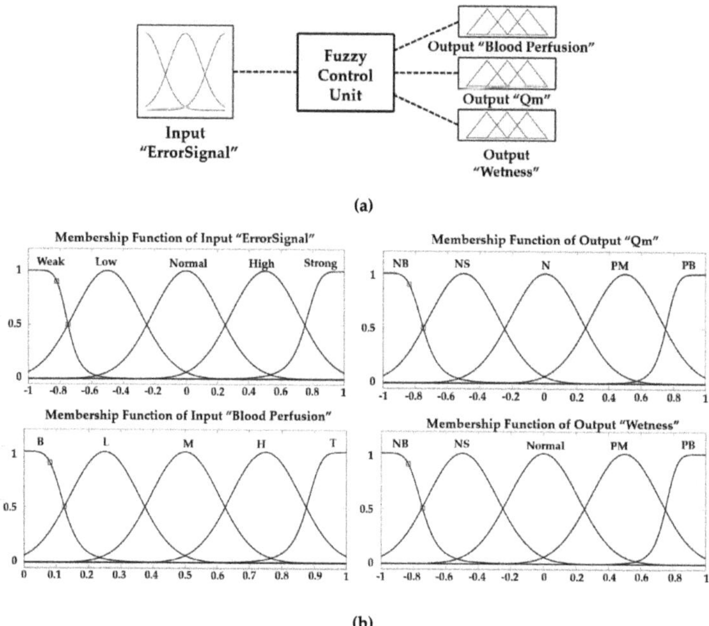


**Figure 1.** (a) The structure of fuzzy control model; (b) membership functions of fuzzy control model.

According to the regulation rules of the human body, the corresponding fuzzy control rules are established. If the input parameter is negative, the body metabolism heat production will increase, and the blood perfusion rate of the skin layer will be decrease. If the input parameters are positive, the blood perfusion rate of the skin layer and the skin wetness will increase. The membership function image of the input and output variables is shown in Figure 1b.

### 2.1.3. Thermal Sensation Index

Thermal sensation is the body's subjective description of its surrounding environment which is highly correlated with the skin and internal tissue temperatures. Although it is difficult to obtain an exact definition of thermal comfort conditions, it is possible to evaluate the body's thermal sensation and comfort by body temperatures. Zhang [49,50] proposed the local sensation index model and local comfort model for each part of the human body—which is known as the UC Berkeley thermal comfort model (UCB model) or Zhang's model—to evaluate the environment, with the aim of predicting "satisfaction" or "dissatisfaction" with the thermal environment. In Zhang's model, the thermal sensation index is defined in a range from −4 to 4. The negative values of the local thermal sensation (LTS) index represent the "cold" sensation while the positive values represent the "hot" sensation. It is an extended ASHRAE 7-point scale, adding "very hot" and "very cold" to accommodate extreme environments: 4, "very hot"; 3, "hot"; 2, "warm"; 1, "slightly warm"; 0, "neutral"; −1, "slightly cool"; −2, "cool"; −3, "cold"; −4, "very cold" [49]. In this paper, the thermal state of the head is evaluated according to Zhang's thermal sensation model in which the LTS index is expressed as follows:

$$LTS = 4 * \left( \frac{2}{1 + EXP\left(-C_1 * \Delta T_{sk,loc} - K_1 * \left(\Delta T_{sk,loc} - \Delta T_{sk,m}\right)\right)} - 1 \right) + C_2 \frac{dT_{sk,loc}}{dt} + C_3 \frac{dT_{core}}{dt} \tag{9}$$

where $LTS$ is the local thermal sensation; $C_1$ is a coefficient that varies for different body parts, (for the head $C_1 = 0.38$ when the $\Delta T_{sk,loc}$ is negative and $C_1 = 1.32$ when the $\Delta T_{sk,loc}$ is equal or greater than zero); $\Delta T_{sk,loc}$ is the difference between the local skin temperature and its set-point temperature (°C); $K_1$ is a coefficient with a value from 0 to 1 that varies for different body parts( for the head, $K_1 = 0.18$); $\Delta T_{sk,m}$ is the difference between mean skin temperature (°C) and the mean set-point temperature (°C); $C_2$ and $C_3$ are the thermal capacities at the skin and core nodes respectively [11].

### 2.2. MRI Analysis of Head Composition and Structure

One of the factors that affect the numerical results is the inappropriate simplification of human body segments. In the study of Yutaka Kobayashi et al. [9], the head segment was simplified into a shape with four layers: the core layer, the first layer, the second layer and the skin layer. Shinichi Tanabe et al. [51] also divided the human body into 16 segments, of which the head segment was simplified as a four layer shape consisting of core, muscle, fat and skin layers. In the circumstances of whole-body thermal response studies, a reasonable simplification is acceptable because the thermal response details within a particular body part is not of important. However, this does not mean that the differences can be neglected. In this study, the micro differences of the thermal state of the head which are affected by internal tissue differences are of interests. Hence, the head cannot be treated as having a simple geometry; the internal structure and the tissue differences need to be considered.

To obtain the detailed parameters of the structure and tissue of the head, the MRI method was used. Muscle and adipose composition within the extracranial part of head have different signals on the image, hence, the proper threshold levels were chosen to sub-select different tissues, and then the threshold values were manually adjusted for excluding other tissues. The internal structure of the head at multiple views and levels are shown in Figure 2a, and the results of the selection of different tissues are shown in Figure 2b. Figure 2c is the comparison of tissue thickness between different individuals at the same section of the head.

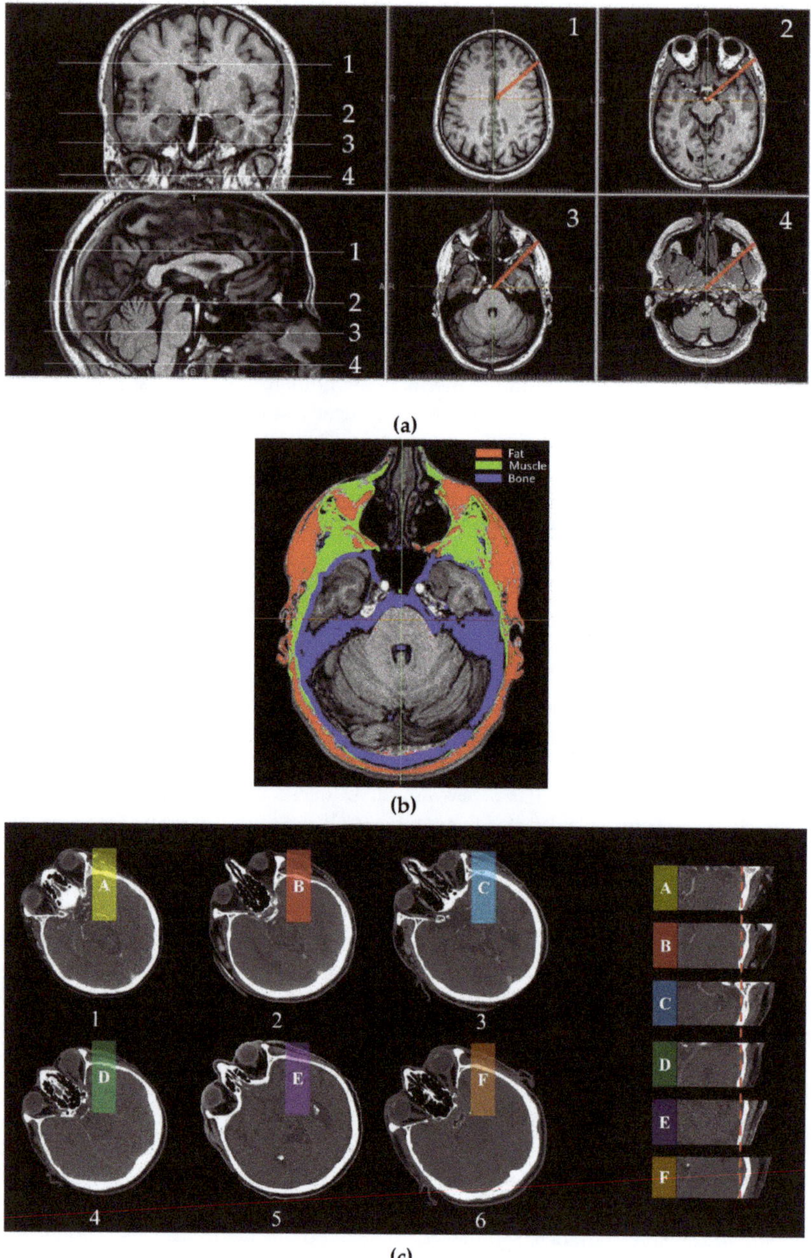

**Figure 2.** The structures of different sections of the head obtained by MRI. In (**a**), the two left images are the front and side view of the head. Sub images of 1, 2, 3, and 4 on the right side show the different sections from the top to bottom of the head. The image (**b**) shows the mask results of different tissues of section (**3**). The image (**c**) shows the comparison of tissue thickness between different individuals at the same section of the head.

As shown in the Figure 2, the head is not a symmetrical entity. The internal structure of the head is complex and the tissue components are diverse. The radial tissue is unevenly distributed on different levels. In general, the head is composed of a skull part and a face part, the skull part is mainly composed of brain tissue, skull, muscle tissue, adipose tissue, and skin from the inside to the outside, and the composition of the face is more complicated. A large amount of muscle tissue is attached to these bones, and adipose tissue and skin layers are located outside the muscle tissue. From the MRI image, it can be found that in different layers, the distribution of muscle tissue, adipose tissue and skin layer is uneven, and the thickness is significantly different. In the skull and facial parts, the radial distribution of the tissues and structures are significantly different. The cavity of the airway is located near the bottom of the skull, and the internal structure of the bilateral sides is complicated. Muscle tissue and skeletal tissue exist alternately in this part.

The physical and thermal properties of tissues are determined according to [35] and listed in Table 1. The measured results of the structure and tissue distribution along the red markers in Figure 2a are for the four sections are shown in Table 2.

**Table 1.** Physical and thermal properties of different tissue [35].

| Tissue | $k$<br>W/m·K | $\rho$<br>kg/m$^3$ | $c$<br>J/kg·K | $w_{bl}$<br>m$^3$/m$^3$·s | $q_m$<br>kg/m$^3$ |
|---|---|---|---|---|---|
| Brain | 0.49 | 1080 | 3850 | 10.132 | 13,400 |
| Bone | 1.16 | 1500 | 1591 | 0 | 0 |
| Muscle | 0.42 | 1085 | 3768 | 0.538 | 684 |
| Fat | 0.16 | 850 | 2300 | 0.0036 | 58 |
| Skin | 0.47 | 1085 | 3680 | 5.48 | 368 |

**Table 2.** Structure and tissue thickness of different sections along the red markers in Figure 2a.

| 1 | | 2 | | 3 | | 4 | |
|---|---|---|---|---|---|---|---|
| Tissue | Thickness (cm) | Tissue | Thickness (cm) | Tissue | Thickness (cm) | Tissue | Thickness (cm) |
| Brain | 8.81 | Brain | 7.37 | Brain | 1.37 | Muscle | 3.18 |
| Bone | 0.46 | Bone | 0.57 | Bone | 1.46 | Bone | 0.48 |
| Muscle | 0.31 | Muscle | 1.71 | Brain | 4.5 | Muscle | 3.56 |
| Fat | 0.78 | Fat | 0.75 | Bone | 0.66 | Bone | 0.58 |
| Skin | 0.27 | Skin | 0.31 | Muscle | 0.69 | Muscle | 0.46 |
| - | - | - | - | Fat | 0.76 | Fat | 0.52 |
| - | - | - | - | Skin | 0.36 | Skin | 0.4 |

### 2.3. Methodology Framework

In the current work, the purpose was to investigate the influence of the difference of the internal structure of the human head and the biological tissue composition on tissue temperature and local thermal sensation through a numerical method. Firstly, a mathematical model of biological heat transfer was established, which included control equation, boundary condition, and dynamic response. By solving the mathematical model, the temperature distribution of the head tissue can be obtained. The boundary condition describes the heat exchange between the human body and the environment. The dynamic response part regulates the heat exchange by adjusting the physiological parameters in boundary conditions and in the governing equation. The internal structure of the head and the physical parameters of the tissue were obtained by MRI. By inputting these different physiological parameters into the model, we were able to obtain the tissue temperature distribution and local thermal sensation value under the corresponding conditions, and then analyze the influence of input parameters on the output. The framework is shown in Figure 3.

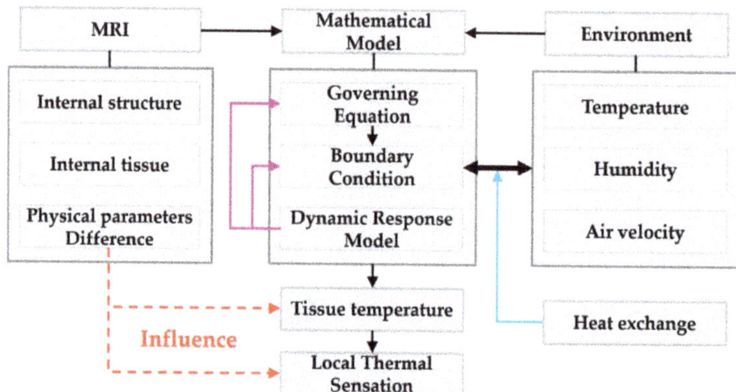



**Figure 3.** The methodology framework of the study

## 3. Results and Discussion

### 3.1. Model Validation

To validate the mathematical model of Section 2.1, two different cases are considered and the results are compared with published experimental and numerical data.

The first case is the comparison to the results reported in [16]. Experimental data are of the core temperatures and skin temperatures of human subjects exposed to the ambient conditions of 28.5 °C, and 31% relative humidity (RH), at a steady state for 4 h. The results are shown in the Figure 4, and it can be seen that the results of the model used in this article are generally consistent with the published data. The reason for the slight difference is that the model uses the actual parameters of the human body's structure and the differences in boundary conditions. However, from the overall trend, the calculation results are acceptable.

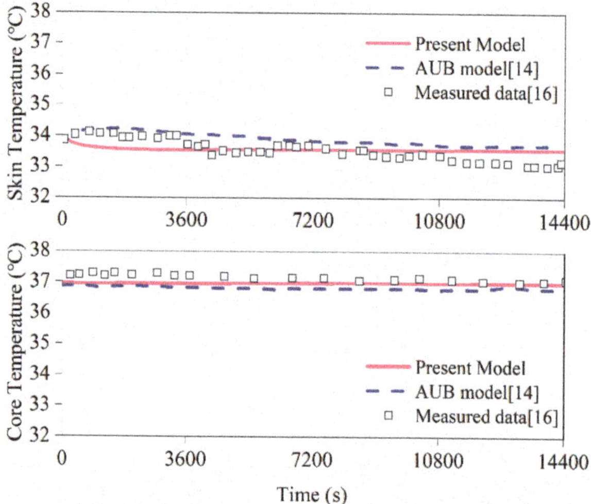



**Figure 4.** Comparison of simulated results of the model used in this study with the published experimental data [16], and simulated results of American University of Beirut model (AUB model) [14] for skin temperature and core temperature at constant ambient conditions of 28.5 °C /31% RH.

In the second case, the ambient temperature and relative humidity change from 30 °C/40% to 48 °C/30% for the first two hours, and then change back to 30 °C/40% in the last hour. Figure 5 shows the simulation results of the model in this paper and the measured data of [16]. It can be found that the temperature of both the skin and the inner tissue are affected by the ambient environment, the increase of air temperature results in the increase of skin temperature. The simulated results of the mathematical model are in agreement with the measured results.

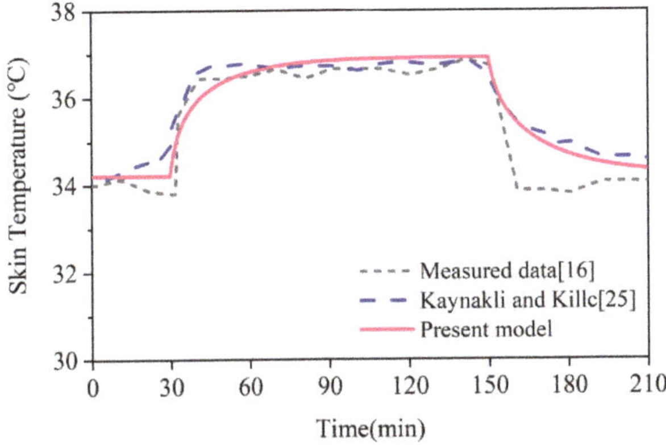


**Figure 5.** Comparison of the simulated results with the measured data [16], and Kaynakli and Kilic [25] as the ambient temperature step change.

One of the most commonly used methods to assess the accuracy of data is the root means square error (RMSE). It describes the degree of coincidence among measured and simulated data.

If we assume that we have a set of measured data $M$ and the simulated result $S$, then the definitions for the RMSE is

$$RMSE = \sqrt{\frac{1}{N}\sum_{i=1}^{N}(S_i - M_i)^2} \tag{10}$$

where $M_i$ is the measured data; $S_i$ is the simulated data; n is number of data pairs.

In general, if the computed values of RMSE are small, the simulated data and the measured data agree well.

According to Equation (10), the RMSE of the present model and other models in the above two cases are calculated. The results are shown in Table 3 below.

**Table 3.** Root mean square error results of different models.

| | | RMSE | |
|---|---|---|---|
| | | **Present Model** | **AUB model** |
| **Case1** | **Core** | 0.2127 | 0.3517 |
| | **Skin** | 0.3080 | 0.3415 |
| | | **RMSE** | |
| | | **Present Model** | **Kaynakli and Killc** |
| **Case2** | **Skin** | 0.5501 | 0.6153 |

From the results of RMSE in the table, it can be seen that the RMSE value of the present model is smaller than that of the corresponding model, indicating that the gap between the results of the present model and the measured values is smaller, and the prediction accuracy of the existing model is improved compared with that of the other models.

### 3.2. Influence of Different Structures on Tissue's Temperature Distribution

The different distribution of head skin temperature has been observed in many studies, the temperature of the forehead, cheek, mouth and nose is not uniform. The skin temperature of the head and face is not only different with the change of the environment, but the stable skin temperature is also different under the same external environment. The difference in the internal structure of the head is the main cause of this result. Based on the MRI analysis and measurement results, the radial temperature distributions of the four sections in Figure 2a were calculated. The exposure time was 2 h, the ambient environment temperature and relative humidity (RH) were 28 °C/30%, and the air velocity was 0.05 m/s. The temperature results of the skin and inner tissues were obtained and are shown in Figure 6.

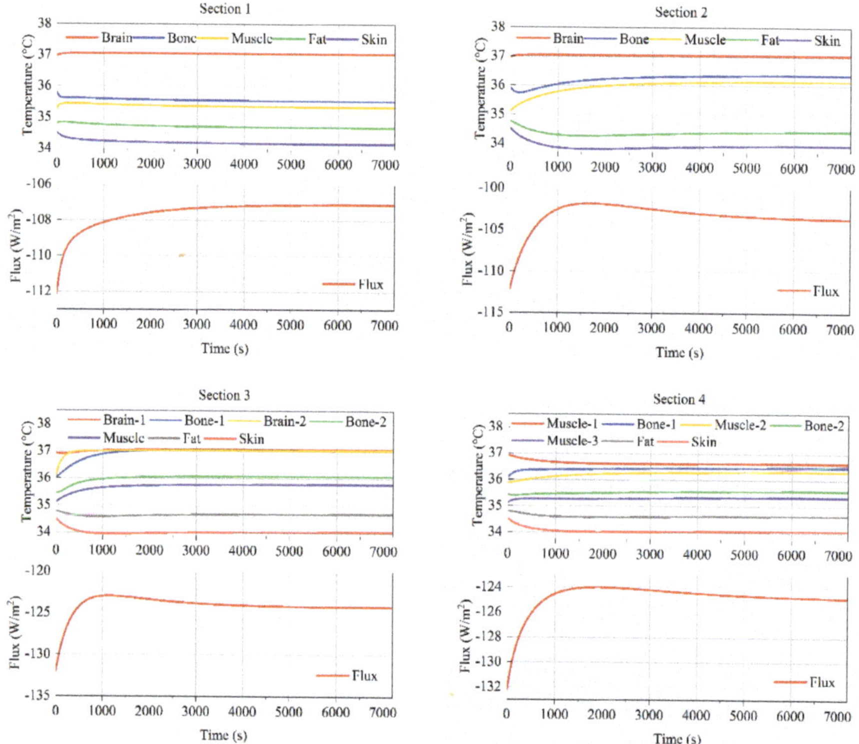

**Figure 6.** Comparison of the temperatures influenced by different compositions and structures of different layers of the head.

From the results in Figure 6, it can be found that there are obvious differences in the temperature distribution in different sections. The four layers in Figure 2a are structures that can roughly represent the forehead, the part around the eyes, the part of the cheeks, and the mouth and nose.

For Sections 1 and 2, although the internal structural composition is similar, the temperature of the core and the skin temperature are different due to the different thickness of the tissues. The skin temperature of Section 2 is significantly lower than that of Section 1. In Section 2, the temperature of the bones and muscles is significantly higher than the corresponding temperature in Section 1. The reason is that the thickness of the muscle tissue in Section 2 is greater than that in Section 1. For Sections 3 and 4, the structure and tissue distribution are both different, resulting in different internal core temperatures and internal tissue temperatures. The core temperature of Section 3 is slightly lower than that of Section 4, but the most stable temperature is almost the same, while the skin temperature of Section 3 is slightly lower than that of Section 4.

From the final stable skin heat flux, the four sections are also quite different. This explains why the head and face have different temperature distributions, internal structure differences, and different tissue thicknesses, leading to differences in internal heat transfer and skin heat loss, which are reflected in uneven skin temperature distribution.

*3.3. Effect of Tissue Thickness Step Change on Temperature Distribution*

It is reported that the percentage of fat in each part of the body will affect the temperature distribution [43]. A higher proportion of fat leads to a lower skin temperature, while the temperature of the tissue inside the fat layer is higher, which indicates that the fat plays a role in heat insulation.

For different individuals, in addition to the difference in fat content, muscle tissue and skin also have different thicknesses and distributions. Although the structure of the same part does not differ much, the thickness of muscle tissue, adipose tissue, and skin varies from person to person. Even in the same person, the parameters of these tissues can change under different physiological conditions. In order to study the effect of these three tissues on head temperature distribution, we assume that the thickness of muscle tissue, adipose tissue, and skin varies from 0.1 to 1.5 cm, 0.1 to 1.5 cm and 0.05 to 0.5 cm, respectively. The air temperature, relative humidity and wind speed of the ambient environment are 28 °C, 30% and 0.05 m/s, respectively. Based on the physiological structure of the forehead, calculations were performed to obtain the effect of different tissue thicknesses on the tissue temperature of each layer. The calculation results are shown in Figures 7–9.

Figure 7 shows the results of the effects of changes in muscle tissue thickness on the temperature of each layer, and Figures 6 and 9 show the results of changes in fat thickness and skin thickness on the temperature of each layer.

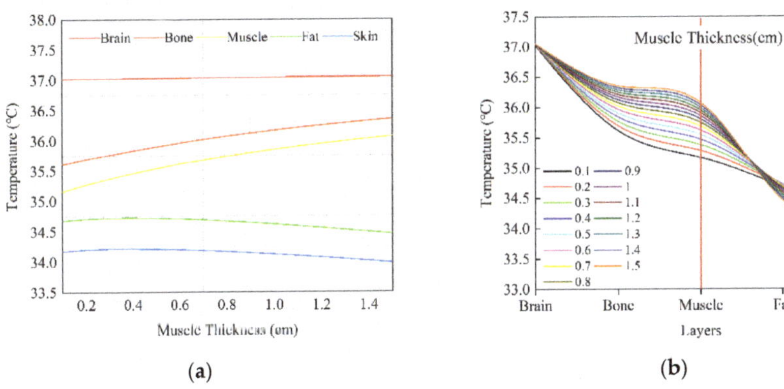



(a)        (b)

**Figure 7.** (**a**) Temperature variation by muscle thickness step change; (**b**) Tissues temperature distribution under different muscle thickness.

As shown in Figure 7a, skin temperature continues to decrease as muscle thickness increases, and the temperature of the fat layer has the same trend as the skin layer. While, on the contrary, the temperature of brain, bone and muscle layers are increased, of which the bone and muscle layers increased more obviously, and the temperature of the brain layer stayed relatively stable. The trends are clearly shown in Figure 7b, showing that not only the inner tissues are affected, but also the muscle layer itself is affected by the increase of its thickness.

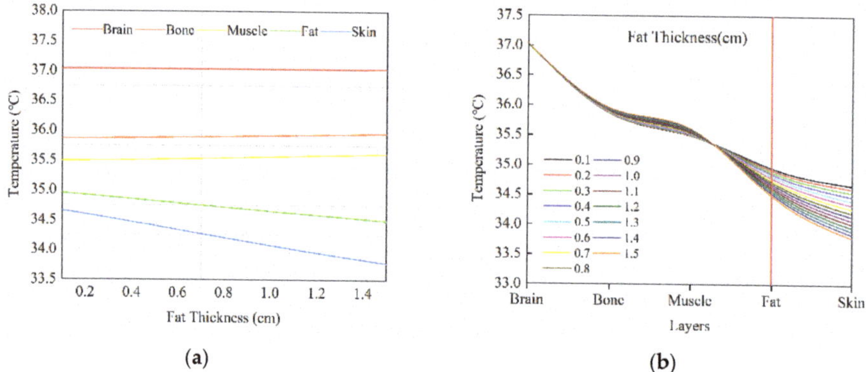

**Figure 8.** (**a**) Temperature variation by fat thickness step change; (**b**) Tissues temperature distribution under different fat thickness.

Figure 8a shows the tissue temperature change with the changes of fat thickness. The temperatures of the brain, bone and muscle are increased, and the fat and skin layers' temperatures decreased as the fat thickness increased. The general trend of the five layers are similar to that in the case of muscle thickness step change, but the absent value of temperature change is different. Temperature changes caused by changes in fat thickness are not as obvious as temperature differences caused by changes in muscle tissue thickness. The temperature changes of muscles and bones are very slow, rising only in a small range, while the temperature of brain tissues is not obvious, and basically remains stable. The temperature of the adipose tissue and the skin drops significantly; a trend of temperature change that is different from that in the case of muscle thickness changes. The increase in fat thickness does not cause the temperature of the fat layer to rise, but rather causes it to decrease, indicating that fat does play a role in blocking heat. This phenomenon is consistent with experimental observations in [42,43].

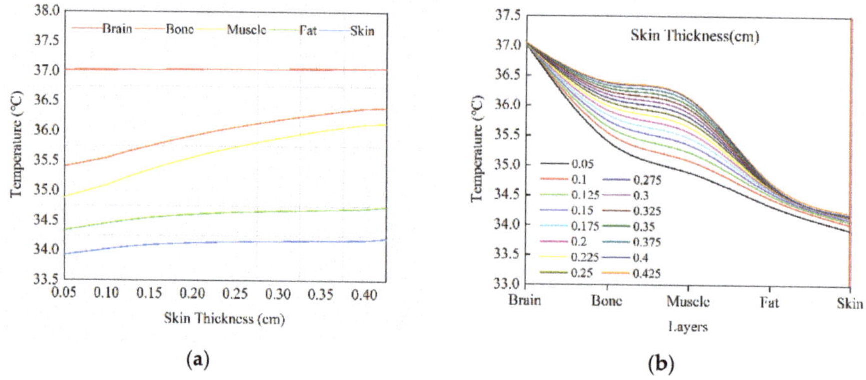

**Figure 9.** (**a**) Temperature variation by skin thickness step change; (**b**) Tissues temperature distribution under different skin thickness.

The skin layer is the outer layer of the body, and the barrier to protect human body, and is also the interface to exchange heat, water and gas with the environment. Although the skin is a thin layer, its function of heat exchange is significant. The results in Figure 9 show that the change in the thickness of the skin layer also has a certain effect on the temperature distribution. As the thickness of the skin increases, the tissue temperature inside the skin layer tends to rise. The temperature of brain tissue is similar to the previous situation, although there is a small change, it remains stable. Similar to the change in muscle thickness, the temperature of bones and muscles showed an upward trend, and reached a higher position. The temperature of the fat layer also increased with the same trend as the skin layer.

From the results of the above three examples, it can be seen that changes in the thickness of muscle tissue, adipose tissue, and skin can all affect the distribution of head temperature. In order to compare the effects of the three tissues on temperature changes, the temperature data of the three examples were normalized and then a regression analysis was performed. The regression data are shown in Table 4.

**Table 4.** Regression results of normalized temperature and tissue thickness change in Figures 7–9.

| Curve | Muscle (Figure 7) | | | Fat (Figure 8) | | | Skin (Figure 9) | | |
|---|---|---|---|---|---|---|---|---|---|
| | Intercept | Slope | r | Intercept | Slope | r | Intercept | Slope | r |
| Brain | 0.9993 | 0.00074 | 0.99104 | 0.9999 | 0.00009 | 0.9962 | 0.999 | 0.0011 | 0.9885 |
| Bone | 0.9798 | 0.02151 | 0.99104 | 0.9971 | 0.0029 | 0.9962 | 0.9709 | 0.0318 | 0.9885 |
| Muscle | 0.9758 | 0.02597 | 0.98721 | 0.9961 | 0.0038 | 0.9962 | 0.9639 | 0.0394 | 0.9862 |
| Fat | 1.0016 | −0.0081 | −0.91704 | 1.001 | −0.0143 | −0.9999 | 0.99 | 0.0107 | 0.9288 |
| Skin | 1.0014 | −0.0071 | −0.91478 | 1.0017 | −0.0274 | −0.9995 | 0.9934 | 0.0071 | 0.8969 |

From the results of the regression analysis, the three tissue thickness changes generally show a linear relationship with the temperature changes, and a high linear fitting degree indicates that there is a strong correlation between temperature changes and tissue thickness changes.

Here we define a coefficient to evaluate the effect of tissue thickness changes on head temperature.

$$C = \frac{1}{n}\sum_{i=1}^{n}|S_i| \tag{11}$$

where, $C$ is the influence coefficient, $S_i$ is the slope of each tissue layer in the regression analysis, $n$ is the total number of tissue layers.

Then, the influence coefficients of muscle, fat and skin are: $C_{muscle} = 0.0127$, $C_{fat} = 0.0097$ and $C_{skin} = 0.0180$, respectively.

At present, research on the influence of body composition on temperature distribution is mainly focused on the effect of the proportion of adipose tissue on temperature. The results of these studies show that the thickness of the fat layer and the temperature of the skin have a negative correlation. This phenomenon has been verified in the calculation results of this paper. Moreover, the influence of different combinations of tissue or biological structures on temperature distribution is also observed in the study. Meanwhile, other tissues of the body also affect the temperature distribution. As mentioned in reference [42,43], the percentage change of muscle tissue will affect the average temperature of the skin, which is also reflected in the calculation results.

The fat layer inside the body parts serves as a heat insulator with a low thermal conductivity of $k$ = 0.16. As the thickness of fat increased, the total heat resistance rises, and it becomes more difficult for inner heat to transfer to outer layers, which results in skin temperature drop, meanwhile the heat accumulates in the inner tissues and results in a temperature increase.

The muscle and the skin layers have a similar effect as the fat layer; a thickness increase of muscle and skin leads to the temperature rise of inner tissue layers and temperature drop of outer layers. However, the difference lies in the effect on the layer itself. The thickness increase in the fat layer does not result in a temperature increase of the fat layer itself, but in the case of thickness change of muscle and skin layers, an increase of thickness in the two layers results in the temperature rise of the muscle layer and the skin layer. The reason for this result is that the fat tissue has a low thermal conductivity ($k = 0.16$), when the thickness of the fat layer increases, the amount of heat transferred to the adipose tissue decreases, and because of its lower metabolic heat production rate ($q_m = 58$) and lower blood perfusion rate ($w_{bl}$ = 0.0036), the overall caloric accumulation of itself decreases, leading to a decrease in the temperature of the fat layer itself, and consequent drop in skin temperature.

On the other hand, the thermophysical parameters of muscle tissue and skin tissue are very close, which leads to very similar laws of temperature changes of muscle tissue and other tissues caused by changes in skin thickness. The temperature of the internal tissue increases, while the temperature of the external tissue decreases. The only difference is that the muscle tissue and skin are in different positions in the physiological structure. The skin is located on the outermost side of the physiological structure. As a result, changes in skin thickness cause the temperature of all internal tissues to rise, including the temperature of the skin layer. The skin layer is located in the outermost layer, so it has a blocking effect on the heat in all the inner layers.

*3.4. Thermal Sensation and Its Sensitivity to Tissue Thickness Vriation*

The purpose of this study was to investigate the effect of tissue differences on the local thermal sensation of the head. Numerical results have shown that the tissue thickness difference has a significant effect on the temperature distribution and also directly affect local thermal sensation. In order to study the relationship between changes in tissue thickness and thermal sensation, or more specifically, to investigate how the thickness change of muscle tissue, adipose tissue and skin tissue affects the thermal sensation of head in an uniformed environment, the thermal sensation index sensitivity analysis was performed, and the results are shown in Figure 10.

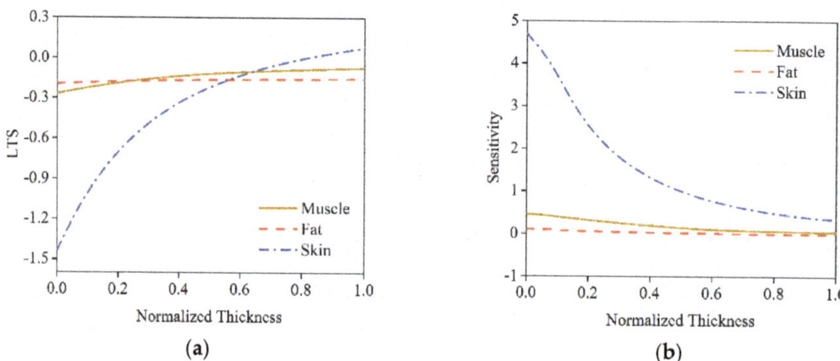


**Figure 10.** (**a**) The variation of the thermal sensation index with tissue thickness change; (**b**) the sensitivity of the thermal sensation index to tissue thickness change.

In Figure 10a, the curve shows the change of the thermal sensation of the head relative to the normalized thickness change of the three tissues (muscle, fat, and skin). During the gradual increase in the thickness of the three tissues (muscle, fat, and skin), it can be clearly seen that the thermal sensation index of the head is gradually increasing. The increase in skin thickness has the greatest effect on the thermal sensation of the head which result in the thermal sensation index increasing from a negative value to a positive value. It can be seen from the figure that when the value of skin thickness is the smallest, the head thermal sensation index LTS is about −1.44, and when the skin thickness is the largest, the thermal sensation index LTS is about 0.076, which has a wide range of variation. For muscle tissue and adipose tissue, there is a definite change in the thermal sensation index of the head, but the change is not as large as the effect of skin on heat sensation. However, from the numerical result when the muscle thickness is the largest, the LTS is about −0.074, and when the fat thickness is the largest, the LTS is about −0.154. This means that increased muscle and fat thickness has less effect on thermal sensation.

Figure 10b is the sensitivity of the head thermal sensation index to the standardized thickness of three tissues (muscle tissue, adipose tissue, and skin tissue). It can be seen that the sensitivity of the thermal sensation index to the change of skin thickness is the highest. That means the thermal sensation of the head is easily affected by skin thickness change. However, this affect is not changed in a linear form. In the initial stage, the increase in skin thickness has a significant effect on improving thermal sensation, but as the skin thickness continues to increase, the skin's effect on increasing thermal sensation becomes slower and the sensitivity of thermal sensation decreases gradually until, in the end, it gradually flattens out, which means that when the skin thickness is high, the thermal sensation of head is less sensitive to skin thickness change.

The sensitivity of thermal sensation to muscle tissue and adipose tissue is at a relatively low level, indicating that the increase in the thickness of these two tissues has a lower effect on changing the thermal sensation of the head. The reason for this phenomenon is that the thermal conductivity of adipose tissue itself is very low, and when the thickness or content is not high, it can play a better thermal insulation effect. As the thickness increases, it does not decrease the temperature of the skin further. For muscle tissue, it has a similar effect to fat tissue, as analyzed for temperature changes in a previous section. This leads to the sensitivity of the results in Figure 10b.

## 4. Conclusions

The purpose of this study was to investigate the influence of the head's internal structure and composition on head temperature distribution and local thermal sensation through numerical calculations. The following conclusions were reached:

(1) In this work, a mathematical model suitable for calculating head temperature was established, in which the active system was established based on the fuzzy control theory, and the usability of this model was proved by comparing it with public data.

(2) The physiological parameters of the head were obtained by MRI, and obvious differences were found in the physiological parameters between individuals, which would have certain influence on the simulation of human thermal comfort.

(3) Based on the mathematical model and the obtained physiological parameters, the temperature distribution under the conditions of different head structures and tissue thickness parameters was calculated. The results show that different internal structures cause differences in skin surface temperature, with the highest skin temperature being 34.17 °C on the forehead, and the lowest 33.94 °C, with a difference of 0.24 °C. The results of the regression analysis showed that skin thickness had the largest influence on head temperature, followed by muscle tissue, and the least influence was adipose tissue, with the influence coefficients of $C_{muscle} = 0.0127$, $C_{fat} = 0.0097$, and $C_{skin} = 0.0180$, respectively.

(4) Local thermal sensation of the head is sensitive to changes in tissue thickness. Local thermal sensation was most sensitive to skin thickness differences, with the highest sensitivity coefficient of 4.69, the lowest of 0.36, and the average of 1.58. The highest sensitivity to muscle tissue was 0.46, the lowest 0.64, and the average 0.2, while the sensitivity coefficient of fat was relatively small, with the highest of 0.11, the lowest of 0.013, and the average of 0.04.

**Author Contributions:** Conceptualization, Y.Z.; investigation, S.H.; methodology, S.H.; resources, Y.G.; visualization, G.Z.; writing—original draft, S.H.; writing—review and editing, Z.H. All authors have read and agreed to the published version of the manuscript.

**Acknowledgments:** The authors appreciate the financial support of project No. 51974015, No. 51904292 and No. 51474017 provided by the National Natural Science Foundation of China, project No. 2017CXNL02 provided by the Fundamental Research Funds for the Central Universities (China University of Mining and Technology), project No. BK20180655 provided by the Natural Science Foundation of Jiangsu Province of China, project No. WS2018B03 provided by the State Key Laboratory Cultivation Base for Gas Geology and Gas Control (Henan Polytechnic University), and project No. E21724 provided by the Work Safety Key Lab on Prevention and Control of Gas and Roof Disasters for the Southern Coal Mines of China (Hunan University of Science and Technology).

**Conflicts of Interest:** The authors declare no conflict of interest.

## References


1. Yang, Y.; Yao, R.; Li, B.; Liu, H.; Jiang, L. Parameterization of Temperature Perception of Ventilation Changes in Full-Face Motorcycle. *Build. Environ.* **2015**, *87*, 1–9. [CrossRef]
2. Johnson, A.T. Respirator masks protect health but impact performance: A review. *J. Biol. Eng.* **2016**, *10*, 4. [CrossRef] [PubMed]
3. Bogerd, C.P.; Aerts, J.M.; Annaheim, S.; Bröde, P.; de Bruyne, G.; Flouris, A.D.; Kuklane, K.; Sotto Mayor, T.; Rossi, R.M. A review on ergonomics of headgear: Thermal effects. *Int. J. Ind. Ergon.* **2015**, *45*, 1–12. [CrossRef]
4. Yang, Y.; Yao, R.; Li, B.; Liu, H.; Jiang, L. Modelling of the microclimate saturation inside a safety helmet. *Build. Environ.* **2015**, *87*, 1–9. [CrossRef]
5. Halimi, M.T.; Dhahri, H.; Khedher, N.B.; Hassen, M.B.; Sakli, F. Thermal Properties of Industrial Safety Helmets. *Build. Environ.* **2009**, *5*, 833–844.
6. Underwood, L.; Vircondelet, C.; Jermy, M. Thermal comfort and drag of a streamlined cycling helmet as a function of ventilation hole placement. *Proc. Inst. Mech. Eng. Part P J. Sports Eng. Technol.* **2018**, *232*, 15–21. [CrossRef]
7. Ghani, S.; ElBialy, E.M.A.A.; Bakochristou, F.; Gamaledin, S.M.A.; Rashwan, M.M. The effect of forced convection and PCM on helmets' thermal performance in hot and arid environments. *Appl. Therm. Eng.* **2017**, *111*, 624–637. [CrossRef]
8. Bogerd, C.P.; Walker, I.; Brühwiler, P.A.; Rossi, R.M. The effect of a helmet on cognitive performance is, at worst, marginal: A controlled laboratory study. *Appl. Ergon.* **2014**, *45*, 671–676. [CrossRef]
9. Kobayashi, Y.; Tanabe, S.I. Development of JOS-2 human thermoregulation model with detailed vascular system. *Build. Environ.* **2013**, *66*, 1–10. [CrossRef]
10. Lv, Y.G.; Liu, J. Effect of transient temperature on thermoreceptor response and thermal sensation. *Build. Environ.* **2007**, *42*, 656–664. [CrossRef]
11. Foda, E.; Almesri, I.; Awbi, H.B.; Sirén, K. Models of human thermoregulation and the prediction of local and overall thermal sensations. *Build. Environ.* **2011**, *46*, 2023–2032. [CrossRef]
12. Zhou, X.; Xiong, J.; Lian, Z. Predication of skin temperature and thermal comfort under two-way transient environments. *J. Therm. Biol.* **2017**, *70*, 15–20. [CrossRef] [PubMed]
13. Lan, L.; Tsuzuki, K.; Liu, Y.F.; Lian, Z.W. Thermal environment and sleep quality: A review. *Energy Build.* **2017**, *149*, 101–113. [CrossRef]
14. Salloum, M.; Ghaddar, N.; Ghali, K. A new transient bioheat model of the human body and its integration to clothing models. *Int. J. Therm. Sci.* **2007**, *46*, 371–384. [CrossRef]
15. Irzmańska, E. Case study of the impact of toecap type on the microclimate in protective footwear. *Int. J. Ind. Ergon.* **2014**, *44*, 706–714. [CrossRef]
16. Zolfaghari, A.; Maerefat, M. A new simplified thermoregulatory bioheat model for evaluating thermal response of the human body to transient environments. *Build. Environ.* **2010**, *45*, 2068–2076. [CrossRef]
17. Ludwig, N.; Trecroci, A.; Caumo, A.; Alberti, G.; Michielon, G.; Formenti, D.; Gargano, M. Dynamics of thermographic skin temperature response during squat exercise at two different speeds. *J. Therm. Biol.* **2016**, *59*, 58–63.
18. Neves, E.B.; Salamunes, A.C.C.; de Oliveira, R.M.; Stadnik, A.M.W. Effect of body fat and gender on body temperature distribution. *J. Therm. Biol.* **2017**, *70*, 1–8. [CrossRef]
19. Lakhssassi, A.; Kengne, E.; Semmaoui, H. Modifed pennes' equation modelling bio-heat transfer in living tissues: Analytical and numerical analysis. *Nat. Sci.* **2010**, *2*, 1375–1385. [CrossRef]
20. Dixit, A.; Gade, U. A case study on human bio-heat transfer and thermal comfort within CFD. *Build. Environ.* **2015**, *94*, 122–130. [CrossRef]
21. Tuzikiewicz, W.; Duda, M. Bioheat transfer equation. *The problem of FDM explicit scheme stability. J. Appl. Math. Comput. Mech.* **2015**, *14*, 139–144. [CrossRef]
22. Lakhssassi, A.; Kengne, E.; Semmaoui, H. Investigation of nonlinear temperature distribution in biological tissues by using bioheat transfer equation of Pennes' type. *Nat. Sci.* **2010**, *2*, 131–138. [CrossRef]
23. Yang, J.-H.; Kato, S.; Seo, J. Evaluation of the Convective Heat Transfer Coefficient of the Human Body Using the Wind Tunnel and Thermal Manikin. *J. Asian Archit. Build. Eng.* **2009**, *8*, 563–569. [CrossRef]

24. Cano, S.P.; Sillero-Quintana, M.; Brito, C.J.; de Andrade Fernandes, A.; Bouzas Marins, J.C.; Costa, C.M.A.; de Azambuja Pussieldi, G.; Moreira, D.G. Daily rhythm of skin temperature of women evaluated by infrared thermal imaging. *J. Therm. Biol.* **2017**, *72*, 1–9.

25. Kaynakli, O.; Kilic, M. Investigation of indoor thermal comfort under transient conditions. *Build. Environ.* **2005**, *40*, 165–174. [CrossRef]

26. Pang, T.Y.; Subic, A.; Takla, M. Thermal comfort of cricket helmets: An experimental study of heat distribution. *Procedia Eng.* **2011**, *13*, 252–257. [CrossRef]

27. Bogerd, C.P.; Rossi, R.M.; Brühwiler, P.A. Thermal perception of ventilation changes in full-face motorcycle helmets: Subject and manikin study. *Ann. Occup. Hyg.* **2011**, *55*, 192–201.

28. Mehnert, P.; Malchaire, J.; Kampmann, B.; Piette, A.; Griefahn, B.; Gebhardt, H. Prediction of the average skin temperature in warm and hot environments. *Eur. J. Appl. Physiol.* **2000**, *82*, 52–60. [CrossRef]

29. Zhou, X.; Lian, Z.; Lan, L. An individualized human thermoregulation model for Chinese adults. *Build. Environ.* **2013**, *70*, 257–265. [CrossRef]

30. Kubota, H.; Kuwabara, K.; Hamada, Y. Prediction of mean skin temperature for use as a heat strain scale by introducing an equation for sweating efficiency. *Int. J. Biometeorol.* **2014**, *58*, 1593–1603. [CrossRef]

31. Mairiaux, P.; Malchaire, J.; Candas, V. Prediction of mean skin temperature in warm environments. *Eur. J. Appl. Physiol. Occup. Physiol.* **1987**, *56*, 686–692. [CrossRef] [PubMed]

32. Katić, K.; Li, R.; Kingma, B.; Zeiler, W. Modelling hand skin temperature in relation to body composition. *J. Therm. Biol.* **2017**, *69*, 139–148. [CrossRef]

33. Abu Bakar, R.; Jusoh, N.; Rasdan Ismail, A.; Zanariah Shamshir Ali, T. Effect on human metabolic rate of skin temperature in an office occupant. *MATEC Web Conf.* **2017**, *90*, 01070. [CrossRef]

34. Laird, I.S.; Goldsmith, R.; Pack, R.J.; Vitalis, A. The effect on heart rate and facial skin temperature of wearing respiratory protection at work. *Ann. Occup. Hyg.* **2002**, *46*, 143–148. [PubMed]

35. Fiala, D.; Lomas, K.J.; Stohrer, M. A computer model of human thermoregulation for a wide range of environmental conditions: The passive system. *J. Appl. Physiol.* **2017**, *87*, 1957–1972. [CrossRef] [PubMed]

36. Dongmei, P.; Mingyin, C.; Shiming, D.; Minglu, Q. A four-node thermoregulation model for predicting the thermal physiological responses of a sleeping person. *Build. Environ.* **2012**, *52*, 88–97. [CrossRef]

37. Katić, K.; Li, R.; Zeiler, W. Thermophysiological models and their applications: A review. *Build. Environ.* **2016**, *106*, 286–300. [CrossRef]

38. Havenith, G.; Fiala, D. Thermal indices and thermophysiological modeling for heat stress. *Compr. Physiol.* **2016**, *6*, 255–302.

39. Zhu He, Z.; Xue, X.; Liu, J. An Effective Finite Difference Method for Simulation of Bioheat Transfer in Irregular Tissues. *J. Heat Transf.* **2013**, *135*, 071003. [CrossRef]

40. Lai, D.; Zhou, X.; Chen, Q. Modelling dynamic thermal sensation of human subjects in outdoor environments. *Energy Build.* **2017**, *149*, 16–25. [CrossRef]

41. Lai, D.; Chen, Q. A two-dimensional model for calculating heat transfer in the human body in a transient and non-uniform thermal environment. *Energy Build.* **2016**, *118*, 114–122. [CrossRef]

42. Weigert, M.; Nitzsche, N.; Kunert, F.; Lösch, C.; Schulz, H. The influence of body composition on exercise-associated skin temperature changes after resistance training. *J. Therm. Biol.* **2018**, *75*, 112–119. [CrossRef] [PubMed]

43. Salamunes, A.C.C.; Stadnik, A.M.W.; Neves, E.B. The effect of body fat percentage and body fat distribution on skin surface temperature with infrared thermography. *J. Therm. Biol.* **2017**, *66*, 1–9. [CrossRef] [PubMed]

44. Davoodi, F.; Hassanzadeh, H.; Zolfaghari, S.A.; Havenith, G.; Maerefat, M. A new individualized thermoregulatory bio-heat model for evaluating the effects of personal characteristics on human body thermal response. *Build. Environ.* **2018**, *136*, 62–76. [CrossRef]

45. Ziade, G.; Semaan, S.; Ghulmiyyah, J.; Kasti, M.; Hamdan, A.L.H. Structural and Anatomic Laryngeal Measurements in Geriatric Population Using MRI. *J. Voice* **2017**, *31*, 359–362. [CrossRef]

46. Chia, C.W.; Ferrucci, L.; Spencer, R.G.; Fishbein, K.W.; Makrogiannis, S.K.; Lukas, V.A.; Okine, M.; Egan, J.M.; Ramachandran, R. Measurement of fat fraction in the human thymus by localized NMR and three-point Dixon MRI techniques. *Magn. Reson. Imaging* **2018**, *50*, 110–118.

47. Lack, C.M.; Lesser, G.J.; Umesi, U.N.; Bowns, J.; Chen, M.Y.; Case, D.; Hightower, R.C.; Johnson, A.J. Making the most of the imaging we have: Using head MRI to estimate body composition. *Clin. Radiol.* **2016**, *71*, e1–e402. [CrossRef]

48. Brooks, T.; Choi, J.E.; Garnich, M.; Hammer, N.; Waddell, J.N.; Duncan, W.; Jermy, M. Finite element models and material data for analysis of infant head impacts. *Heliyon* **2018**, *4*, e01010. [CrossRef] [PubMed]
49. Zhang, H.; Arens, E.; Huizenga, C.; Han, T. Thermal sensation and comfort models for non-uniform and transient environments: Part I: Local sensation of individual body parts. *Build. Environ.* **2010**, *45*, 380–388. [CrossRef]
50. Zhang, H.; Arens, E.; Huizenga, C.; Han, T. Thermal sensation and comfort models for non-uniform and transient environments, part II: Local comfort of individual body parts. *Build. Environ.* **2010**, *45*, 389–398. [CrossRef]
51. Tanabe, S.I.; Kobayashi, K.; Nakano, J.; Ozeki, Y.; Konishi, M. Evaluation of thermal comfort using combined multi-node thermoregulation (65 MN) and radiation models and computational fluid dynamics (CFD). *Energy Build.* **2002**, *34*, 637–646. [CrossRef]

MDPI

St. Alban-Anlage 66

4052 Basel

Switzerland

Tel. +41 61 683 77 34

Fax +41 61 302 89 18

www.mdpi.com

*Atmosphere* Editorial Office

E-mail: atmosphere@mdpi.com

www.mdpi.com/journal/atmosphere